ANTIVIRAL NUCLEOSIDES:

CHIRAL SYNTHESIS AND CHEMOTHERAPY

ANTIVIRAL NUCLEOSIDES:

CHIRAL SYNTHESIS AND CHEMOTHERAPY

Edited by

C.K. CHU

2003

ELSEVIER

Amsterdam – Boston – Heidelberg – London – New York – Oxford
Paris – San Diego – San Francisco – Singapore – Sydney – Tokyo

ELSEVIER B.V.
Sara Burgerhartstraat 25
P.O. Box 211, 1000 AE Amsterdam, The Netherlands

First edition 2003

Library of Congress Cataloging in Publication Data
A catalog record from the Library of Congress has been applied for.

British Library Cataloguing in Publication Data
A catalogue record from the British Library has been applied for.

ISBN: 0-444-51319-1

∞ The paper used in this publication meets the requirements of ANSI/NISO Z39.48-1992 (Permanence of Paper).
Printed in Hungary.

Dedicated to these nucleoside chemists and biologists.

(Photographed at the 15th International Round Table held in Leuven, Belgium, September 10-14, 2002)

PREFACE

Since the discovery of AZT in 1985, soon after the discovery of the human immunodeficiency virus (HIV), a number of laboratories of nucleosides and viral pharmacology intensified their efforts in order to come-up with more safe and potent anti-HIV agents. From these efforts, scientists further discovered ddI, ddC and d4T as potent and clinically useful agents, although these agents also possess different profiles of antiviral activity as well as side-effects. With these agents in hand we opened a new era of combination chemotherapy of viral diseases. However, retrospectively, the major breakthrough came when 3TC (lamivudine) was discovered as a potent anti-HIV agent. The discovery of 3TC was almost a shock to conventional nucleoside chemists as 3TC was an oxathiolane derivative, and furthermore it was an L-nucleoside. L-Nucleosides have previously never been demonstrated any significant biological activity. Of course, we know now that more than 70-80 % of AIDS patients are currently taking 3TC as part of combination therapy. Furthermore, the discovery of 3TC fueled the chemistry of L-nucleosides for the last ten years. Thanks to 3TC, we are now witnessing at least six L-nucleosides (FTC, L-FMAU, L-d4FC, L-OddC, L-dT and L-dC) currently undergoing clinical trials. This all happened in the 90s, thus nucleoside chemists call it the "L-nucleosides decade"

Parts of this book reflect the development of the L-nucleosides chemistry and pharmacology, plus other conventional antiviral nucleosides discovered during the last ten years. This book covers experimental antiviral agents discovered up to the middle of 2002. At The University of Georgia, I am using the chapters in this book as the text for our graduate students in medicinal chemistry. It works out well because these chapters not only cover modern carbohydrate and nucleoside chemistry, but also deal with biology and chemotherapy of antiviral agents. Thus, these chapters encompass the so called "from bench to beside." That is what our aspiring graduate students in medicinal chemistry need to be exposed to due to the interdisciplinary nature of drug discovery and development. I hope this book will be useful to those who are already involved in the field for a quick review as well as graduate students who are entering the field of antiviral chemistry and chemotherapy.

Finally, I would like to dedicate this book to scientists who have contributed to the field of nucleoside chemistry and biology. Some of them participated as part of the Gordon Conference in New Port, Rhode Island in 1997. Keep-up your good work!

C. K. Chu
Athens, Georgia USA
August, 2002

CONTENTS

CHAPTER 1

RECENT ADVANCES IN ANTIVIRAL NUCLEOSIDES

GIUSEPPE GUMINA, YONGSEOK CHOI and CHUNG K. CHU

1.1. Introduction

During the last two decades, treatment of viral infections has advanced remarkably, thanks to the heroic efforts of chemists and pharmacologists, the rapid progress of molecular virology as well as the cumulative knowledge of more detailed mechanism of action of antiviral agents.[1] In recent years, we are facing an outburst of new and emerging viral diseases, such as new strains of hepatitis and herpes viruses, Ebola virus, West-Nile virus, plus a number of exotic viruses which, although still isolated in small areas of the world, have the potential for pandemic outbreak. Besides, the threat that viruses and other microorganisms could be used as biological weapons in warfare or bioterrorism has become a reality. Although vaccination is a valuable tool to fight viral diseases and in some cases is available and successful, the difficulty associated with state- or world-wide vaccination programs makes antiviral chemotherapy a more practical approach in the fight to epidemic viral infections.

Among the most successful antiviral agents, nucleoside analogs have been the drugs of choice in the treatment of a number of deseases caused by herpes simplex virus (HSV), human cytomegalovirus (HCMV), varicella zoster virus (VZV), human immunodeficiency virus type 1 (HIV-1) and human hepatitis B (HBV) and C (HCV) virus. Since 1980, a variety of biologically interesting and promising nucleosides have been discovered, some of which are being used clinically or are undergoing preclinical or clinical development. Currently, eighteen nucleosides are clinically being used for the treatment of HIV-1, herpes virus, HBV, RSV and HCV infections (Table 1). Despite these achievements, continued discoveries of novel nucleoside analogs are needed in order to overcome common problems in antiviral chemotherapy, such as toxicity, metabolic instability and, above all, the emergence of resistant viral strains as well as of new and emerging viral diseases. So far, a number of reviews have been published, regarding specific nucleoside classes,[2] general aspects of nucleosides[3] and their chemistry[4] as well as their antiviral activity spectrum and target of actions.[5] In view of these reviews, the purpose of this chapter is to give a brief overview on the most recent advances in antiviral nucleosides focusing on the structure-activity relationships with particular regard to the biochemical mode of action of the most promising nucleosides.

Antiviral Nucleosides: Chiral Synthesis and Chemotherapy, Ed. by C.K. Chu. 1 — 76

Table 1. Antiviral nucleosides used in clinics

Generic name	Acronyms	Target viruses	Mode of action
Anti-HIV agents			
Zidovudine	AZT	HIV-1	Reverse transcriptase inhibitor/ chain terminator
Didanosine	ddI	HIV-1	
Zalcitabine	ddC	HIV-1	
Stavudine	d4T	HIV-1	
Lamivudine	3TC	HIV-1	
Abacavir	1596U89	HIV-1	
Tenofovir disoproxil	PMPA	HIV-1	
Anti-HBV agent			
Lamivudine	3TC	HBV	Reverse transcriptase inhibitor/ chain terminator
Anti-Herpetic agents			
Idoxuridine	IdU	HSV-1/2	DNA polymerase inhibitor; topical use
Trifluridine	TFT	HSV-1/2	
Acedurid	EdU	HSV-1/2	
Vidarabine	araA	HSV-1/2	
Acyclovir	ACV	HSV-1/2, VZV	Selective viral DNA polymerase inhibitor
Valaciclovir	val-ACV	HSV-1/2, VZV	Valine ester prodrug of acyclovir
Penciclovir	PCV	HSV-1/2, VZV	Selective viral DNA polymerase inhibitor; topical use
Famciclovir	FCV	HSV-1/2, VZV	Oral prodrug of PCV
Ganciclovir	DHPG	HCMV	Selective viral DNA polymerase inhibitor
Cidofovir	(S)-HPMPC	HCMV	
Virazole	Ribavirin	RSV, HCV	Viral RNA polymerase inhibitor

1.2. Structural features of nucleosides as antiviral agents

Nucleoside analogs as inhibitors of viral replications usually act by interaction of their triphosphates with viral polymerases. As structural units of nucleic acids, the nucleoside triphosphates (NTPs) are the substrates for polymerase enzymes, which catalyze the polymerization of the NTPs. The biosynthesis of the NTPs is controlled by nucleoside kinases. The structural requirements of nucleosides to interact with kinases and polymerases have important implications in the design of potential antiviral nucleosides (Figure 1). The 5'-hydroxymethyl group and the base moiety of nucleosides interact with kinases and their complementary nucleotides on the DNA template. The sugar moiety of the nucleoside can be considered as a spacer to connect the hydroxymethyl group and the base moiety.[6] Therefore, modification of the sugar moiety has provided opportunities in the design of biologically active nucleosides.

Some viruses, such as herpes viruses, encode their own nucleoside-phosphorylating enzymes, which offers the potential for a therapeutic target.[6] Nucleosides, which are preferably phosphorylated by viral enzymes rather than by the cellular homologue, are only activated in infected cells and can have high selectivity against these viruses. This is, for example, the main factor in the success of acyclovir (ACV). However, other viruses, such as HIV and HBV, do not encode nucleoside kinases. In order to be active against these viruses, nucleoside analogs have to be phosphorylated by cellular kinases. Thus, the selectivity between antiviral activity and cellular toxicity depends on the substrate specificity of the NTPs for viral and host polymerases, and often the therapeutic exploitation of active nucleosides is compromised by the toxicity resulting from inhibition of the host enzymes or incorporation in the host nucleic acids.

In general, enzymes act on one enantiomer of a chiral substrate, the specificity of which is related to the unique structure of the enzymes.[7] However, recent findings have indicated that there are some exceptions to this rule among enzymes involved in the phosphorylation of nucleosides.[7,8,9,10] For instance, herpes virus thymidine kinases (TKs) phosphorylate both D- and L-enantiomeric forms of several uracil analogs as well as acyclic nucleosides, cellular deoxycytidine (dCyd) kinase phosphorylates both enantiomeric forms of several dCyd analogs, and some viral DNA polymerases, such as herpes viruses, HIV-1 RT and HBV DNA polymerase, are inhibited by the triphosphates of a number of L-nucleosides. These findings offer new opportunities for antiviral chemotherapy, although, at the molecular level, it is not completely understood how kinases phosphorylate both D- and L-nucleosides.

In recent years, a growing number of nucleoside analogs have been discovered which exert their antiviral activity by inhibiting enzymes different from polymerases, such as inosine monophosphate dehydrogenase, *S*-adenosylhomocysteine hydrolase, orotidine 5'-monophosphate decarboxylase and CTP synthetase.[11] Such compounds may prove useful because, by targeting different enzymes, they may offer synergistic action with classic polymerase inhibitors.

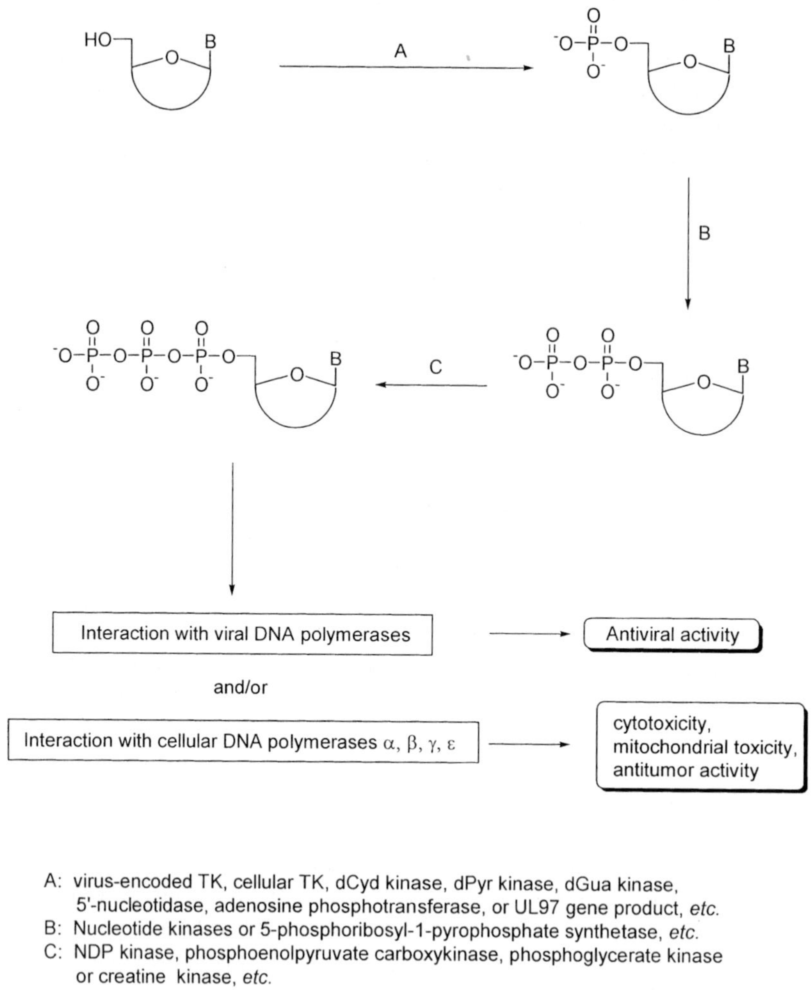

Figure 1. General mode of action of nucleoside analogs.

1.3. 2'-Deoxy nucleosides and related analogs

2'-Deoxy nucleoside analogs have proved effective against DNA viruses, such as HSV, VZV, EBV and HBV. Some nucleosides of this class show poor selectivity between the viral polymerases and the host polymerases due to their structural resemblance to natural substrates. However, modification of the base moiety or the sugar portion as well as the use of the unnatural L-enantiomer have been shown to reduce cellular toxicity, as in the case of 2'-fluoro-5-methyl-β-L-arabinofuranosyluracil (L-FMAU).[12]

Since the discovery of the first antiherpes compound, 5-iodo-2'-deoxyuridine (**1**, IdU),[13a] a modifications of the 5-position of the pyrimidine moiety have produced a number of active antiviral compounds (Figure 2).[13] Several 5-substituted 2'-deoxy-uridines, such as 5-trifluoromethyl-2'-deoxyuridine (**2**, TFT)[13b] and 5-ethyl-2'-deoxyuridine (**3**, EdU),[13c] have been approved for the treatment of herpetic keratitis. IdU and TFT are phosphorylated to their triphosphates by the virus-encoded TK. The triphosphates inhibit HSV DNA polymerase as well as, even though to a lesser extent, cellular polymerases. EdU has higher affinity for the herpesvirus-induced TK than for the cellular TK, and its triphosphate is incorporated to a large extent into the viral DNA.[3a,13d]

BVdU (**4**, brivudin) was originally synthesized by Walker and co-workers and shown to be a potent and selective anti-herpes agent.[14] It is specifically phosphorylated by virus-encoded TK and nucleoside diphosphate (NDP) kinase to give BVdUTP, which may act as either an inhibitor of or a substrate for viral DNA polymerase. However, BVdU is cleaved by pyrimidine nucleoside phosphorylases to (*E*)-5-(2-bromovinyl)uracil (BVU), which is cytotoxic.[15] The marked loss of activity of BVdU in thymidine kinase-deficient (TK^{-}) HSV-1 or VZV strains has been bypassed by its incorporation into phosphoramidate prodrugs (*vide infra*).[16] The inhibitory effects of several 5-alkynyl-2'-deoxyuridine analogs on virus replication, host cell metabolism and tumor cell proliferation have been investigated, among which 5-ethynyl-2'-deoxyuridine (**5**) is the most cytotoxic against L1210 cells.[17] 5-Heteroaryl-substituted 2'-deoxyuridines,[18,19] i.e. 5-(3-Bromoisoxazol-5-yl)-2'-deoxyuridine (**6**), 5-(5-bromothien-2-yl)-2'-deoxyuridine (**7**) and 5-(5-chloro-thien-2-yl)-2'-deoxyuridine (**8**) also share with BVdU a common antiviral spectrum against various strains of HSV-1 and VZV, but not HSV-2, HCMV or TK^{-} HSV-1.[5c]

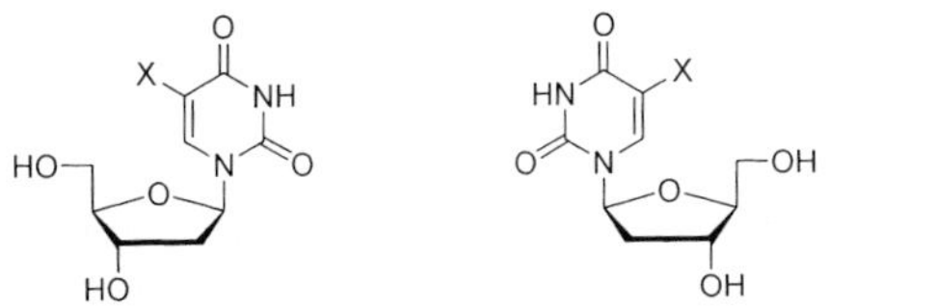

1 (IdU, X = I)	**9** (L-IdU, X = I)
2 (TFT, X = CF_3)	**10** (L-BVdU, X = (*E*)-bromovinyl)
3 (EdU, X = Et)	**11** (L-dT, X = CH_3)
4 (BVdU, X = (*E*)-bromovinyl)	
5 (X = C≡CH)	
6 (X = 3-Br-isoxazol-5-yl)	
7 (X = 5-Br-thien-2-yl)	
8 (X = 5-Cl-thien-2-yl)	

Figure 2. 2'-Deoxyuridine analogs.

Focher *et al.* have demonstrated that L-IdU (**9**), L-BVDU (**10**) and L-thymidine (**11**, L-dT) (Figure 2) are not recognized by human cytosolic TK *in vitro*, but function as a substrate for HSV-1 TK and inhibit HSV-1 proliferation in infected cells.[8] L-dT is selectively phosphorylated *in vivo* to L-dTMP by HSV-1 TK. L-dTMP is further phosphorylated to the di- and triphosphate forms by non-stereospecific cellular kinases. L-dTTP not only inhibits HSV-1 DNA polymerases *in vitro*, but also human DNA polymerases α, γ, δ and ε, HIV-1 RT, *E. coli* DNA polymerase I and calf thymus terminal transferase, although DNA polymerase β is resistant. Spadari *et al.* have also reported that HSV-1 TK shows no stereoselectivity and phosphorylates both D- and L-dT to their corresponding monophosphates with identical efficiency, with a K_i value of 2 μM, almost identical to the K_M for the natural substrate thymidine (2.8 μM).[9] L-IdU and L-BVdU inhibit HSV-1 TK with activities comparable to those of their corresponding D-enantiomers. In addition, the L-isomers of IdU and BVdU have no effect on human thymidylate synthase and are fully resistant to hydrolysis by nucleoside phosphorylase.[7] However, Chu and co-workers reported that L-BVdU and L-BVaraU show no activity against herpes viruses.[20] Furthermore, L-dT and L-2'-deoxycytidine (L-dC) do not show any inhibitory effect against HIV, HSV-1, HSV-2, EBV, VZV and vaccinia virus, whereas they show selectively potent anti-HBV activity with an EC_{50} value of 0.05-0.26 μM in 2.2.15 cells and duck HBV and with an EC_{50} value of 0.05 μM in primary duck hepatocytes without any cellular toxicity (CC_{50} >2000 μM).[10,21,22] L-dT and L-dC do not inhibit the growth of human bone marrow progenitor cells, although L-dT is a substrate of cytosolic TK and mitochondrial TK, and L-dC is phosphorylated by dCyd kinase and mitochondrial TK (Figure 3).[10,21] L-dC is not a substrate of dCyd deaminase, but its L-dCMP is deaminated to form L-dUMP, which is further converted to L-dUTP. L-dTTP and L-dCTP inhibit woodchuck hepatitis virus DNA polymerase with an IC_{50} value of 0.34 and 2.0 μM, respectively, whereas none of them is a substrate for the HIV RT or for human DNA polymerases α, β or ε up to 100 μM. Moreover, L-dT and L-dC do not cause any reduction in mitochondrial DNA content, any lactic acid accumulation nor the alteration in mitochondrial morphology or function up to 10 μM.[10,21] L-dT and L-dC are currently undergoing clinical trials as anti-HBV agents.

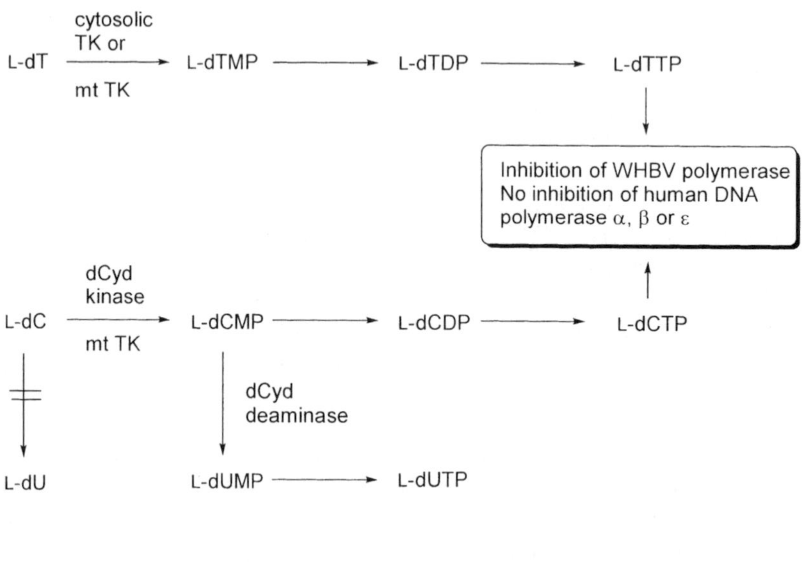

Figure 3. Metabolic pathway of L-dT and L-dC.[21]

During the synthesis of 5-alkynyluridine analogs (**12**), 3-glycosyl-6-substituted-furano[2,3-*d*]pyrimidine-2-one derivatives (**13**) have been obtained as cyclic by-products (Scheme 1). These furanopyrimidine derivatives exhibit potent and selective *in vitro* inhibition against VZV.[23] In this series, the 6-octyl derivative is the most potent, followed by the 6-decyl and 6-nonyl derivatives. Shorter chains (<C_6) led to antiviral activity similar to that of ACV and C_8-C_{10} led to higher antiviral activity. Longer chains (>C_{11}) reduced the potency against VZV, probably due to low water solubility. Interestingly, the furanopyrimidine derivatives (**13**) exhibited only anti-VZV activity. It seems that these analogs may be phosphorylated by VZV TK, as the complete loss of antiviral activity in the VZV TK$^-$ assays seems to support.

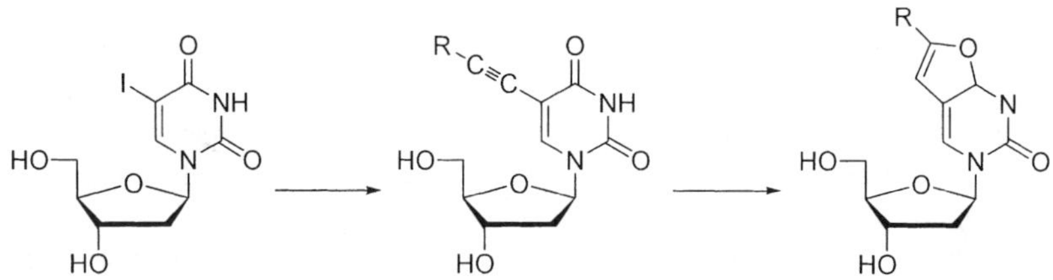

Scheme 1. Synthesis of furanopyrimidine derivatives.

Substitution of the oxygen on the furanose ring by a sulfur or methylene group also retains comparable antiviral activity with increased stability of the glycosidic bond. Among 2'-deoxy-4'-thiouridine analogs, 4'-thio BVdU (**14**, S-BVdU) is the most interesting compound (Figure 4), showing potent activity against HSV-1, HSV-2 and VZV (EC_{50} 0.6, 10 and 0.08 μM, respectively) with no cytotoxicity and improved *in vivo* stability.[24] Other 5-substituted analogs in this series also have good activity against HSV-1 and VZV *in vitro* without any apparent cytotoxicity (*e.g.* 5-ethyl, 5-vinyl and 5-chloroethyl).[25] Among them, the 5-ethyl analog has the broadest antiherpetic spectrum, being active against HSV-1, HSV-2 and VZV. Furthermore, the isopropyl and cyclopropyl analogs have significant activity *in vitro* against HSV-1 and VZV, whereas no activity is observed for their oxygen counterparts.[26] Among 2'-deoxy-4'-thio-*ribo* purine analogs, the 2-amino-6-(cyclopropylamino)purine derivative (**15**) is the most potent and selective agent against HCMV and HBV replication *in vitro* (EC_{50} 0.2 and 0.0072 μM in 2.2.15 cells, respectively), but it is also nephrotoxic *in vivo*.[27] Uenishi *et al.* have reported the synthesis and biological activity of D- and L-2'-deoxy-4'-thiouridines and their 5-trifluoromethyluridine analogs, D- and L-4'-thiothymidine, and D- and L-2'-deoxy-4'-thiocytidine.[28] D-Thymine, D-cytosine and D-5-trifluoromethyluracil derivatives are also potent inhibitors of the growth of L1210 cells. On the other hand, none of the L-nucleosides showed any cytotoxicity toward L1210 and KB cells except for the L-thymidine analog, which was slightly toxic toward L1210 cells.

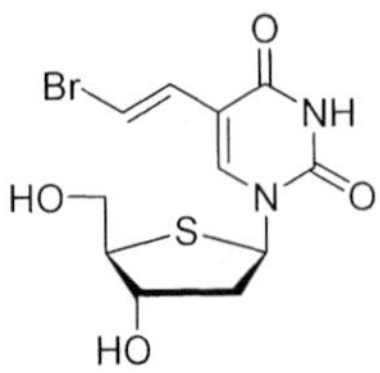

14 (S-BVdU)

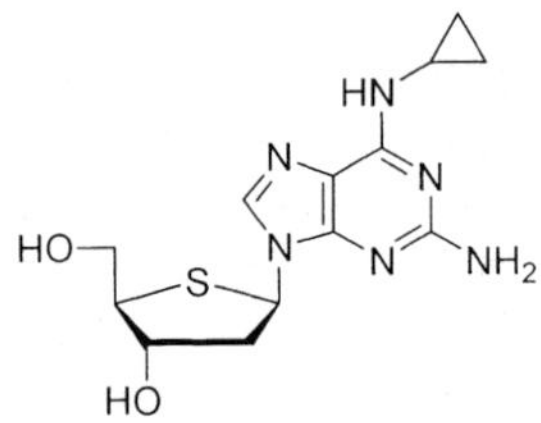

15

Figure 4. Biologically active 2'-deoxy-4'-thionucleosides.

The racemic carbocyclic analogs of several 2'-deoxyribofuranosides are also active against HSV-1 and HSV-2 replication in cell culture. Among them, the carbocyclic analogs of IdU (**16**, C-IdU) and BVdU (**17**, C-BVdU) show similar selectivity and potency to their parent compounds (Figure 5).[29] Racemic C-BVdU and its analogs as well as C-IdU are equally selective, albeit slightly less potent in their antiherpes action than their parent compounds. Although resistant to degradation by pyrimidine nucleoside phosphorylases, C-BVdU is no more effective than BVdU in systemic (oral, intraperitoneal) or topical treatment of HSV-1 infections in mice. However, both (-)- and (+)-enantiomers of C-IdU (**18**) and C-BVdU (**19**) are active against HSV-1, which indicates that both may act as substrates for HSV-1 TK.[30]

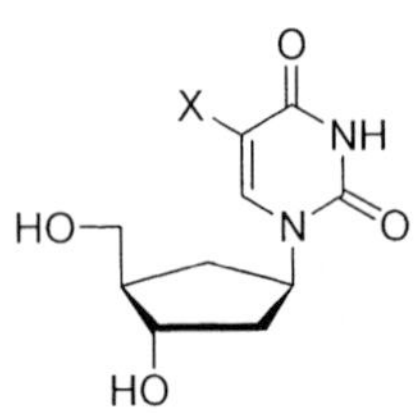

16 (C-IdU, X = I)
17 [C-BVdU, X = (*E*)-bromovinyl]

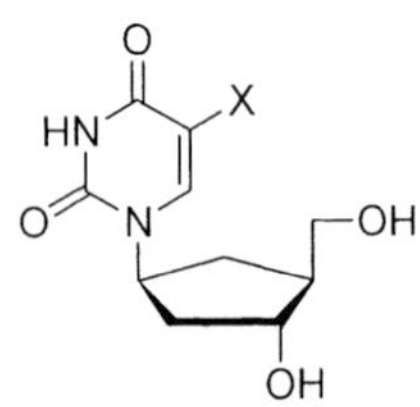

18 (L-C-IdU, X = I)
19 [L-C-BVdU, X = (*E*)-bromovinyl]

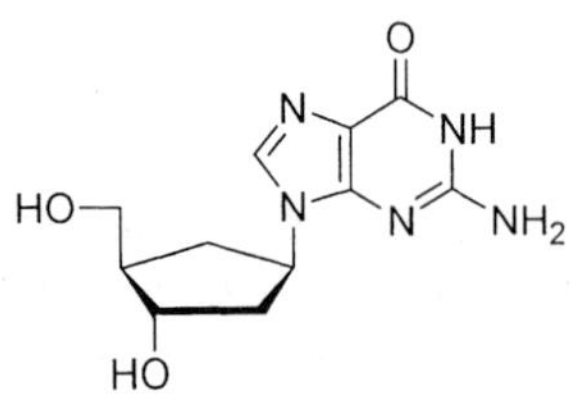

20 (CdG)

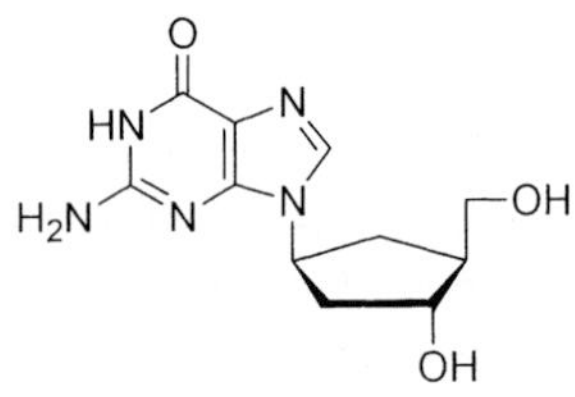

21 (L-CdG)

Figure 5. Biologically active carbocyclic 2'-deoxy nucleosides.

Among purine analogs, (±)-carbocyclic 2'-deoxyguanine (2'-CdG, **20**) shows the most potent antiherpetic activity.[31] Secrist *et al.* resolved the enantiomers, and reported that D-2'-CdG is as active and potent as (±)-2'-CdG against HSV-1 and HSV-2, whereas L-2'-CdG displays only modest activity against HSV-1.[32] According to the metabolic study of (±)-2'-CdG, D-2'-CdG (**20**) and L-2'-CdG (**21**, Figure 5), D-2'-CdG seems to be a good substrate for the virus-encoded kinase and a very poor substrate for cellular phosphorylating enzymes.[33] Besides, both D- and L-2'-CdG are phosphorylated by dCyd kinase from MOLT-4 cells, 5'-nucleotidase from Hep-2 cells, and mitochondrial deoxyguanosine (mt-dGua) kinase from human platelets and CEM cells.[34] For both dCyd kinase and mt-dGua kinase, L-CdG is a better substrate with K_M values of 0.63 and 4.9 mM, respectively *vs.* 1.98 and 1.2 mM for D-CdG. In the case of 5'-nucleotidase, D-CdG is a better substrate with a V_{max}/K_M value of 0.02 for L-CdG and 0.05 for D-CdG.[34] In addition, D-CdG shows a 50% inhibition of HBV DNA polymerase activity at 5 ng/mL in 2.2.15 cells, and at 25 ng/mL the complete disappearance of HBV replication has been observed.[35] D-CdG is phosphorylated to its triphosphate (although the exact identity of the enzymes responsible for this phosphorylation is not clear), which can be efficiently incorporated into HBV DNA. D-CdGTP is a competitive inhibitor of dGTP for both HBV DNA polymerase and eukaryotic DNA polymerase δ, with a 6-fold lower K_i for the viral enzyme.[36] Unfortunately, D-CdG is toxic with a 50% inhibition of cell growth (HepG2 2.2.15 cells at 32 μM).[37]

Replacement of the oxygen with an ethenyl group produces compounds with potent and selective anti-HBV activity in 2.2.15 cells.[38] Entecavir (BMS-200475, **22**), originally synthesized as an anti-herpesvirus agent, displayed also moderate activity against HSV-1, HSV-2 and VZV (Figure 6).

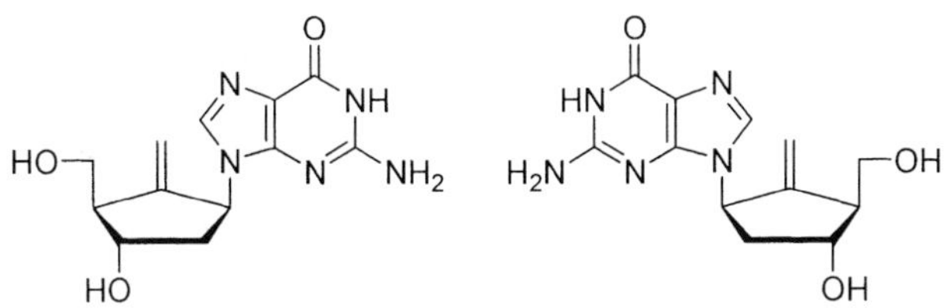

22 (entecavir) **23**

Figure 6. BMS-200475 and its L-enantiomer.

Activity was also seen with HCMV, a herpes virus lacking TK, but no activity was detected against RNA viruses such as HIV or influenza. Further studies have established that entecavir is one of the most potent anti-HBV nucleosides discovered *in vitro* as well as *in vivo*, (EC_{50} 3 nM, IC_{50} 30 μM) in 2.2.15 cells. Treatment with entecavir results in no apparent inhibitory effects on mt-DNA content.[34] Furthermore, daily oral treatment at doses ranging from 0.02 to 0.5 mg/kg of body weight for 1 to 3 months effectively reduces the level of woodchuck hepatitis virus (WHV) viremia in chronically infected woodchucks as measured by reductions in serum WHV DNA levels and endogenous hepadnaviral polymerase activity. However, WHV viremia in BMS-200475-treated WHV carriers eventually returns to pretreatment levels after therapy is discontinued.[39] *In vitro* biochemical studies indicate that entecavir can be efficiently phosphorylated by cellular enzymes to its triphosphate, which is a potent inhibitor of HBV DNA polymerase, inhibiting both priming and elongation steps of HBV DNA replication.[40] The enantiomer (**23**) of entecavir as well as the adenine, thymine and iodouracil analogs, are much less active against HBV.[34] Entecavir is currently undergoing clinical trials as an anti-HBV agent.

9-β-D-Arabinofuranosyl adenine (**24**, ara-A, vidarabine) has been known to have significant antiviral activity *in vitro* against herpes and vaccinia virus[41] and is also a potent inhibitor of HBV DNA polymerase (Figure 7).[42] Due to its low water solubility, its 5'-monophosphate (ara-AMP) is administered intramuscularly, and ara-AMP has been extensively studied for treating chronic HBV infections in humans.[5e] Although a 8-week treatment has been shown to effect the loss of HbeAg and HBV DNA, in many cases serious neurotoxicity is evident after 4 weeks. Ara-A is phosphorylated by cellular enzymes to its triphosphate, which interfere with viral nucleic acid replication. Unfortunately, vidarabine is deaminated rapidly by adenine deaminase to arabinosyl hypoxanthine, which has weak antiviral activity.[5e] Its carbocyclic analog, cyclaradine (**25**), synthesized in efforts to develop deaminase-resistant ara-A derivatives, exhibits significant anti-HSV-1 and anti-HSV-2 activity.[3a]

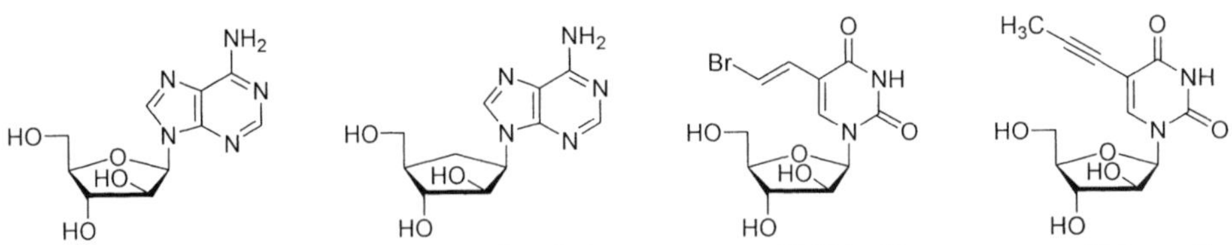

24 (ara-A) **25** (cyclaradine) **26** (BVaraU) **27** (zonavir)

Figure 7. Biologically interesting arabinofuranosyl nucleosides.

Substitution of an arabinofuranose for the ribose moiety of BVdU leads to BVaraU (**26**), the most potent anti-VZV agent discovered so far which, for this reason, had been registered in Japan for the treatment of herpes zoster (shingles).[43] However, several patients who had been treated with BVaraU along with 5-fluorouracil (5-FU) died because of the drugs interaction. In fact, 5-bromovinyluracil, released by phosphorolytic cleavage of the glycosylic bond, is a potent inhibitor of dihydropyrimidine dehydrogenase, whose inhibition results in elevating 5-FU to lethal levels.[44] The L-enantiomer of BVaraU does not exhibit antiviral activity against any herpes viruses, including VZV.[20,45]

Introduction of an alkenyl or alkynyl group at 5-position of the uracil base also produces potent anti-herpetic activity.[46] β-D-Arabinofuranosyl 5-propynyluracil (**27**, zonavir) is a good substrate for viral kinases, particularly for VZV TK, which converts it to the monophosphate whereas cellular cytosolic thymidine kinase is not effective.[47]

The monophosphate is then specifically converted to the diphosphate by the thymidylate kinase activity of VZV TK. The triphosphate of zonavir is a potent inhibitor of the VZV-specific DNA polymerase and this inhibition is probably the major mechanism of the antiviral activity.

4'-Thio derivatives of BVaraU and related analogs have selective antiviral activities against HSV-1, HSV-2 and VZV, but not superior to the 4'-oxo nucleosides.[48] 4'-Thioarabinofuranosyl guanine and diaminopurine have the most potent anti-HCMV and anti-proliferative activities, whereas arabinosyl guanine and diaminopurine show only marginal antiviral activity.[48a] The L-enantiomer of 4'-thioarabinofuranosyl cytosine does not exhibit significant antiviral activity.[48b]

Introduction of a fluorine atom at the 2'-position of nucleoside analogs has produced a variety of interesting antiviral agents (Figure 8). 2'-Fluoro-β-D-arabinofuranosyl pyrimidine nucleosides are potent agents against herpes virus.[49] 1-β-D-Arabinofuranosyl-5-iodocytosine (FIAC) and 1-β-D-arabinofuranosyl-5-iodouridine (FIAU, **28**) are phosphorylated in HSV-1 infected cells by virus-encoded TKs.[50] The 5-alkenyl analogs were also found to be active against HSV-1, HSV-2 and VZV.[49c] Furthermore, FIAC, FIAU, FMAU (**29**) and FEAU (**30**) have significant anti-HBV activity.[5e] Studies in HepG2 cells indicate that FIAU is activated by host cell enzymes including cellular TK, thymidylate kinase and pyrimidine diphosphate kinase.

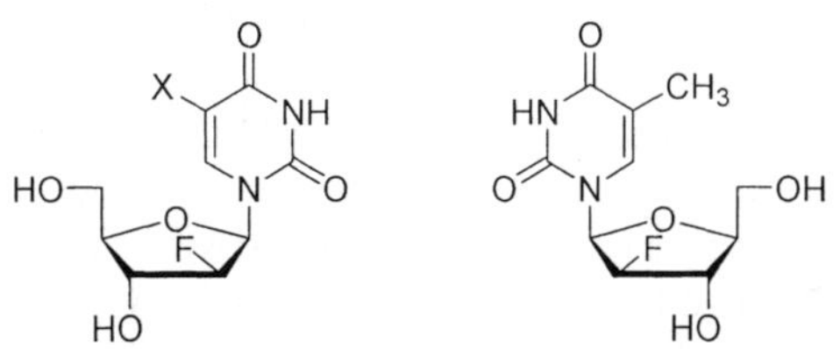

28 (FIAU, X = I)
29 (FMAU, X = CH_3)
30 (FEAU, X = Et)

31 (L-FMAU)

Figure 8. 2'-Deoxy-2'-fluoro-arabinofuranosyl nucleosides.

Unfortunately, problems associated with the toxicity of potential therapeutic compounds have been demonstrated by the results of the FMAU[51] and FIAU (fialuridine) clinical trials (Figure 9).[6,52] Although initial trials in humans showed very good efficacy in terms of reducing the plasma levels of HBV as measured by viral DNA concentrations or viral polymerase activity, longer trials, in which the duration of drug treatment was extended, had to be curtailed when serious toxic effects became apparent.[6] These included myopathy, lactic acidosis, peripheral neuropathy, pancreatitis and liver failure, and the severity of the toxic effects was such that several patients died.[6] The primary cause of this delayed toxicity is due to the incorporation of the drug into mitochondrial DNA (mt-DNA), which causes damage to the mitochondrial function.[53] Studies in HepG2 2.2.15 cells indicate that FIAU is activated by host cell enzymes including cellular TK, thymidylate kinase and pyrimidine diphosphate kinase to FIAUTP, which inhibits the viral DNA polymerase. However, FIAUTP is also efficiently used as a substrate by DNA polymerase γ, which incorporates it into mt-DNA, causing disruption in the replication of DNA, resulting in either decreased production of proteins or the production of defective proteins.[53]

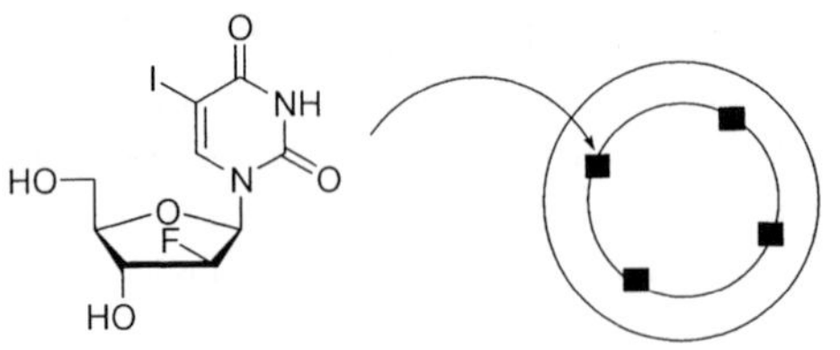

Figure 9. Mitochondrial toxicity caused by FIAU: internalization of FIAU into mt-DNA, which cannot be repaired by exonucleases.[53]

Since the broad spectrum of biological activity of 2'-F-arabinofuranosyl nucleosides was discovered, a number of structural modifications of these analogs have been carried out. Chu and co-workers have demonstrated that the enantiomer of FMAU, L-FMAU (clevudine, **31**, Figure 8) has potent anti-HBV as well as anti-EBV activity.[12] Most importantly, L-FMAU has low cytotoxicity in a variety of cell lines, including MT2, CEM, H1 and 2.2.15 and bone marrow progenitor cells. L-FMAU is phosphorylated stepwise to L-FMAUMP, L-FMAUDP and L-FMAUTP in 2.2.15 cells by cytosolic TK, dCyd kinase or mt-dPyd kinase, respectively (Figure 10), acting as a potent inhibitor of HBV DNA polymerase.[54] However, it is not utilized as a substrate by human DNA polymerase α, β, γ or δ. In addition, L-FMAU exhibits potent anti-EBV activity. The metabolic studies suggest that EBV-specific TK in H1 cells can phosphorylate L-FMAU to its mono, di- and triphosphates.[54] Interestingly, L-FMAUTP is not a substrate for HBV or EBV DNA polymerases unlike other antiviral nucleosides, which suggests that the anti-HBV and anti-EBV activity of L-FMAU may not be due to its incorporation into HBV and EBV DNA.[54] Currently undergoing clinical trials against chronic hepatitis B virus infection, L-FMAU has been found to be one of the most potent anti-HBV agents so far in woodchuck as well as in humans, since no significant viral rebound was observed in woodchucks or humans.

The 4'-thio substitution of 2'-fluoro nucleosides also retains their antiviral activity (Figure 11). Machida *et al.* have reported that 2'-fluoro-4'-thioarabinofuranosyl nucleosides (4'-thio-F-araNs) are active against HSV-1, HSV-2, VZV and HCMV.[48] 4'-Thio-F-araG (**32**) and 4'-thio-F-araDAP (**33**) have particularly potent activity against all herpes viruses tested, equipotent to arabinosyl guanine and diaminopurine. These compounds also have a 6-fold lower EC_{50} value than ganciclovir against clinical isolates of HCMV. In addition, 4'-thio-F-araA (**34**) shows biological activities similar to that of araA. 2'-Fluoro-5-methyl-4'-thio-β-L-arabinofuranosyluracil (**35**, L-SFMAU), the enantiomer of 4'-thio-FMAU (**36**), also has moderate activity against HSV-1 and HSV-2.[55]

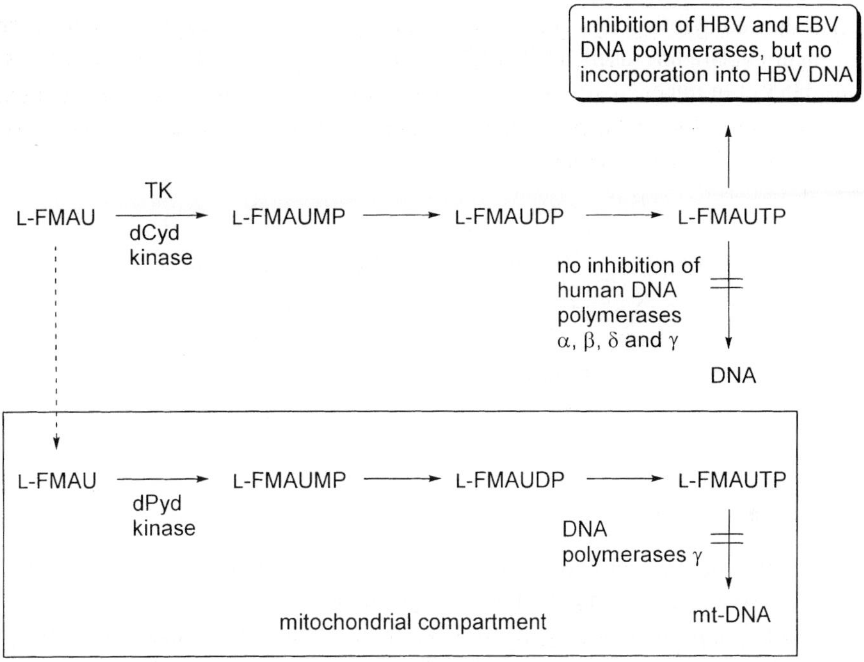

Figure 10. Proposed metabolism of L-FMAU.[54]

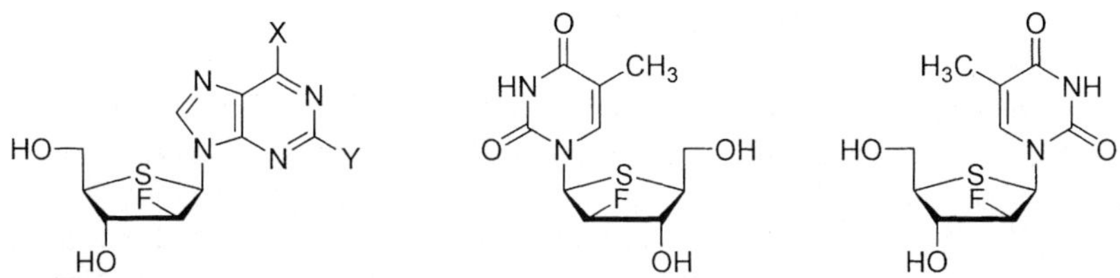

32 (S-FaraG, X = OH, Y = NH_2)
33 (S-FaraDAP, X = Y = NH_2)
34 (S-FaraA, X = NH_2, Y = H)

35 (L-S-FMAU)

36 (S-FMAU)

Figure 11. 2'-Deoxy-2'-fluoro-4'-thioarabinofuranosyl nucleosides.

Among carbocyclic derivatives of FMAU (Figure 12), C-FMAU (**37**) showed moderate anti-HSV-1 activity although it is less active than FMAU.[56] The 2'-ara-fluoroguanosine derivative[57] is potent against HSV-1 and HSV-2 and poorly active against VZV, but it has also showed cytotoxicity. The 2’-fluoro analog of cyclaradine (**38**) is 10 times more active than cyclaradine itself against HSV-1 and HSV-2, and more active than ACV against HSV-2 in mice.[58]

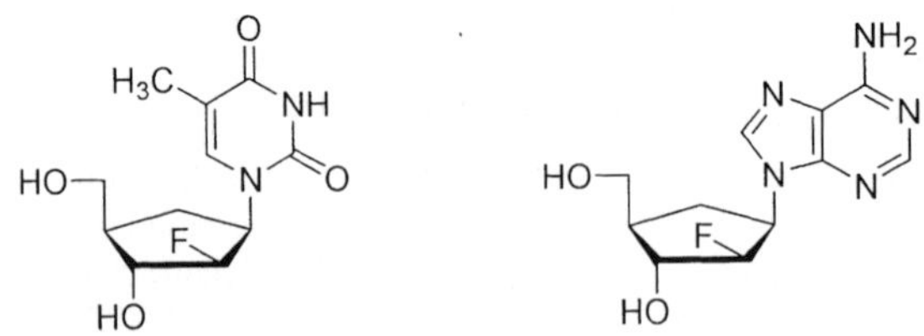

37 (C-FMAU) **38** (C-F-araA)

Figure 12. Carbocyclic 2’-deoxy’2’-fluoro-arabinofuranosyl nucleosides.

Introduction of geminal fluorine atoms at the 2’-position has resulted in the discovery of 2’-deoxy-2’,2’-difluorocytidine (**39**, gemcitabine),[59] which has been approved by the FDA for the treatment of pancreatic cancer (Figure 13). Gemcitabine shows a complex mechanism of action, inhibiting the synthesis of DNA and RNA as well as inhibiting ribonucleotide reductase.[60] Its guanosine analog also shows similar activity.[61] Among the series of L-enantiomers of gemcitabine including L-gemcitabine (**40**), only the adenine analog shows marginal anti-HIV-1 activity without cytotoxicity (EC_{50} 3.4 μM in PBM cells).[62] The 4’-thio analog of gemcitabine (**41**, 4’-thiogemcitabine)[63] also shows moderate antineoplastic activity. However, its enantiomer, L-4’-thiogemcitabine (**42**), shows neither antiviral nor antitumor activity against a panel of five different tumor cell lines.[64]

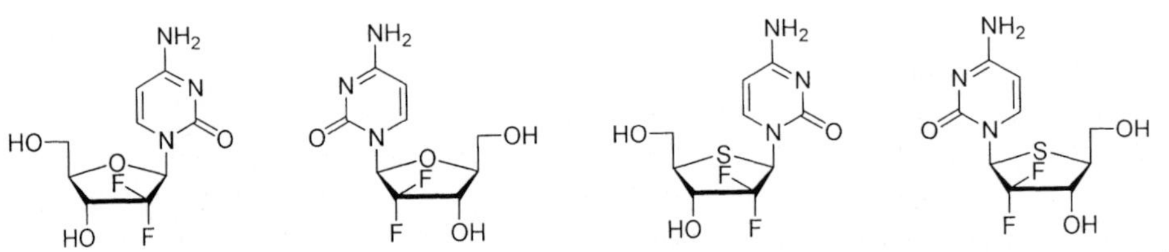

39 (gemcitabine) **40** (L-gemcitabine) **40** (S-gemcitabine) **42** (L-S-gemcitabine)

Figure 13. 2’-Deoxy-2’-fluorocytidine derivatives.

The methylidene substitution of 2’-deoxycytidine provides 2’-deoxy-2’-methylidenecytidine (**43**, DMDC, Figure 14),[65] endowed with anti-neoplastic activity against several solid tumor cell lines as well as leukemia. DMDC is resistant to cytidine deaminase[66] and its diphosphate is a potent inhibitor of ribonucleotide reductase.[67] DMDC and 2'-deoxy-2'-methylidene-5-fluorocytidine (DMDFC) (**44**) are potent inhibitors of HSV-1, HSV-2, VZV and HCMV with significant anti-proliferative activity.[68] The (*E*)-5-(2-bromovinyl)uracil analog (BV-DMDU) has moderate antiviral activity against HSV-1, HSV-2, VZV and HCMV. Among the 2'-deoxy-2'-methylidene pyrimidine nucleoside analogs, BV-DMDU showed the most potent and selective anti-VZV activity, which was more potent than ACV, but less active than BVaraU.[68]

Also the 4'-thio analog of DMDC (**45**, Figure 14) has potent antineoplastic properties *in vitro* with an IC_{50} value of 0.0091 and 0.12 μg/mL in CCRF-HSB-2 and KB cells, respectively.[63] Its enantiomer, L-4'-thio-DMDC (S-DMDC, **46**), does not show antitumor activity against different cell lines.[64] In addition, various 5-substituted 4'-thio-DMDUs show potent anti-HSV-1 activity (EC_{50} 0.016-0.096 μg/mL). 5-Ethyl- and 5-iodo-4'-thio-DMDUs are also active against HSV-2 (EC_{50} 0.17 and 0.86 μg/mL, respectively), and 5-bromovinyl-4'-thio-DMDU has efficacy against VZV with an EC_{50} value of 0.013 μg/mL without significant cytotoxicity.[69]

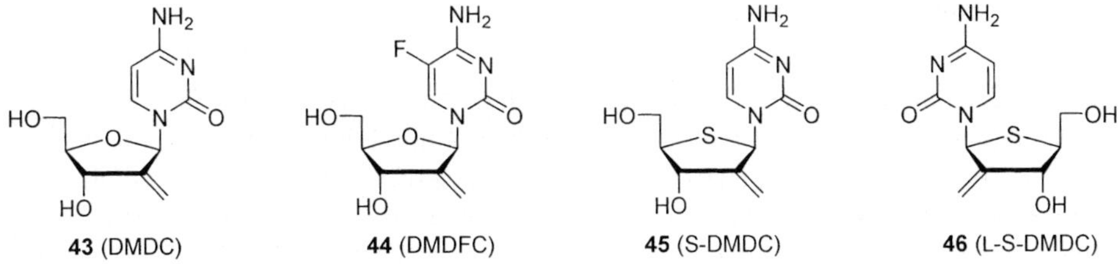

Figure 14. 2'-Deoxy-2'-methylidenecytidine derivatives.

Introduction of a functional group at the 4'-α-position of 2'-deoxynucleosides elicits potent antiviral activity. For example, 4'-azido analogs exhibit potent anti-HIV activity in A3.01 cell cultures with EC_{50} values of 0.003-0.8 μM.[70] The guanine analog is the most potent compound, but it is also cytotoxic. Further evaluations of 4'-azidothymidine (**47**, ADRT) in H9, PBL and MT-2 cells infected with HIV have demonstrated a similar inhibitory profile to that of AZT. Interestingly, ADRT retains its activity against HIV mutants that are resistant to AZT. In the 4'-methoxy series, adenosine, thymidine and guanosine analogs are also inhibitors of HIV, but 2-3 orders of magnitude less active than their azido counterparts.[70] A metabolic study has revealed that ADRT is not a substrate for thymidine phosphorylase, but is metabolized by kinases (Figure 15).[71] Thymidine kinase (TK) phosphorylates ADRT to its monophosphate with a K_i value of 5.2 μM, and a K_M value of 8.3 μM, in comparison to a K_M value of 0.7 μM for thymidine. ADRTMP has a low affinity toward thymidylate kinase and thymidylate synthase, which suggests that ADRT can be activated effectively by other cellular kinases without significant interference of normal thymidine metabolism. In cultured human lymphocytes (A3.01, H9 and U937 cells), ADRT is phosphorylated efficiently to ADRTTP, which is a poor competitive inhibitor against dTTP toward DNA polymerases α and β with K_i values of 62.5 and 150 μM, respectively. However, ADRTMP is incorporated into cellular DNA, which can lead to mutations.[72] Because of these toxicity issues, the development of ADRT as an anti-HIV agent was discontinued.

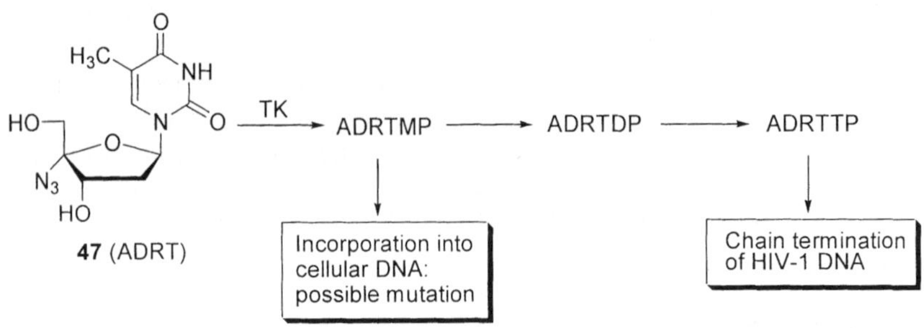

Figure 15. Metabolism of 4'-azidothymidine (ADRT).[71]

4'-α-Substituted 2'-deoxycytidines also exhibit potent anti-HIV-1 activity (Figure 16), although they are cytotoxic to the host cells (MT-4).[73] In the series, the cyano derivative (**48**) is the most potent against HIV-1 with an EC_{50} value of 1.2 nM in MT-4 cells followed by ethynyl, ethenyl, ethyl, methyl and chloroethenyl derivatives. The methyl and cyano derivatives show moderate activity against HSV-1 and HSV-2. 2'-Deoxy-4'-*C*-methyl pyrimidine nucleosides show potent anti-HSV-1 and anti-VZV activity *in vitro*.[74] 4'-Methyl-BVdU (**49**, Figure 16) displays more potent activity than BVdU against VZV, and it is cytotoxic against human T-cell leukemia, CCRF-HSB-2. The *arabino* analog, 4'-methyl-BVaraU, shows a weaker antiviral activity than that of 4'-methyl-BVdU without any cytotoxicity. The 4'-hydroxy *C*-deoxyguanosine analog (**50**, 4'-OH-CdG) has antiviral activity equivalent to ACV against HSV-1 and HSV-2, but is inactive against VZV.[3b,75]

4'-*C*-Ethynyl-substituted 2'-deoxyribonucleosides (Figure 16) have shown impressive anti-HIV activity, with EC_{50} in the nano- or subnanomolar range, but many also display severe toxicity, with IC_{50} in the micromolare range.[76] Thus, the diaminopurine (**51**), guanosine (**52**) and cytosine (**53**) derivatives show an EC_{50} of 0.3, 1.4 and 4.8 nM, respectively, and an IC_{50} of 0.82, 1.5 and 0.92 μM, respectively, in MT-4 cells. The thymidine (**54**), 5-bromouridine (**55**), 5-iodouridine (**56**), 5-fluorocytidine (**57**) and guanidine (**58**) analogs, although less active, do not show toxicity in MT-4 cells, with the best therapeutic index found in the cytidine analog **54** (EC_{50} 0.030 μM, IC_{50} >100 μM, TI >3333).[76]

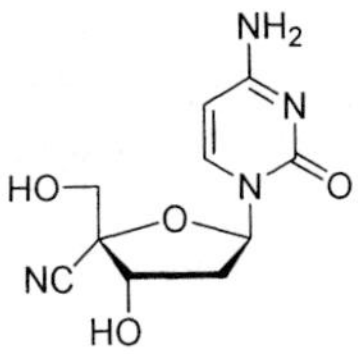

48

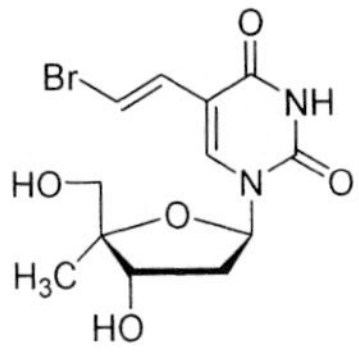

49

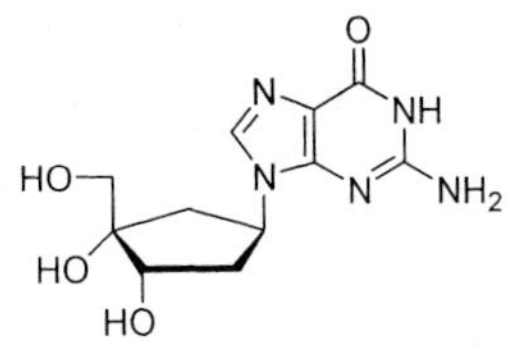

50 (4-OH-CdG)

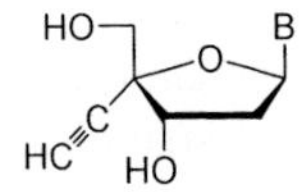

51 (B = 2,6-di-aminopurine)
52 (B = guanosine)
53 (B = cytosine)
54 (B = thymine)
55 (B = 5-Br-uracil)
56 (B = 5-I-uracil)
57 (B = 5-F-cytosine)
58 (B = guanine)

Figure 16. 4'-Substituted-2'-deoxy nucleoside analogs.

Marquez and co-workers have synthesized conformationally locked nucleosides with bicyclo[3.1.0]hexane templates to investigate the correlation between puckering of the sugar ring and biological activity (Figure 17).[77] The antiherpetic activity of these nucleosides is associated with the northern conformation of the thymidine analog, (N)-methanocarba-T (**59**), which is more active than ACV against HSV-1 and HSV-2 with EC_{50} values of 0.03 and 0.09 μg/mL, respectively. The cytosine analog has anti-HSV-1 activity (EC_{50} 0.14 μg/mL) and the adenine analog is active against HCMV. On the other hand, the southern conformation of the thymidine analog, (S)-methanocarba-T (**60**) does not show any antiherpetic activities. Unlike other nucleoside analogs, it seems that the diphosphorylation of **59** in HSV-1-infected cells is the rate-limiting step in the activation to the active triphosphate.[78]

Conformationally restricted nucleosides have also been synthesized by Chu and co-workers, who reported the complete D- and L-series of 2',3'-dideoxy-2',3'-*endo*-methylene nucleosides (**61** and **62**, respectively).[79] None of the synthesized compounds, however, showed anti-HIV activity.

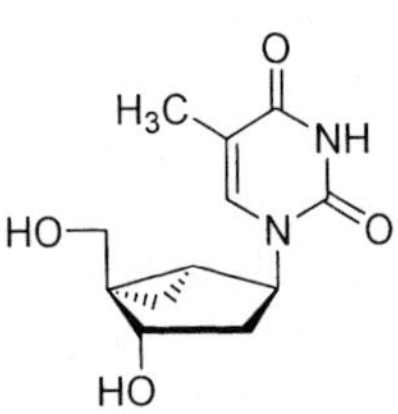

59
[(N)-methanocarba-T]

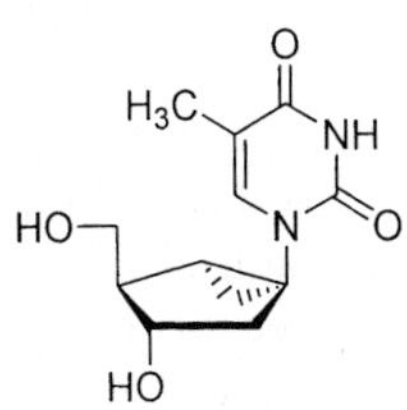

60
[(S)-methanocarba-T]

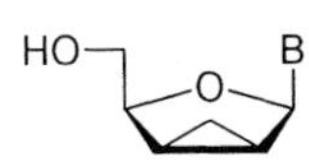

61

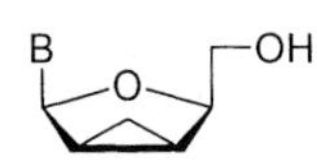

62

Figure 17. Conformationally locked carbocyclic thymidine analogs.

Computer-based conformational studies and biological evaluation of cyclohexenyl nucleosides have demonstrated that a cyclohexene and a furanose ring can be considered as bioisosters.[80,81] Thus, both enantiomers of cyclohexenyl-G (**63** and **64**, Figure 18) show potent antiviral activity against HSV-1, HSV-2, VZV, HCMV and HBV. The antiviral activities of the two isomers are comparable, although the D-isomer is slightly more potent in all the tested systems. The fact that both isomers show reduced activity in TK$^-$ HSV-1 suggests that intracellular phosphorylation plays an important role in the bioactivation of these compounds.[80,81]

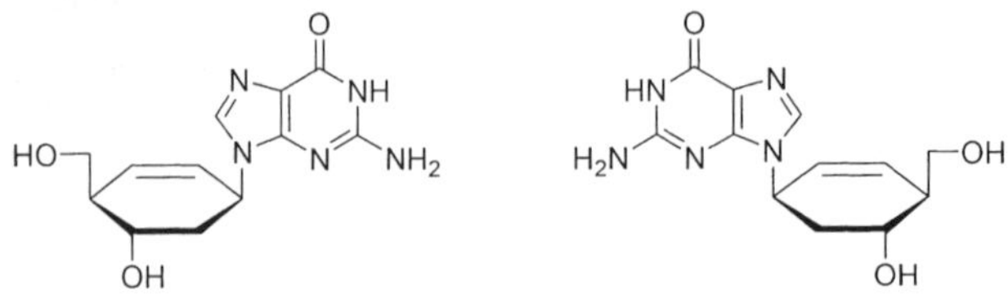

63
D-cyclohexenyl-G

64
L-cyclohexenyl-G

Figure 18. D- And L- cyclohexenyl-G.

1.4. 2',3'-Dideoxy nucleosides and related analog

The discovery of dideoxynucleosides, such as 2',3'-dideoxycytidine (ddC),[82] 2',3'-dideoxyinosine (ddI)[82] and 3'-azido-3'-deoxythymidine (AZT)[83] as potential therapeutic agents for the treatment of acquired immunodeficiency syndrome (AIDS) has triggered an extensive development of this class of compounds to identify active anti-HIV agents, which inhibit the virus-associated RT reaction by terminating DNA chain elongation. A number of selective HIV-1 inhibitors have now been approved for the treatment of HIV infections, such as AZT, ddC, ddI, d4T,[84] 3TC,[85] abacavir[86] and tenofovir disoproxil.[87,88,89,90] Some of these, particularly 3TC, are also active against HBV. Given that the DNA polymerase of HBV is also a reverse transcriptase, it is not surprising that such degree of overlap exists as a chain terminator of DNA synthesis. The finding that (±)-dioxolane-thymine[91] and (±)-BCH-189[92] (see below) are potent anti-HIV agents and that the L-isomer of BCH-189 is more potent and less toxic than its D-isomer have opened the new era of L-nucleosides. Since then, a number of L-nucleoside analogs have been synthesized and biologically evaluated, and the importance of chirality and its influence on the antiviral activity of the L-nucleosides has been recognized.[2f,2i]

ddI (**65**, Figure 19) is a potent and selective anti-HIV agent in ATH8 cells.[82] It is phosphorylated to ddIMP by cytosolic 5'-nucleotidase, then aminated to ddAMP by adenylosuccinate synthase/lyase enzymes, and converted to ddADP and ddATP by cellular nucleotide kinases.[93] ddATP is the active agent against HIV-RT.[94] Peripheral neuropathy and pancreatitis are the major side effects of ddI.

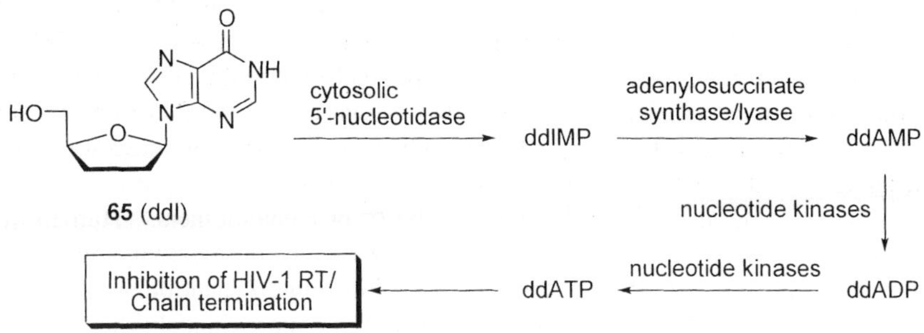

65 (ddI)

Figure 19. Metabolism of the anti-HIV-1 agent ddI (didanosine).[93,94]

In various cell lines, ddC (**66**, Figure 20) is phosphorylated to ddCMP, ddCDP and ddCTP by dCyd kinase, CMP/dCMP kinase and NDP kinase, respectively.[5f] The affinity of ddCTP for DNA polymerase α is poor, and intermediate for DNA polymerase β and high for DNA polymerase γ.[95a] ddC exerts delayed cytotoxicity and reduces the cellular content of mt-DNA[95b] and, at higher concentrations, causes a delayed distortion of mitochondrial ultrastructure.[96] It also exhibits a significant inhibitory effect on the replication of HBV DNA.

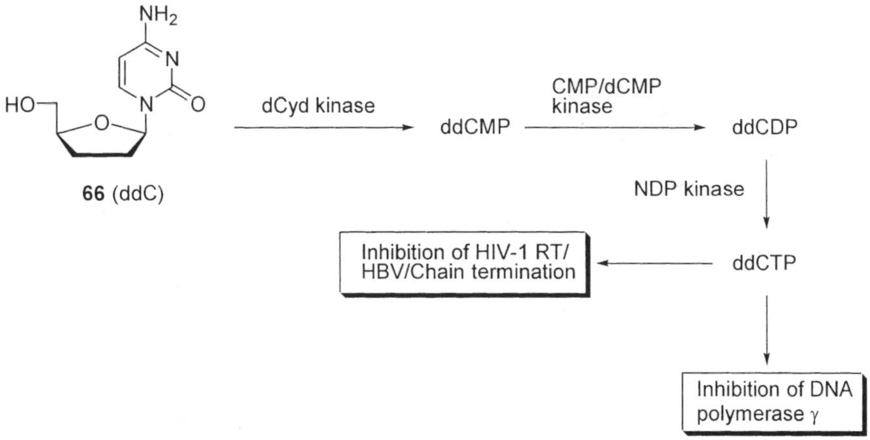

66 (ddC)

Figure 20. Metabolism of the anti-HIV-1 agent ddC (zalcitabine).

A number of L-2',3'-dideoxy nucleosides also show moderate to potent anti-HIV and anti-HBV activities (Figure 21).[45,97,98] Among these analogs, β-L-2',3'-dideoxy-5-fluorocytidine (**67**, L-FddC) is the most active against HIV-1, approximately 3- to 4-fold more potent than ddC *in vitro*.[97] In addition, L-FddC and L-ddC (**68**) are potent anti-HBV agents with an EC_{50} value of 0.01 μM without any toxicity up to 100 μM against host mt-DNA synthesis. L-ddA shows moderate anti-HIV-1 and anti-HBV activity in PBM and 2.2.15 cells, respectively.[98] An enzymatic study of D-ddA (**69**) and L-ddA (**70**) has been performed with respect to adenosine kinase, dCyd kinase, adenosine deaminase (ADA) and purine nucleoside phosphorylase (Figure 22).[98] Adenosine deaminase was strictly enantioselective and favored D-ddA, whereas adenosine kinase and purine nucleoside phosphorylase had no apparent preference for the D- or L-enantiomers.[98] Human dCyd kinase showed a remarkable inversion of the expected enantioselectivity, with L-ddA having better substrate efficiencies than its corresponding D-enantiomer.[98] β-2',3'-L-Dideoxy-5-azacytidine shows potent anti-HIV activity at approximately the same level as ddC. However, unlike ddC, it has no antiviral activity against HBV.

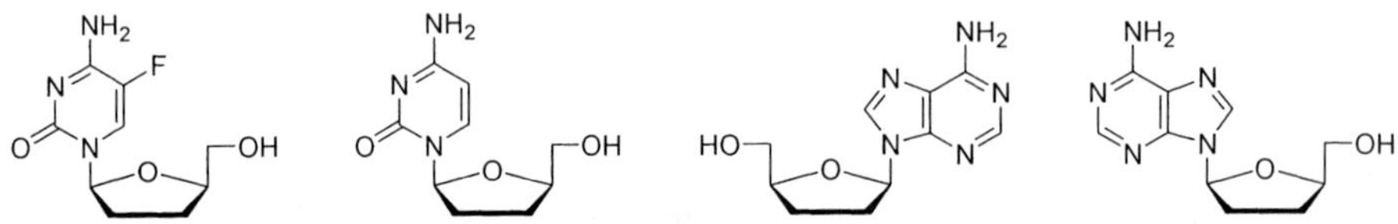

67 (L-FddC) **68** (L-ddC) **69** (ddA) **70** (L-ddA)

Figure 21. 2',3'-Dideoxy nucleosides with antiviral activity.

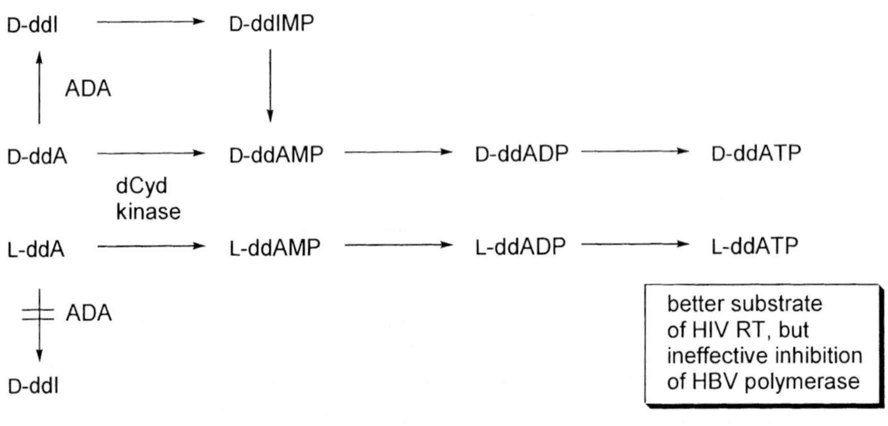

Figure 22. Comparative metabolism of D-ddA and L-ddA.[98]

Replacement of a 4'-oxygen atom of both D- and L-2',3'-dideoxynucleosides with a methylene group, a sulfur atom or amino groups fails to elicit any significant antiviral activity except for 2',3'-dideoxy-4'-thiocytidine (**71**) (Figure 23), which displays modest activity *in vitro* against HIV (EC_{50} 1.0 and 38 μg/mL in CEM and MT-2 cells, respectively).[99]

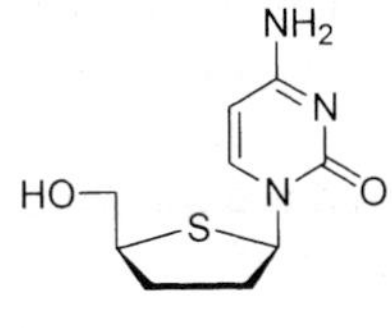

71 (S-ddC)

Figure 23. 4'-Thio-ddC (S-ddC).

Among the 3'-azido-2',3'-dideoxy nucleosides, AZT (**72**, Figure 24), which is the first anti-HIV-1 agent approved for use against HIV-1, has remained one of the most potent and selective anti-HIV agents.[83] The uracil analog of AZT (**73**, AZdU)[100] and the 5-methylcytosine analog (**74**, AZdMeC)[100] are less potent but also less toxic than AZT. The guanine (AZdG) and diaminopurine (AZdDAP) analogs of AZT have also been reported to be more effective against HIV than ddA.[101,102]

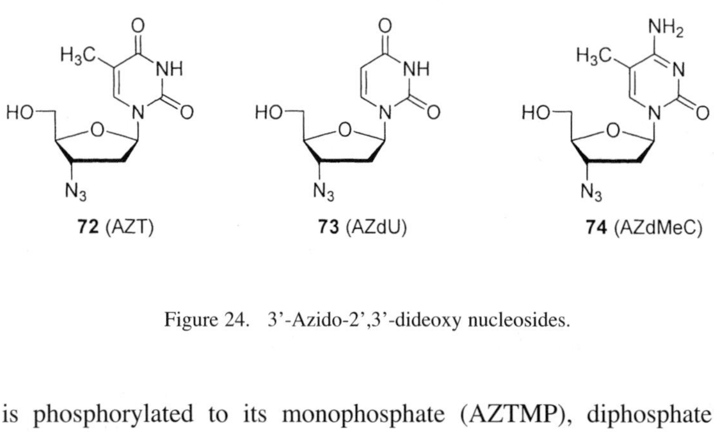

72 (AZT) **73** (AZdU) **74** (AZdMeC)

Figure 24. 3'-Azido-2',3'-dideoxy nucleosides.

AZT is phosphorylated to its monophosphate (AZTMP), diphosphate (AZTDP) and triphosphate (AZTTP) by TK, TmdK and NDP kinase, respectively (Figure 25).[5f] AZTTP acts as a competitive inhibitor or alternate substrate of HIV RT leading to viral DNA chain termination and has much less affinity to cellular DNA polymerase α.[103,104] In the presence of 2 μM AZTTP, the activities of HIV-RT and cellular DNA polymerase γ were inhibited by more than 80 and 90%, respectively.[105] However, AZT is associated with several toxicities, particularly bone marrow suppression including anemia and leukopenia. The hematopoietic toxicity of AZT is generally due to high intracellular levels of AZTMP.[106] Moreover, a number of AZT-resistant HIV strains have been isolated from AIDS patients, which stimulated an extensive search for new anti-AIDS drugs.[3a,107,108]

Among "unnatural" enantiomers of this class of compounds, L-AZT is about 10,000 times less active than its D-counterpart.[109] The finding that L-AZTTP inhibits HIV reverse transcriptase (RT), as well as HBV DNA polymerase, at sub-micromolar concentration, suggests that L-AZT is devoid of antiviral activity because it is not efficiently phosphorylated intracellularly in lymphocytes or hepatocytes.[110]

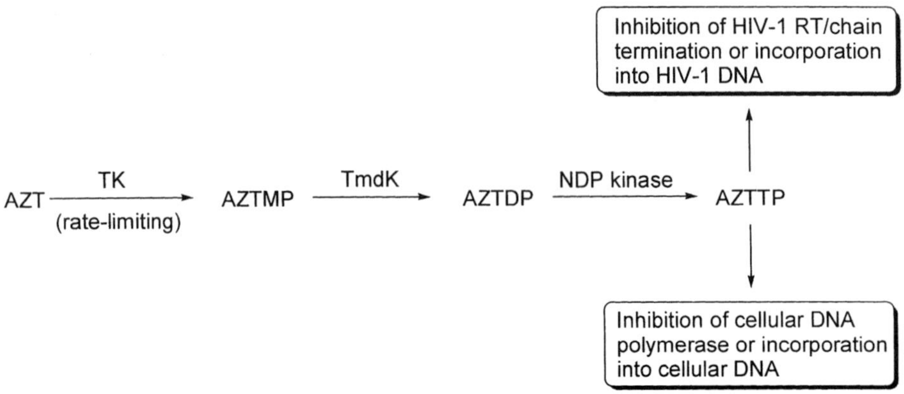

Figure 25. Cellular metabolism of AZT.[103]

AZdU was the first nucleoside analog with a uracil base found to have anti-HIV activity at submicromolar concentrations. It is phosphorylated to AZdUTP, which acts as both an HIV-RT inhibitor and a proviral DNA chain terminator.[5b,111,112] Although its anti-HIV activity is less potent than that of AZT, its toxicity on bone marrow cells is also significantly lower than that of AZT.[113] In human PBM cells, AZdUMP is the predominant intracellular metabolite and levels of AZdUMP are two orders of magnitude greater than AZdUTP. AZdUMP is also converted to its 5'-*O*-diphosphohexose and 5'-*O*-diphospho-*N*-acetyl-glucosamine. This unique metabolism may explain the lower toxicity of AZdU.[114] However, AZdU shows cross-resistance with AZT-resistant HIV virus.[5b] The clinical trials of AZdU were discontinued due to its extensive metabolism to 5'-glururonide.

AZdMeC shows potent anti-HIV activity in human PBM cells and macrophages and is less toxic than AZT in human bone marrow cells.[5b,115] Metabolic studies indicate that this compound is slowly converted intracellularly to AZT.[115] The major metabolite of AZdMeC is AZTMP with no formation of AZdMeCMP. The low toxicity of this compound is related to the lack of formation of AZTTP in human bone marrow cells. AZdMeCTP efficiently inhibits HIV-RT, competing with dCTP while binds to human DNA polymerase α with much lower affinity (< 6000-fold).[5f] AzdMeC is deaminated to AZT in monkeys, but not in humans.

Additional modifications of the sugar moiety of AZT have failed to produce any compounds with antiviral activity, including 2'-β-fluoro-AZT,[116] 4'-thio-AZT[117] and carbocyclic AZT.[118]

Conformationally locked AZT analogs, (N)-*methano*-carba-AZT (**75**) and (S)-*methano*-carba-AZT (**76**) (Figure 26) have been reported.[119] The chemically synthesized 5'-triphosphates of the two isomers have been evaluated as RT inhibitors using both a recombinant enzyme and an enzyme purified from wild-type viruses.[119] Inhibition of RT occurs only with the conformationally locked $_2E$ (N)-*methano*-carba-AZT (**75**) triphosphate. This inhibition is equipotent to and kinetically indistinguishable from that produced by AZTTP. On the other hand, the antipodal $_3E$ (S)-*methano*-carba-AZT (**76**) triphosphate does not inhibit RT.

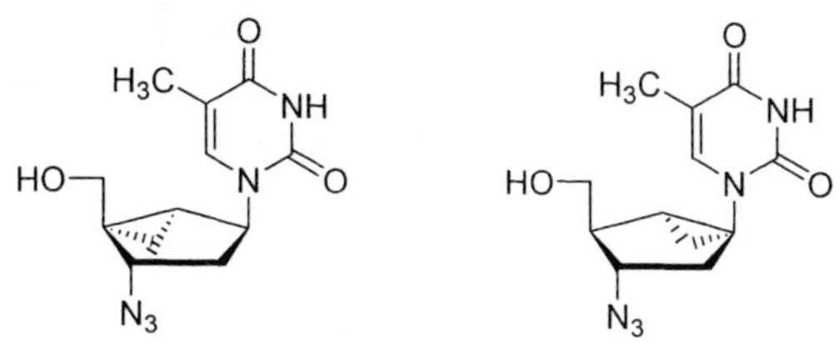

75 [(N)-methanocarba-AZT] **76** [(S)-methanocarba-AZT]

Figure 26. Conformationally locked carbocyclic AZT analogs.

Introduction of a fluorine atom at the 3'-α-position of 2',3'-dideoxy nucleosides (ddNs) increases their anti-HIV activity (Figure 27), whereas 3'-β-fluoro derivatives do not show significant anti-HIV activity.[5a] Particularly, in MT-4 and CH3 cells 3'-α-fluoro-2',3'-dideoxyuridine (FddU) has significantly increased anti-HIV activity than ddU,[120] and in MT-4 cells 3'-fluoro substitution of ddG has greater anti-HIV activity than the parent nucleoside. The fluoro-substituted diaminopurine derivative (FddDAP) has higher anti-HIV activity than ddG.[102] Analogously, FLT (**77**) shows potent anti-HIV activity in various cell lines.[121,122] FLT is phosphorylated intracellularly to FLTTP,[123] which is one of the most potent inhibitors of HIV RT *in vitro*.[124] Unfortunately, FLT produces toxic effects similarly to AZT in cultures of normal human hematopoietic progenitor cells.[125] Thus, despite the promising *in vitro* anti-HIV profiles of FLT, clinical trials with this nucleoside analog failed due to its severe hematological toxicity.[5f,126]

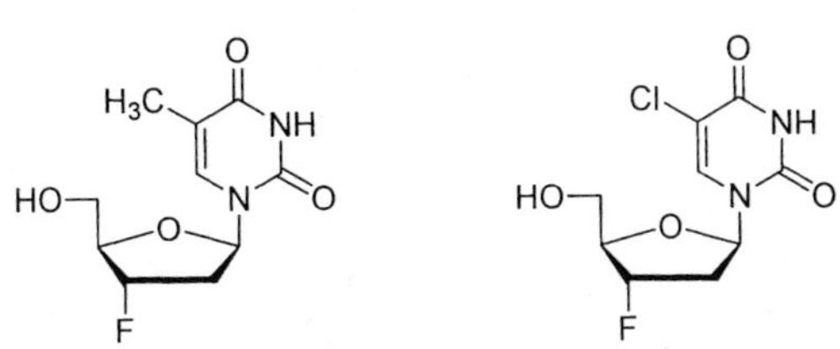

77 (FLT) **78** (FddClU)

Figure 27. 3'α-F-2',3'-dideoxy nucleosides.

FddClU (**78**) also shows significant antiviral activity against HIV and remarkably low cytotoxicity in human leukemic cells and bone marrow progenitor cells.[5f,127] Furthermore, FddClU induces very little resistance in HIV-1 and is active against strains of HIV which are resistant to AZT, ddI, ddC and 3TC as well as many non-nucleoside RT inhibitors.[128] FddClU is metabolized to its mono-, di- and triphosphate in human cells, and the monophosphate is the predominant metabolite. The triphosphate selectively inhibited HIV-RT and DNA polymerase γ while it had little effects on DNA polymerases α and β.[128,129]

Among other L-3'-α-fluoro-ddNs, the cytidine analog shows moderate anti-HBV activity in 2.2.15 cells, but none of them possess anti-HIV activity.[129] Also carba-2',3'-dideoxy-3'-fluorothymidine, the carbocyclic analog of FLT,[130] and related carbocyclic nucleosides[131] failed to elicit significant antiviral activity against HIV.

Introduction of a fluorine atom at the 2'-β-position of 2',3'-dideoxy purine nucleosides retains their anti-HIV activity. Moreover, both 2'-F-ara-ddA (**79**, Figure 28) and 2'-F-ara-ddI are stable in acidic conditions under which ddA and ddI decompose instantaneously by acid-catalyzed glycosylic bond cleavage.[5f,132] Both 2'-F-ara-ddA and 2'-F-ara-ddI retain the same anti-HIV activity as their parent drugs in ATH8 cells but seem to be slightly more cytotoxic.[132] Phosphorylation of 2'-F-ara-ddA by dCyd kinase forms 2'-F-ara-ddAMP, which is then sequentially phosphorylated to 2'-F-ara-ddADP and 2'-F-ara-ddATP. In MT-4 cells, the levels of 2'-F-ara-ddADP and 2'-F-ara-ddATP are 20- and 5-fold higher than the levels of ddADP and ddATP under the same incubation conditions.[133] As a potent ADA inhibitor, 2'-deoxycoformycin significantly increased the levels of 2'-F-ara-ddATP when 2'-F-ara-ddA was incubated in ATH8 cells.[134]

2',3'-Dideoxy-2'-fluoro-β-D-arabinofuranosyl cytosine (**80**, 2'-F-ara-ddC) is effective against several strains of HIV in a number of different cell lines, but the *in vitro* therapeutic index of this compound is considerably lower than that of AZT.[116,135,136] For this reason, initial clinical trials of 2'-F-ara-ddC were discontinued. On the other hand, the *ribo* derivatives of 2'-F-ara-ddNs including the cytidine analog do not retain the antiviral activity of the parent compounds.[137] Additional introduction of a fluorine atom at the 3'-position either in the *ribo* or in the *arabino* configuration also abolishes the anti-HIV activity of the parent drug.[137]

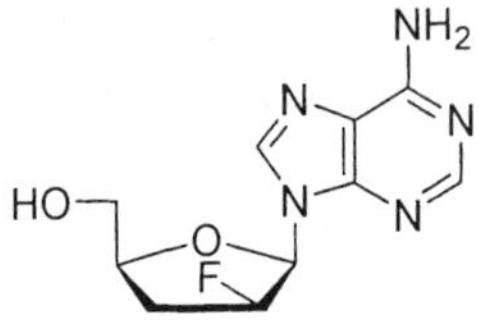

79 (2'-F-ara-ddA)

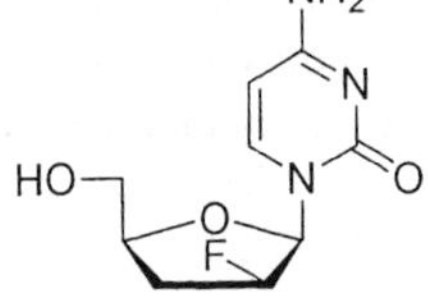

80 (2'-F-ara-ddC)

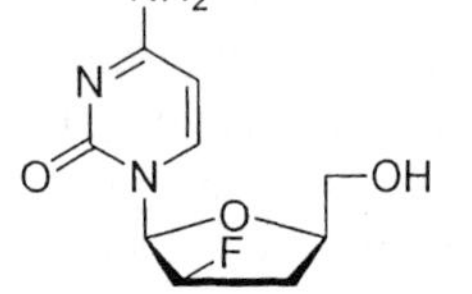

81 (L-2'-F-ara-ddC)

Figure 28. Biologically active 2',3'-dideoxy-2'-fluoro nucleosides.

Chu and co-workers have reported synthesis and antiviral activity of L-2'-F-ara-ddNs, against HIV and HBV.[138] Among the synthesized compounds, the cytosine analog (**81**) shows moderate anti-HIV activity (Figure 28). 2',3'-Dideoxy-2',2'-difluoro-β-L-ribofuranosyl nucleosides have also been reported, but none of them showed significant activity or toxicity.[139]

Fluorination at the 2' or 3'-position of 2',3'-dideoxy-4'-thionucleosides provides fluorinated nucleosides of four different configurations, of which only the cytidine analog displays weak activity against HIV in ATH8 cells.[140]

Substitution of a hydroxymethyl group in ddNs has also produced some anti-HIV activity. Originally, the 3'-hydroxymethyl branched nucleosides of 2-deoxyribofuranose were synthesized as anti-tumor agents.[141] Both α- and β-thioguanine analogs with 2,3-dideoxy-3-(hydroxymethyl)-D-*erythro*-pentofuranose show an inhibitory effect on the growth of WI-L2 human lymphoblastoid cells and are phosphorylated and incorporated into the DNA to the same extent of Mecca lymphosarcoma in mice, proving more effective than the parent analog, α-2'-deoxythioguanosine. Since the pandemic of AIDS, a number of 2',3'-dideoxy-3'-*C*-hydroxymethyl nucleosides have been prepared and evaluated against HIV.[142] Among them, the adenine derivative (**82**, Figure 29) appears to be the most effective in inhibiting viral replication in H9 cells with an activity comparable to ddI and AZT. The cytosine derivative is also a potent inhibitor of HIV-1 *in vitro* and a broad-spectrum antiviral agent. The 4'-thio-[143a] and carbocyclic[143b] analogs of 2',3'-dideoxy-3'-*C*-hydroxymethyl nucleosides are devoid of anti-HIV activity.

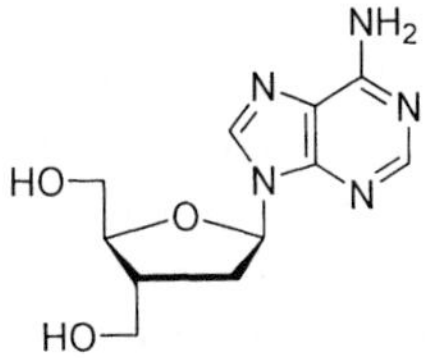

82

Figure 29. 2',3'-Dideoxy-3'-*C*-hydroxymethyladenosine.

Various *C*-branched functionalities, such as hydroxymethyl, fluoromethyl, azidomethyl and aminomethyl, have also been introduced to the 2,3-dideoxy-*erythro*-pentofuranose, 3-deoxyribofuranose and 3-deoxyarabinofuranose moieties.[144] Among the derived nucleosides, 2',3'-dideoxy-3'-*C*-hydroxymethylthymidine has significant anticancer activity against L1210, P388, S-180 and CCRF-CEM cells with ED_{50} values of 50, 5, 10 and 1 μM, respectively. However, none of these compounds show any significant antiviral activity against HSV-1, HSV-2 or HIV. Evaluation of these compounds against thymidine kinases derived from HSV-1 (strain KOS), HSV-2 (strain 333) and mammalian (K562) cells shows that TK from HSV-1 is inhibited significantly by both 3'-deoxy-3'-*C*-(hydroxymethyl) and 3'-deoxy-3'-*C*-(fluoromethyl)thymidines.

Introduction of a fluorine atom in the 2'-β-position of 2',3'-dideoxy-3'-*C*-hydroxymethyl nucleosides has also been considered. The thymine, 5-iodouracil and cytosine analogs showed weak anti-HSV-1 activity.[145] Their carbocyclic analogs are inactive against HIV-1 and HSV-1.[146]

1.5. 2',3'-Unsaturated nucleosides and related analog

Although for nucleoside analogs it has not been possible to elaborate a pharmacophore, 2',3'-unsaturated sugars are probably the most effective moieties for the inhibition of HIV and HBV replication. Among compounds with this feature, 2',3'-didehydro-2',3'-dideoxythymidine (**83**, d4T, stavudine)[84] and its carbocyclic 2-amino-6-cyclopropylaminopurine analog (**84**, 1592U89, abacavir)[86] have been approved for the treatment of HIV infection (Figure 30).

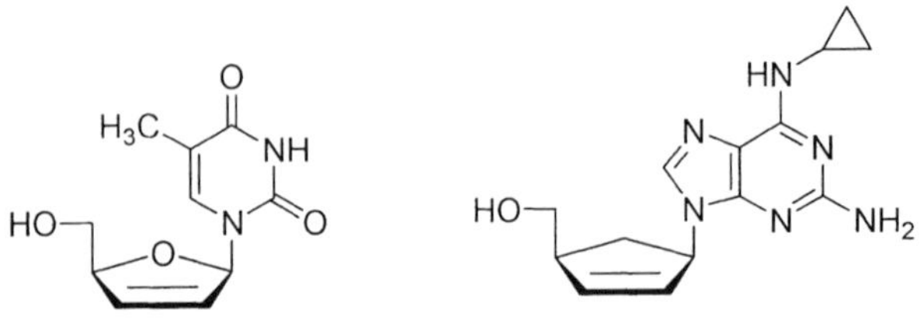

Figure 30. Anti-HIV-1 agents d4T and abacavir.

The phosphorylation of d4T to its monophosphate is the rate-limiting step of the sequential conversion to d4TTP, which inhibits HIV-RT equipotently with AZTTP.[5f,147] d4TTP inhibits DNA polymerase γ and is incorporated into the viral DNA, thus terminating DNA synthesis at the incorporation site (Figure 31). The 3',5'-exonuclease ε cannot remove d4TMP from the 3'-end of DNA once it is incorporated into cellular DNA, whereas, in the case of AZTTP, the enzyme maintains about 20% of its normal deoxynucleotide excision capability.[148] It has been shown that d4T causes peripheral neuropathy. d4T, however, shows 10-fold less toxicity to human hematopoietic progenitor cells compared to AZT.[149] After exposure of human bone marrow cells to similar extracellular levels of parent drugs, steady-state level of d4TMP incorporated into cellular DNA was 10- to 50-fold less than that of AZTMP.[150] In CEM cells, d4T decreased mt-DNA synthesis with higher potency than that of AZT.[151]

L-d4FC (**85**) and L-d4C (**86**, Figure 32) have also been reported to have potent antiviral activities.[152] L-d4FC showed potent anti-HBV activity (EC_{50} 2 nM in 2.2.15 cells) and anti-HIV activity (EC_{50} 0.09 μM in CEM cells), whereas L-d4C was less potent against both viruses (8 nM and 1.0 μM for HBV and HIV, respectively). However, both compounds inhibited cell growth at concentrations below 20 μM. Nevertheless, L-d4FC did not exhibit significant inhibition of mt-DNA at 100 μM.

The fluorinated derivative D-d4FC (**87**, Figure 32) has potent anti-HIV activity *in vitro* with an EC_{50} value of 0.05 μM in PBM cell and anti-HBV activity with an EC_{50} value of 3 nM in 2.2.15 cells without cytotoxicity up to 100 μM in both cell lines.[153,154] A comparison of the antiviral activity of D-d4FC and L-d4FC shows that the latter is active against HIV (EC_{50} 0.034 μM) and HBV replication (EC_{50} 0.01 μM), but has significant cytotoxicity in various cell lines.[153] Another comparison of the antiviral activity of the two enantiomers has been reported, in which L-d4FC was active against HBV in 2.2.15 cells and HIV in MT-2/IIIB cell line with an EC_{50} value of 0.008 and 0.2 μM, respectively, while D-d4FC showed anti-HBV and anti-HIV activity with an EC_{50} value of >0.3 and 0.2 μM, respectively.[155] These results are different from those previously published.[153] In any case, as reported, L-d4FC appears to be more toxic than D-d4FC.[155] An important feature of D-d4FC is its activity against 3TC- and AZT-resistant viral strains.[156] The combination of its resistance profile, rapid uptake and conversion to the active triphosphate, and intracellular half-life of 13 to 17 h[156] make D-d4FC a promising anti-HIV candidate. Both D- and L-d4FC are currently undergoing clinical trials.

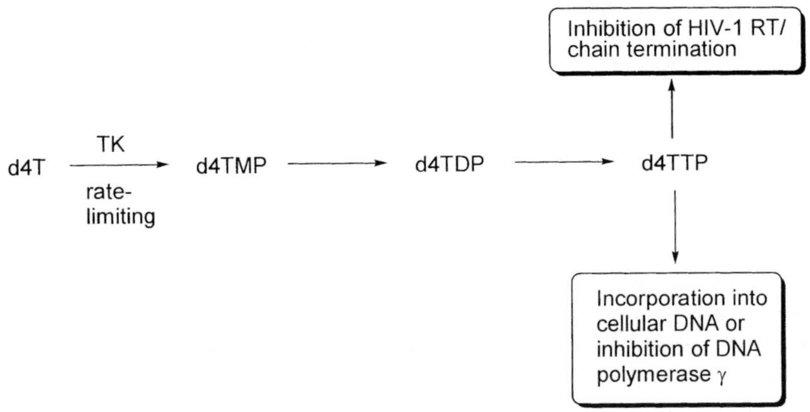

Figure 31. Cellular metabolism of d4T.[147]

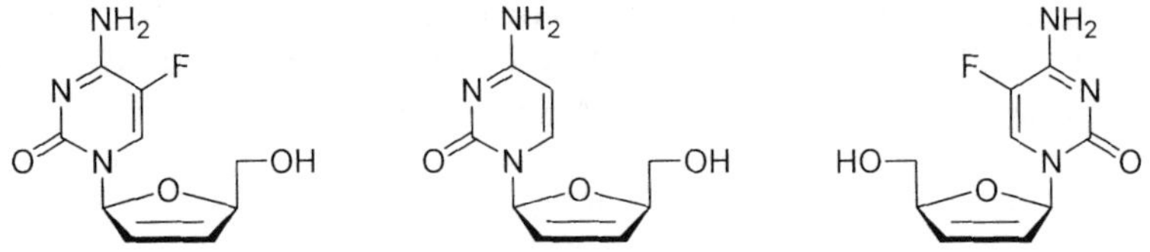

85 (L-d4FC) **86** (L-d4C) **87** (D-d4FC)

Figure 32. Anti-HIV-1 and anti-HBV agents L-d4FC, L-d4C and D-d4FC.

In a series of purine analogs, L-d4A has shown potent anti-HIV-1 activity with an EC_{50} value of 0.38 and 0.54 μM in PBM and CEM cells, respectively, and moderate anti-HBV activity with an EC_{50} value of 1.2 μM in 2.2.15 cells.[157] L-d4I and L-d4G exhibit moderate anti-HIV-1 activity with EC_{50} of 5.5 and 14.1 μM in PBM cells, respectively.[157]

Although D-2',3'-didehydro-2',3'-dideoxyguanosine d4G (**88**, Figure 33) was found to be inactive against HIV-1,[158] recent transient kinetic studies with HIV-1 RT showed that its triphosphate could potentially be an inhibitor of the viral enzyme.[159] The reason for the lack of activity of d4G was found to be its solution instability. In fact, the stable prodrug cyclo-d4G (**89**) is active against HIV-1 (EC_{50} 8.6 μM in MT-2 cells), with increased stability, lipophilicity and solubility, as well as decreased toxicity compared to d4G.[159]

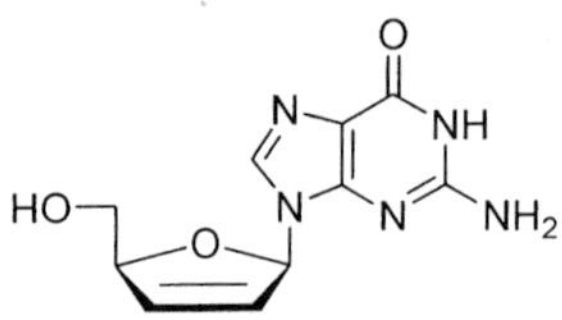

88 (d4G)

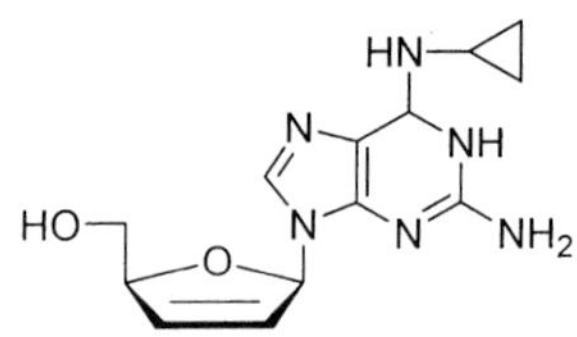

89 (cyclo-d4G)

Figure 33. d4G and its prodrug cyclo-d4G.

2'- or 3'-Substituted d4N analogs also retain their antiviral activity. In particular, introduction of a fluorine atom at the 2'-position produces compounds with anti-HIV and anti-HBV activity (Figure 34). Among the 2'-fluorinated d4N analogs, cytosine derivative D-Fd4C (**90**) shows moderate anti-HIV activity with significant toxicity.[136] A complete SAR study of the D-series has also shown that the 5-F-cytosine analog (**91**) possesses potent anti-HIV and anti-HBV activity.[160,161] Most of the purine analogs have moderate to potent anti-HIV activity. In addition, the adenosine and inosine derivatives show no cross-resistance against 3TC/FTC-resistant strains.[160] Chu and co-workers demonstrated that a series of L-Fd4N has an interesting biological profile.[162] Among these compounds, cytosine (**92**) and 5-F-cytosine (**93**) exhibit potent anti-HIV-1 and anti-HBV activity. In addition, L-Fd4A is moderately active against HIV and HBV. Further study of this series revealed that L-Fd4C and L-Fd4FC are among the most potent anti-HBV agents (EC_{50} 2 and 4 nM in 2.2.15 cells, respectively).[163]

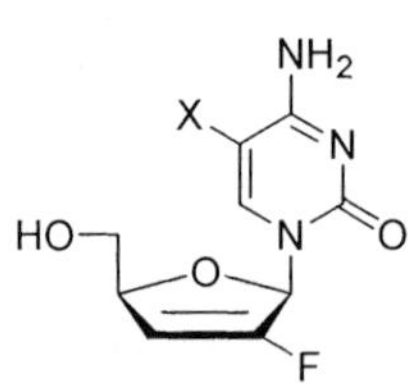

90 (D-Fd4C, X = H)
91 (D-Fd4FC, X = F)

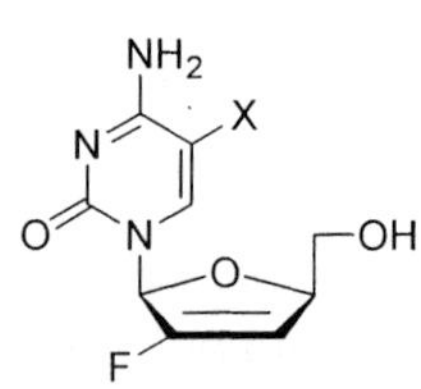

92 (L-Fd4C, X = H)
93 (L-Fd4FC, X = F)

Figure 34. D- and L-2'-fluorinated-2',3'-unsaturated cytidines.

Among 3'-fluoro-2',3'-unsaturated D-nucleosides (Figure 35), 3'-Fd4C (**94**) and 3'-Fd4A (**95**) show modest anti-HIV activity in H9 cells.[164,165] The thymidine analog is marginally active against HIV in MT-4 cells.[136] The 3'-fluoro-2',3'-unsaturated-L-cytosine (**96**) is a potent anti-HIV agent, with EC_{50} of 0.03 μM in PBM cells with little or no significant toxicity.[166]

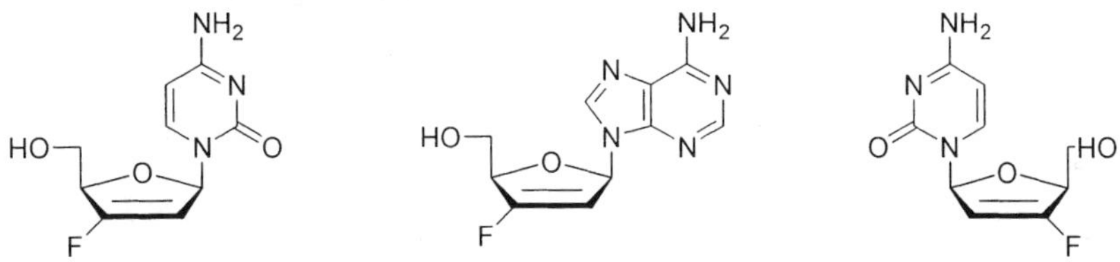

94 (D-3'-F-d4C) **95** (D-3'-F-d4A) **96** (L-3'-F-d4C)

Figure 35. 3'-Fluorinated d4Ns.

The 4'-thio analogs of d4C also show marked anti-HBV and anti-HIV activity (Figure 36).[99b] Particularly, L-4'-thio-d4C (**97**) and L-4'-thio-d4FC (**98**) exhibit significant anti-HIV (EC_{50} 0.8 and 0.4 μM in HeLa CD4 cells, respectively) and anti-HBV (EC_{50} 0.8 and 3.5 μM in HepG2 cells, respectively) activity without toxicity. No other antiviral activity of these compounds has been detected up to 100 μM against HSV-1, HSV-2, VZV, HCMV and influenza.[99b]

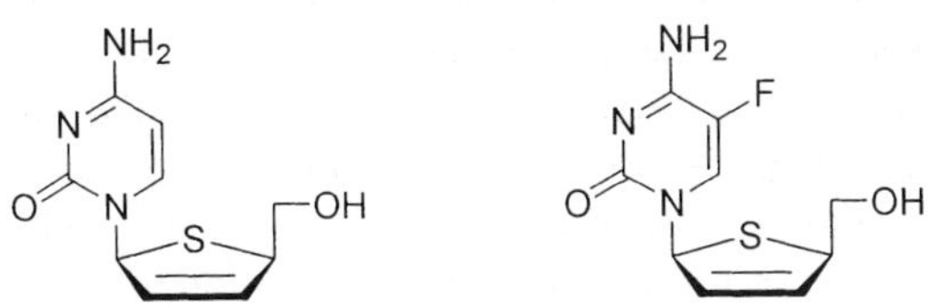

97 (S-d4C) **98** (S-d4FC)

Figure 36. Biologically active L-4'-thio-d4Ns.

The recently reported D- (**99**) and L- (**100**) 2'-fluorinated 4'-thio-2',3'-unsaturated cytidines (Figure 37) also show potent anti-HIV activity, with EC_{50} values of 0.37 and 0.47 μM, respectively, and no significant toxicity up to 100 μM.[167,168,169]

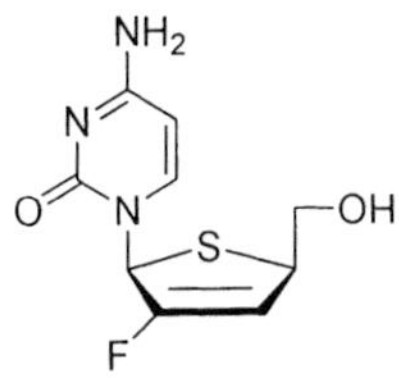

99 (D-2'-F-*S*-d4C)

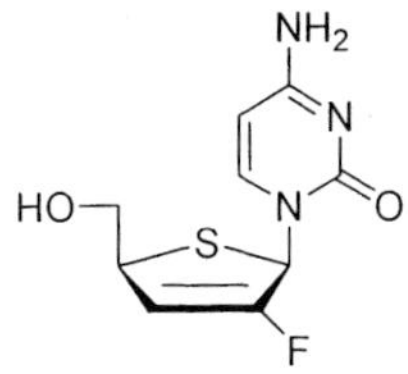

100 (L-2'-F-*S*-d4FC)

Figure 37. D- And L-2'-fluorinated-4'-thio-2',3'-unsaturated cytidines.

Replacement of the furanose ring with a cyclopentyl ring yields selective HIV inhibitors. Structure-activity relationship studies indicate that the optimal anti-HIV activity requires a 2-amino-6-substituted purine and a 2',3'-unsaturated carbocyclic sugar moiety.[5f,170] Racemic carbovir (CBV) was first reported to show anti-HIV activity with low toxicity in H9 cells. (-)-CBV (**101**) was found to be the biologically active isomer against HIV, having a 75-fold higher activity than its (+)-counterpart (**102**, Figure 38), although HIV-RT is equally sensitive to (-)-CBVTP and (+)-CBVTP.[171] The difference in the anti-HIV activity of CBV enantiomers appears to result from the preferential stereoselective phosphorylation of (-)-CBV over its (+)-counterpart. CBV is anabolized intracellularly to its mono-, di- and tri- phosphates rather inefficiently. The enzyme mediating the monophosphorylation is cytosolic 5'-nucleotidase, while the diphosphorylation is catalyzed by GMP kinase (Figure 39).[172] Both these enzymes show preferential selectivity for (-)-CBV over (+)-CBV.[173] In contrast with AZTTP, CBVTP is an inhibitor of HIV-RT, but essentially has no effect on DNA polymerase α, β and γ.[171,174] (-)-CBVTP inhibits HIV-1 RT with an apparent K_i similar to that of AZTTP and, in addition, (-)-CBVMP is also incorporated into the proviral DNA and acts as a chain terminator.[175] CBV and AZT do not affect each other with regard to their intracellular anabolism. The cytotoxicity of CBV may be due to the inhibition of the DNA synthesis, and DNA polymerase α would be responsible for the majority of the incorporation of CBV into DNA in CEM cells.[176] (-)-CBV has no delayed adverse effect on mt-DNA synthesis,[177] however, its poor solubility has prevented the development of CBV as a drug. For this reason, abacavir was synthesized to improve its solubility and pharmacokinetic profile.

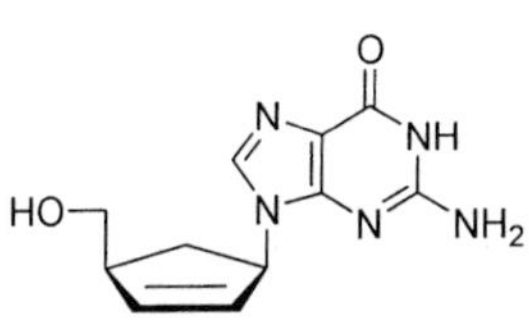

101 [(-)-CBV]

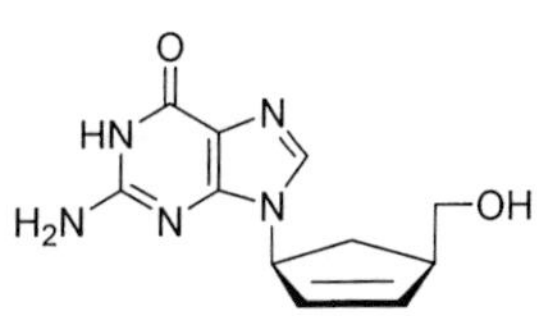

102 [(+)-CBV]

Figure 38. Carbovir and its (+) isomer.

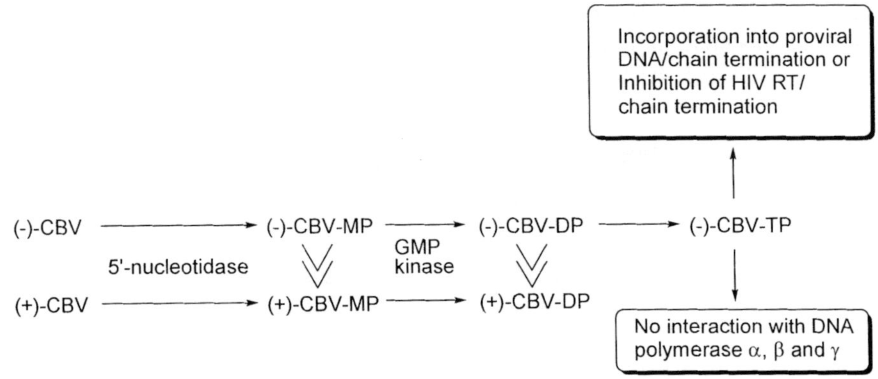

Figure 39. Cellular metabolism of (-)- and (+)-CBV.[172]

Abacavir (**84**, Figure 40), the 6-cyclopropylamino analog of CBV, shows significant inhibition of HIV in PBL cultures with the potency equivalent to AZT, and has synergistic anti-HIV activity in combination with AZT.[86] Its cytotoxicity is low in various human T-cells and bone marrow cells. In addition, toxicity common to other dideoxynucleosides such as peripheral neuropathy and hematopoietic toxicity has not been detected during preclinical studies. The intracellular activation of abacavir sequentially included its monophosphorylation by adenosine phosphotransferase, deamination to (-)-CBVMP by cytosolic deaminase, and two further phosphorylation steps to form (-)- CBVDP and (-)-CBVTP (Figure 40).[178] Therefore, abacavir overcomes the pharmacokinetic and toxicological deficiencies of CBV while maintaining potent and selective anti-HIV activity.[178] For this reason, it has been approved by the FDA for the treatment of HIV infection, and has also been incorporated in a combination with AZT and 3TC (trizivir®).[179,180]

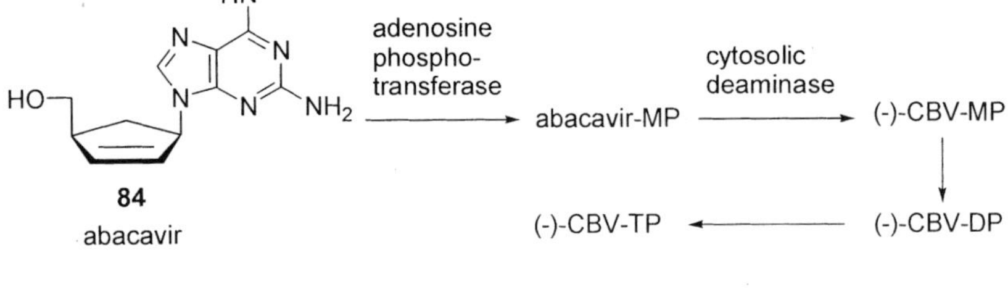

Figure 40. Cellular metabolism of abacavir.[178]

Among L-carbocyclic 2',3'-didehydro-2',3'-dideoxy nucleosides, only the adenine analog (-)-BCA (**103**) has potent *in vitro* anti-HBV activity (EC_{50} 0.9 µM in 2.2.15 cells) as well as moderate anti-HIV activity (EC_{50} 2.4 µM in PBM cells) without cytotoxicity up to 100 µM. Its D-counterpart, (+)-BCA (**104**) is devoid of anti-HIV activity (Figure 41)[181,182]

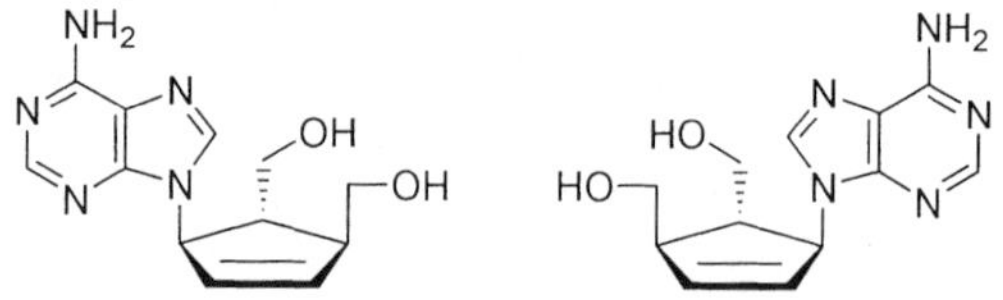

103 [(-)-BCA] **104** [(+)-BCA]

Figure 41. 2',3'-Didehydro-2',3'-dideoxy-4'-*C*-hydroxymethyl carbocyclic nucleosides.

In order to obtain more potent antiviral compounds by increasing the bioavailability of the drug or bypassing critical steps such as the first phosphorylation, a number of prodrugs of 2'-deoxy and 2',3'-unsaturated nucleosides have been prepared.[2j] In the attempt of obtaining higher intracellular levels of d4T, a number of prodrugs have been developed which can deliver d4TMP. Phosphoramidate derivatives of d4TMP (**105**)[183] efficiently deliver the monophosphate (**106**), according to the mechanism showed in Figure 42. The same approach has proved successful in delivering 3TCMP,[184] AZTMP,[185] ddAMP[186] and d4AMP.[186,187]

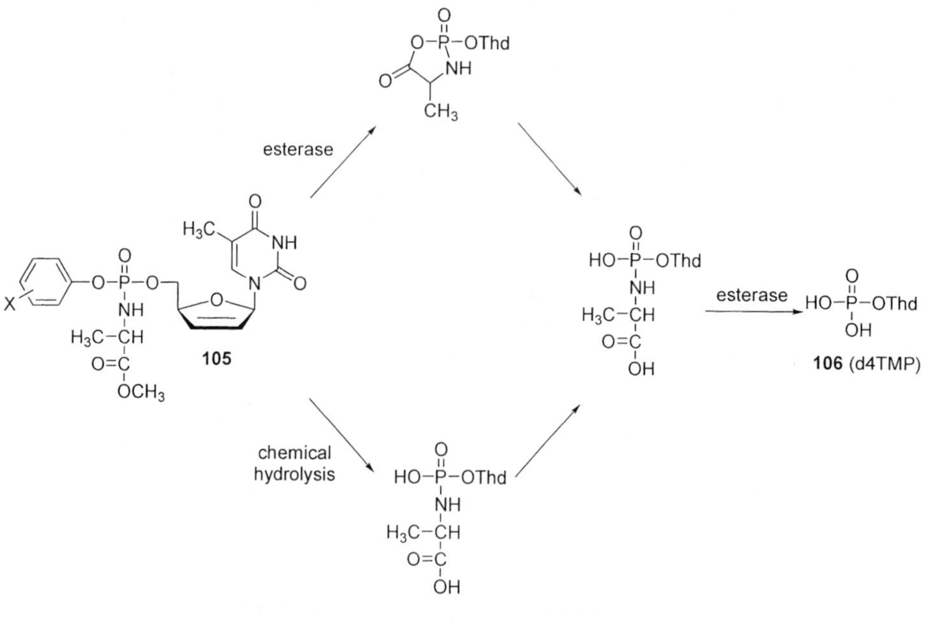

Figure 42. Phosporamidate prodrugs of d4T and their mechanism of action.

In the *cyclo*Sal-pronucleotide approach, nucleotides are delivered intracellularly thanks to pH-driven selective chemical hydrolysis of the prodrug (Figure 43).[188,189,190,] The tandem cleavage originates with the hydrolysis of a phenyl ester (**107**) followed by hydrolysis of a benzyl ester in the resulting phosphotriester (**108**) with liberation of the nucleotide (**109**). This concept is based upon the principle that selection of phenyl, benzyl and alkyl phosphate esters can influence the hydrolysis steps of the tripartate approach. The phenyl ester is cleaved first, because of stabilization caused by delocalization of the negative charge in the aromatic ring, affording the 2-hydroxybenzylphosphodiester. This concept has been applied to anti-HIV and antitumor agents such as d4T,[191,192,193] 5-FU,[194] AZT,[195,196,197] 2',3'-dideoxyadenosine (ddA),[198,199] d4A[199] and 2'-fluoro-2',3'-dideoxyadenosines (F-*ara*-ddA and F-*ribo*-ddA).[200]

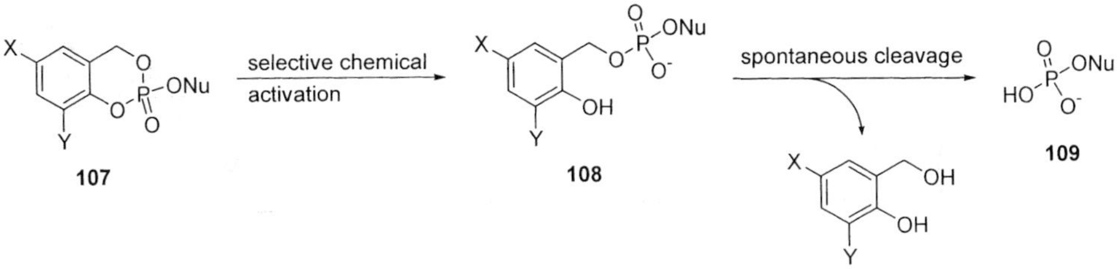

Figure 43. Proposed decomposition of *cyclo*Sal-pronucleotides

1.6. Nucleosides with a heterocyclic sugar ring moiety

This important class of nucleosides will be extensively discussed in Chapter 3. Since the discovery of (±)-dioxolanethymine (**110**)[91] and (±)-BCH-189 (**111**)[92] as potent anti-HIV agents as racemic mixtures, all the four possible diastereomers have been synthesized (Figure 44). From extensive structure-activity relationships studies of these isomers, a number of compounds have made considerable impact on HIV and HBV chemotherapy.[201,202] Currently, among this series of nucleosides, 3TC (lamivudine) has been approved by the FDA for the treatment of HIV-1 and chronic HBV infections, and 2',3'-dideoxy-5-fluoro-3'-thiacytidine (FTC)[201] and β-D-2,6-diaminopurine dioxolane (DAPD)[202c] are under clinical development as anti-HIV-1 and anti-HBV agents, and β-L-dioxolane cytosine (L-OddC)[202d] as an anti-cancer agent.

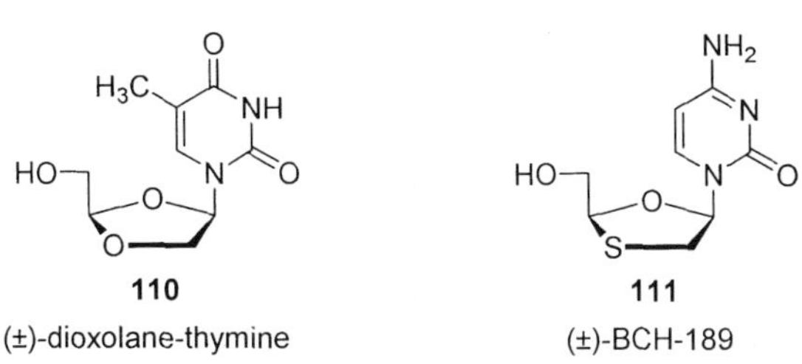

Figure 44. Racemic dioxolane-thymine and BCH-189.

Belleau and co-workers originally reported that (±)-BCH-189 exhibited potent *in vitro* anti-HIV activity in T-cells as well as human peripheral lymphocytes. (±)-BCH-189 was found to be less toxic than AZT, inhibited AZT-resistant virus, and was non-toxic at 100 mg/kg given orally over 14 days in rats.[85] Moreover, it showed potent anti-HBV activity in 2.2.15 cells.[203] Since then, various approaches for the asymmetric synthesis of optically active compounds have been reported including enzymatic resolution.[93,204] Of significance was that while both enantiomers have potent anti-HIV activity, cytotoxicity resides mainly with the natural (+)-D-isomer (Table 2).[205]

Table 2. Comparison of antiviral activities and cytotoxicities of DL- and LL-oxathiolane cytosine analogs.[205]

	Anti-HIV-1 (EC_{50}, μM)		Anti-HBV (EC_{50}, μM)	Cytotoxicity (IC_{50}, μM)	
	PBM	CEM	2.2.15	PBM	CEM
(+)-D-BCH-189	0.2	0.1	0.5	2.7	>100
(-)-L-BCH-189 (3TC)	0.002	0.007	0.001	>100	>100

The first asymmetric syntheses of enantiomers of (±)-BCH-189 and their *trans*-isomers were described by Chu and co-workers from D-mannose, D-galactose or L-gulose as starting materials.[206] An extensive study of structure-activity relationships has made clear that the unnatural L-2',3-dideoxy-3'-thiacytidine (**112**, 3TC, Figure 45) is more potent against HIV-1 in human PBM cells as well as against hepatitis B virus (HBV) in 2.2.15 cells than its racemate or its D-enantiomer (**113**). Most importantly, the comprehensive SAR of the enantiomerically pure D- and L-isomers revealed that most of the nucleosides, among which the 5-fluoro analogs **114** and **115**, exhibited not only good to excellent anti-HIV-1 activity, but also low toxicity in PBM as well as Vero cells.[206]

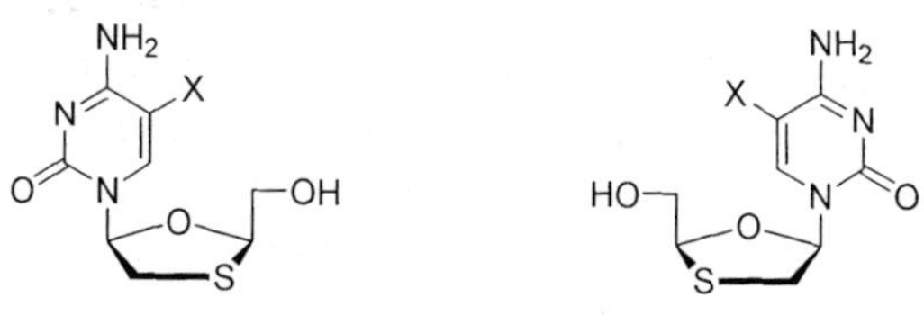

112 [(-)-3TC, X = H)]
114 [(-)-FTC, X = F]

113 [(+)-BCH189, X = H]
115 [(+)-FTC, X = F]

Figure 45. 3TC, FTC and their enantiomers.

Regarding the cellular metabolism of the optically pure isomers of (±)-BCH-189, 3TC is resistant to deamination or enzymatic hydrolysis,[207] whereas the D-isomer is deaminated to 2'-deoxy-3'-thiauridine, although no hydrolysis of glycosyl bond is observed (Figure 46). dCyd kinase is the enzyme responsible for the monophosphorylation of 3TC,[208] which is a better substrate than the D-counterpart.[209] Further phosphorylation of 3TC or (+)-BCH-189 to its di- and triphosphate is accomplished by deoxycytidylate and NDP kinase, respectively.

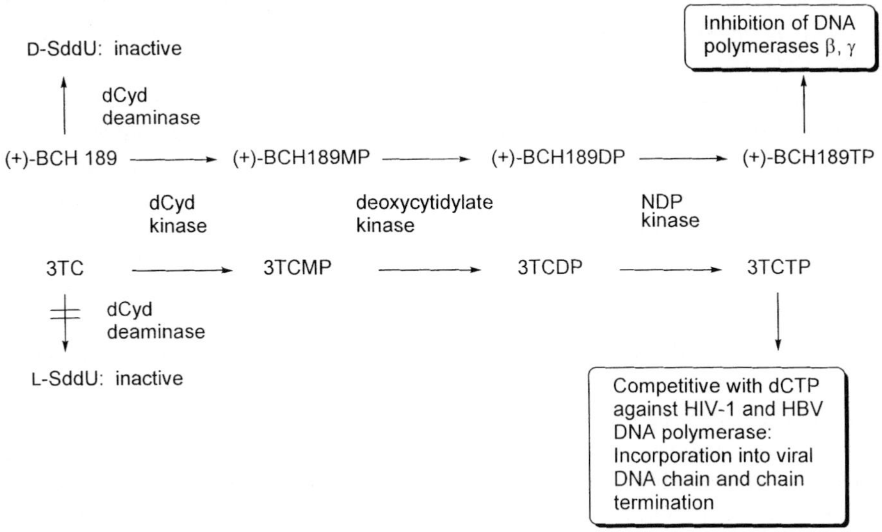

Figure 46. Metabolism of (+)-BCH-189 and 3TC.

3TCTP is a competitive inhibitor (with respect to dCTP) of the RNA-dependent DNA polymerase activity with apparent K_i = 10.6 ± 1.0 to 12.4 ± 5.1 μM, depending on the template and primer used.[210] DNA-dependent DNA polymerase activity is inhibited by 50% by a 3TCTP concentration of 23.4 ± 2.5 μM when dCTP is present at a concentration equal to its K_M value. 3TCTP is a rather weak inhibitor of DNA polymerase γ, but (±)-BCH-189 is about 650 times more inhibitory than 3TC, due to the activity of the D-isomer. This observation might explain why 3TC is more potent and less toxic than its D-counterpart against HIV-1 *in vitro*.[2f] Furthermore, 3TCTP is not an inhibitor of DNA polymerase β, whereas (±)-BCH-189 has a significant inhibitory effect on this enzyme (Figure 46).[210] Chain elongation studies with 3TC show that 3TCTP is incorporated into newly synthesized DNA and that transcription is terminated in similar fashion as seen with ddCTP.

As mentioned before, 3TCTP also shows a potent inhibitory effect against HBV-associated DNA polymerase, and is a better inhibitor of HBV DNA polymerase than its D-counterpart.[207]

Schinazi *et al.* have reported the anti-HIV activity of the racemates as well as the single enantiomers of the 5-fluoro congener of 3TC, FTC (emtricitabine, **114**,

Figure 45).[201,211,212,213] FTC shows potent *in vitro* anti-HIV-1, HIV-2, SIV, and FIV activity in various cell cultures. Like 3TC, FTC exhibits 20-fold more potency against HIV-1 in human PBM cells and less toxicity in myeloid progenitor cells than its D-enantiomer (**115**). It also shows anti-HBV activity in hepatoma cell lines (HepG2 cells), whereas its D-enantiomer is significantly less potent.[212] However, both enantiomers do not show significant cytotoxicity in human bone marrow progenitor cell assays and any detectable hepatotoxic effects at concentrations above their antiviral activities. Currently, (-)-FTC (coviracil) is undergoing phase III clinical trials against HIV and HBV infection and racemic FTC (racevir) is in phase I clinical trials as an anti-HIV agent.

Enzymatic studies of FTC show a similar profile to 3TC, in which the D-enantiomer of FTC is a substrate for dCyd deaminase, and FTC is resistant to deamination by the same enzyme.[213a] dCyd kinase and NDP kinase phosphorylate FTC to its triphosphate, which functions as chain terminator of viral DNA synthesis, similarly to 3TC. Schinazi *et al.* have also reported that highly 3TC/FTC-resistant HIV-1 variants dominate the replicating virus population after two or more cycles of infection in the presence of 3TC or FTC.[213b] These variants are cross-resistant to 3TC and FTC but are susceptible to ddC, AZT, and ddI. DNA sequence analysis of the RT gene amplified from resistant viruses consistently identified a mutation at codon 184 from Met (ATG) to Val (GTG or GTA) or Ile (ATA).

Synthesis of the 3'-deoxy-3'-oxa-thymidine analog (±)-dioxolane-T and other natural pyrimidine base analogs has given promising leads for the inhibition of HIV-1 as well as HBV and herpes virus replication. Originally, (±)-dioxolane cytosine was reported by Belleau and co-workers as an anti-HIV agent and subsequently, (±)-dioxolane thymidine was also reported as moderately active against HIV-1.[91,92] Extensive studies of the structure-activity relationships has led to the synthesis of β-L-dioxolane cytosine (**116**, L-OddC) and its 5-fluoro congener (**117**, L-F-OddC), which exhibit potent *in vitro* anti-HIV-1 and HBV activities (Figure 47).[202] However, L-OddC is quite toxic and stable to degradation by cytidine deaminase and deoxycytidine deaminase.

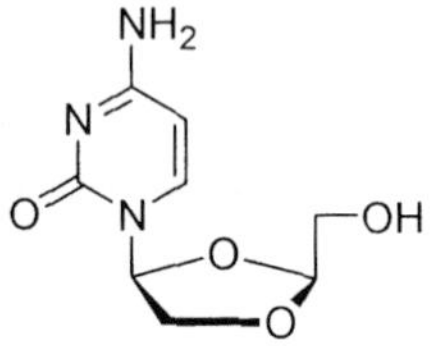

116 (L-OddC)

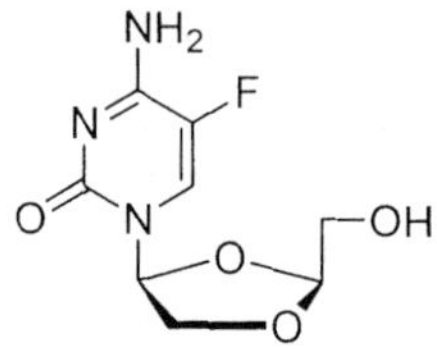

117 (L-F-OddC)

Figure 47. β-L-Dioxolane cytosine and 5-fluorocytosine.

L-OddC is metabolized in cells by dCyd kinase to its monophosphate, and subsequently to the di- and triphosphate, which inhibits DNA polymerase α, β, and γ (Figure 48).[214] L-OddC exhibits potent antitumor activity against various solid tumor cell lines, including prostate, renal, hepatoma, and colon.[214] Thus, L-OddC is the first

L-nucleoside analog ever shown to have anticancer activity, and also the first true chain terminator capable of inhibiting tumor growth.[215] L-OddC is currently undergoing phase II clinical trials against leukemia and solid tumors.

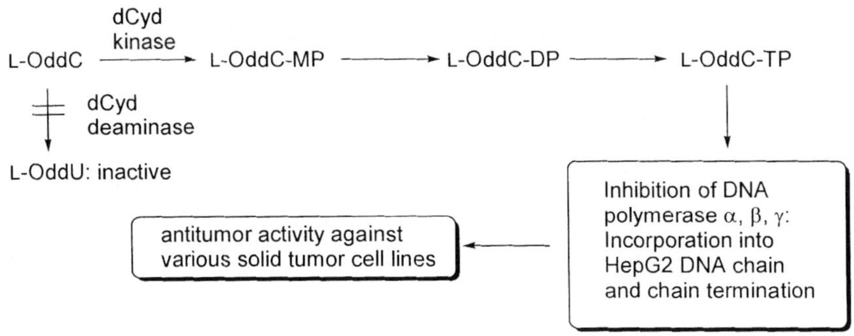

Figure 48. Metabolism of L-OddC.[214b]

Among purine derivatives, β-2,6-diaminopurine dioxolanes, DAPD (**118**) and its enantiomer, L-DAPD (**119**), display potent anti-HIV and anti-HBV activities (Figure 49).[202c,d] Interestingly, L-DAPD is more potent against HIV-1 (EC_{50} 0.014 μM) than DAPD (EC_{50} 0.7 μM) in human PBM cells, while DAPD shows more potent anti-HBV activity (EC_{50} 0.009 μM) than its L-isomer (EC_{50} 8.3 μM) with a favorable toxicity profile.

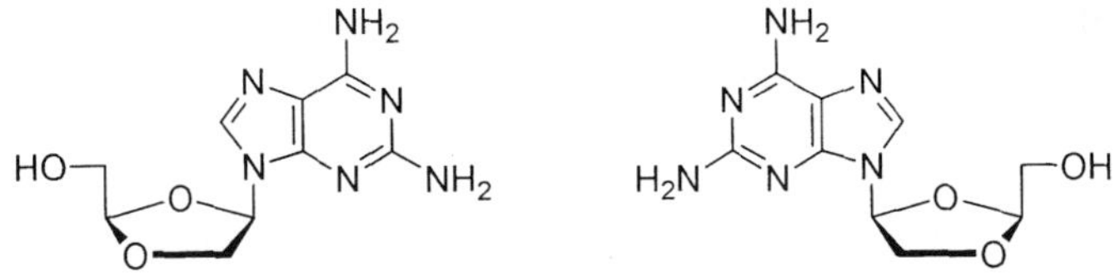

118 (DAPD) **119** (L-DAPD)

Figure 49. Enantiomers of β-2,6-diaminopurine dioxolane.

Pharmacokinetic studies suggest that DAPD is the prodrug of the corresponding guanine derivative, dioxolane-guanine (**120**, DXG). DAPD and β-D-2-amino-6-chloropurine dioxolane (**121**, ACPD) are converted to DXG by ADA, and β-D-2-aminopurine dioxolane (**122**, APD) is converted to DXG by xanthine oxidase (Figure 50).[216] As discussed extensively in Chapter 3, DAPD and DXG are active against 3TC-resistant HIV and HBV strains, which provides DAPD a promising therapeutic potential.

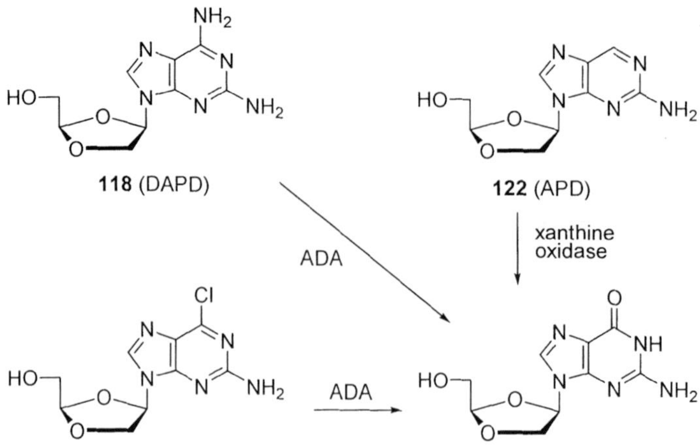

Figure 50. Pharmacokinetics of DAPD, APD, and ACPD.[216]

Mansour and co-workers reported synthesis and activity of BVU analogs with a dioxolane moiety, among which the β-L-dioxolane nucleoside shows significant activity against HSV-1 (EC_{50} 0.3 μg/mL) and HCMV (EC_{50} 5 μg/mL). The β-D-oxathiolane nucleoside demonstrated potent activity against HSV-2 (EC_{50} 2.9 μg/mL).[217] In an extensive SAR study, Chu and co-workers recently reported the activity of a series of dioxolane and oxathiolane (*E*)-5-(2-halovinyl)uracil nucleosides (Figure 51) against a number of viruses.[20] The β-L-dioxolane nucleosides show potent anti-VZV and anti-EBV activities, which can be related to the size of the halogen atoms [chlorovinyl (**123**) < bromovinyl (**124**) < iodovinyl (**125**) against VZV and iodovinyl < bromovinyl < chlorovinyl against EBV]. β-L-(*E*)-5-(2-Iodovinyl)uracil dioxolane (**125**, L-IV-OddU) is 60-fold more potent against VZV than ACV. No inhibition of CEM cell growth or mt-DNA synthesis is observed for any compounds at concentrations up to 200 μM. This selectivity has been explained, in the case of β-L-(*E*)-5-(2-bromovinyl)uracil dioxolane (**124**, L-BV-OddU), with selective phosphorylation by viral TK, but not human TK.[218] Unlike other D-configuration BVU analogs, such as BVdU and BVaraU, L-BV-OddU is metabolized only to its corresponding monophosphate instead of the di- or triphosphate, which suggests a unique inhibitory mechanism other than DNA chain termination. As mentioned above, L-dioxolane derivatives with 5-substituted uracil show potent anti-EBV activities (Figure 51).[219] β-L-5-Iodouracil dioxolane (**128**, L-I-OddU) is the most potent anti-EBV agent with an EC_{50} value of 0.03 μM without any cytotoxicity up to 100 μM. Also in this series, their activities can be related to the size of the halogens [EC_{50} Cl (**126**) 0.15; Br (**127**) 0.07; I (**128**) 0.033 μM]. L-I-OddU is an efficient substrate for EBV TK, but not for human cytoplasmic dThy or mt-dPyd kinases, with L-I-OddUMP being the major metabolite.[220] L-I-OddU and L-Br-OddU are currently undergoing preclinical studies as potential anti-EBV agents.

123 (L-CV-OddU, X = Cl)
124 (L-BV-OddU, X = Br)
125 (L-IV-OddU, X = I)

126 (L-Cl-OddU, X = Cl)
127 (L-Br-OddU, X = Br)
128 (L-I-OddU, X = I)

Figure 51. Structure of L-I-OddU, L-BV-OddU and related nucleosides.

Reversed oxathiolane nucleosides, such as 2'-deoxy-3'-oxa-4'-thiocytidine (dOTC, Figure 52), have also been found active against HSV-1, HSV-2, HBV and HIV-1 in a panel of cell lines. The BVU analog has demonstrated potent activity against HSV-2 and the cytosine and 5-F-cytosine derivatives have exhibited appreciable antiviral activity in cord blood mononuclear cells (CBMCs) and U937 (human monocyte) cell lines.[217,221] (+)-dOTC is moderately active against HBV in 2.2.15 cells. (±)-dOTC (**129**) is phosphorylated within cells *via* the dCyd kinase pathway and approximately 2 to 5% is converted into the racemic triphosphate derivatives (Figure 52).[222] Both 5'-triphosphate derivatives (TP) of (±)-dOTC are more potent than 3TCTP at inhibiting HIV-1 RT *in vitro*. In cell culture experiments, (±)-dOTC is a potent inhibitor of primary isolates of HIV-1 with an IC_{50} for viruses resistant to 3TC and viruses resistant to 3TC and AZT of 2.53 and 2.5 μM, respectively.[222] After 14 days of continuous culture, at concentrations up to 10 μM, no measurable toxic effect on HepG2 cells or mitochondrial DNA replication within these cells has been observed. dOTC is currently undergoing phase I clinical trials as an anti-HIV agent.

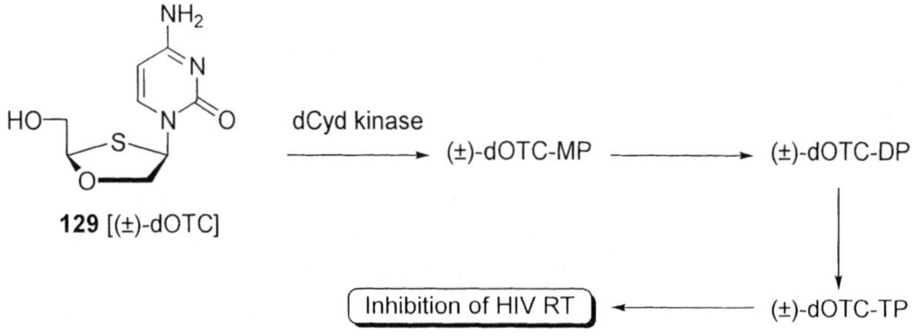

Figure 52. Metabolism of (±)-dOTC.[222]

Prepared as bioisosters of 3TC and FTC, oxaseleno compounds (±)-Se-ddC (**130**) and (±)-F-Se-ddC (**131**, Figure 53), have also been found to exhibit potent anti-HIV-1

(EC_{50} 2.7 and 0.73 μM, respectively) and anti-HBV (EC_{50} 1.2 and 1.2 μM) activities.[223] Resolution of the racemic mixtures showed that most of the anti-HIV activity of **130** and **131** resides with the (-)-isomers (EC_{50} 0.9 μM for **130** and 0.2 μM for **131**).[224]

Substitution of 3'-sulfur in 3TC with an amino group does not produce antiviral activity against HIV.[225]

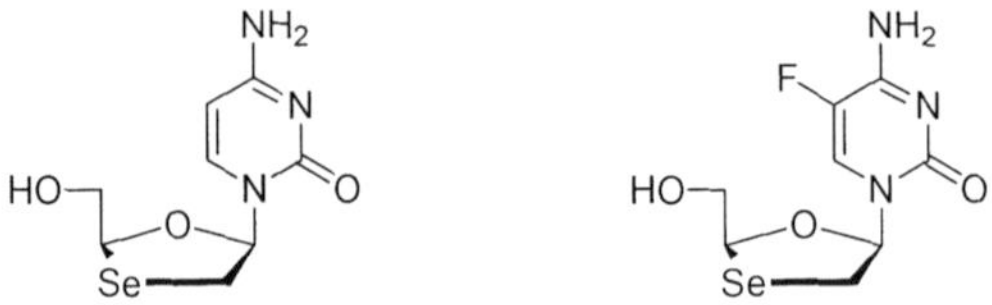

130 [(±)-Se-ddC] **131** [(±)-F-Se-ddC]

Figure 53. Oxaselenonucleosides.

Isodideoxynucleosides,[226] in which the base is transposed from the 1' to the 2' position of the sugar, have been reported by Huryn *et al.*[227] and Nair *et al.*[228] (Figure 54). The isomeric form of ddA, (*R*,*R*)-iso-ddA (**132**), is as active against HIV-1 as ddA, with a better hydrolytic stability.[227] The anabolism of (*R*,*R*)-iso-ddA has been studied in CEM cell cultures.[229] The formation of (*R*,*R*)-iso-ddATP is significant and increases almost linearly upon incubation for 24 h. In comparison, phosphorylation of ddA yields three to four times the amount of ddATP as (*R*,*R*)-iso-ddATP, but the amount of ddATP does not increase much after 10 h. (*R*,*R*)-iso-ddATP competitively inhibits the incorporation of dATP into a synthetic polynucleotide primer. Nitrogen or sulfur analogs of iso-ddA do not show significant antiviral activity.[230,231]

(*S*,*S*)-Iso-ddA (**133**), which can be viewed as an L-related ddN, also shows significant antiviral activity against HIV in MT-4 and PBL cell cultures and is also active against AZT-resistant HIV strains.[228] Synergistic effects are observed in combination with AZT, ddI or FTC. (*S*,*S*)-IsoddA has little cytotoxicity in leukemic cell lines and lower inhibition on human bone marrow cells than AZT. In CEM cells, it is metabolized, although rather inefficiently, to (*S*,*S*)-isoddATP, which is a potent inhibitor of HIV-RT. The metabolism of (*S*,*S*)-IsoddA is unique. The nucleoside is neither phosphorylated by adenosine kinase nor oxidized by adenosine deaminase. Instead, the first step in its metabolic activation seems to be phosphorylation by dCyd Kinase.[232] Compared to ddATP, (*S*,*S*)-isoddATP is a weaker inhibitor of DNA polymerase β and γ, but a stronger inhibitor of DNA polymerase α.[228]

Another series of isoddNs is related to the natural sugar, D-apiose, and its enantiomer, L-apiose, which can be considered as regioisomers of the natural nucleosides through transposition of the hydroxymethyl group from the normal 4'-position to the 3'-position (**134** and **135**, Figure 54). A comprehensive study of these dideoxynucleosides has been accomplished by Sells and Nair[233] However, these compounds do not show any antiviral activities, and even further modifications of the apiosyl moiety, with the introduction of ring substituents such as 4'-hydroxymethyl,[234] 3'-fluoro,[235] 3'-azido[236] and 3'-amino,[236] have failed to induce significant antiviral activity.

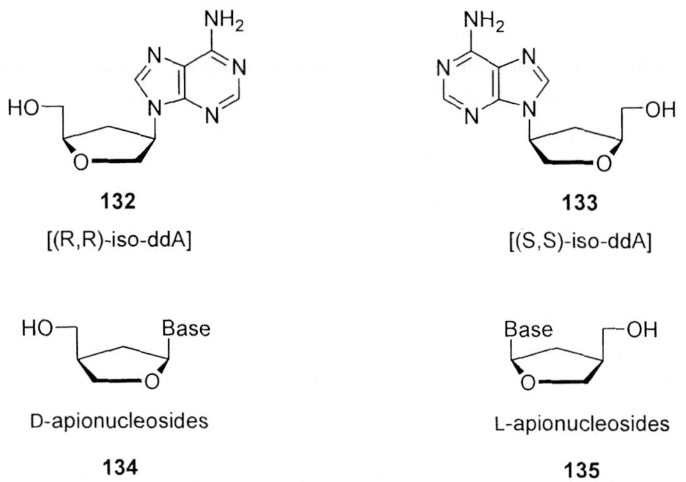

Figure 54. Enantiomers of iso-ddA and apionucleosides.

A series of branched-chain sugar isonucleosides has been reported to exhibit significant antiviral activity against herpes viruses (Figure 55).[237] The hydroxymethylguanosine analog (**136**) displays potent and selective anti-HSV-1 and anti-HSV-2 activity. Although the antiherpetic activity *in vitro* of this compound is lower than that of ACV, it displays superior efficacy in mouse infections. The BVU analog (**137**) also shows selective activity against HSV-1 and VZV, with no cytostatic effect on WI-38 cell growth at >800 μM.

Several other substituted isonucleosides have been prepared, including 3'-hydroxymethyl and 3'-azidomethyl derivatives, but none of them has shown any significant antiviral activity.[238]

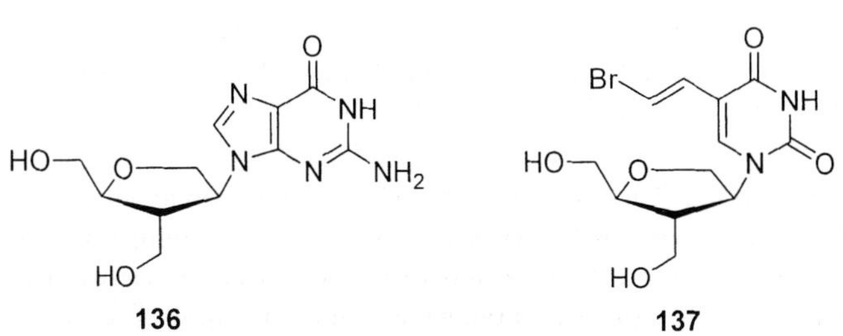

Figure 55. 3'-*C*-Hydroxymethylisonucleosides.

1.7. 3- or 4-Membered ring nucleosides

Transformation of the furanose ring to a 3- or 4-membered ring produces compounds with interesting biological activity. Oxetanocin A, 9-(2-deoxy-2-hydroxymethyl-β-D-erythro-oxetanosyl)adenine (**138**) (Figure 56), is an antibiotic produced by *Bacillus megaterium*, which inhibits infection of T cells by HIV-1 *in vitro*.[239] Chemical and enzymatic modifications of oxetanocin A (OXT-A) have afforded 2,6-diamino (2-amino-OXT-A), guanine (OXT-G), hypoxanthine (OXT-H) and xanthine (OXT-X) analogs.[240] In MT-4 cells, OXT-A markedly reduces the expression of HIV antigens, and 2-amino-OXT-A, OXT-G and OXT-H also show significant anti-HIV activity. In addition, OXT-G is very potent and selective in inhibiting the replication of HCMV *in vitro* (EC_{50} 0.1 μg/mL) and against HSV-2 (EC_{50} 3.5 μg/mL).[241]

The thymidine analog of oxetanocin, A-73209 (**139**), is a potent *in vitro* and *in vivo* inhibitor of HSV-1, HSV-2 and VZV.[242] A-73209 is two logs more potent against five TK^+ strains of VZV *in vitro* and one log more potent against TK^+ HSV-1 strains than ACV. A-73209 is more effective than ACV against lethal systemic or intracerebral HSV-1 infections in mice. L-oxetanocin (L-OXT-A, **140**) is inactive against HIV.[243]

Early reports of racemic carbocyclic analogs of OXT-A and OXT-G have described the protective effect of both carbocyclic analogs on $CD4^+$ ATH8 cells against the infectivity and cytopathic effect of HIV-1, suppressing proviral DNA synthesis.[244,245,246] In addition, carbocyclic OXT-G showed excellent activity against HSV and it was suggested that it is phosphorylated by virus-encoded TK prior to exerting its antiviral effect. In contrast, the adenine congener, carbocyclic OXT-A, is a good inhibitor of HCMV *in vitro* and *in vivo*. However, severe cytotoxicity to host cells has prevented further development of this compound as an anti-HCMV agent.

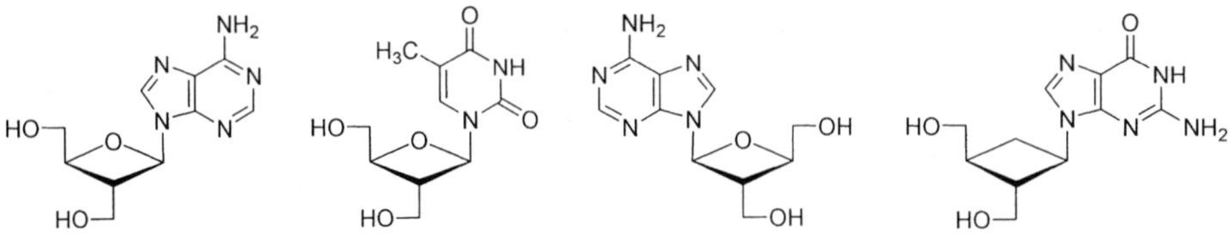

138 (OXT-A) **139** (A-73209) **140** (L-OXT-A) **141** (LBV)

Figure 56. Oxetanocin A (OXT-A), its thymine analog A-73209 and lobucavir (LBV).

The active enantiomer of carbocyclic OXT-G (**141**, lobucavir, LBV, Figure 56) displays an impressive broad-spectrum antiviral activity against a wide variety of herpesviruses and HBV as well as HIV.[247] The mechanism of action of LBV against HSV-1, HSV-2 and VZV consists in the inhibition of the viral polymerases after phosphorylation by the virally encoded TK (Figure 57).[247] However, HCMV, HBV and HIV do not encode enzymes which are capable of mediating LBV phosphorylation.

It is known that HCMV has homologs of a herpesvirus-encoded protein kinase (UL97 gene), which mediates the phosphorylation of ganciclovir (GCV). In the case of VZV, both the herpesvirus TK and protein kinase may independently enable the phosphorylation of LBV. Furthermore, LBV is phosphorylated to its triphosphate intracellularly in both HCMV-infected and uninfected cells, with phosphorylated metabolites levels 2- to 30-fold higher in infected cells. These studies[247] suggest that LBVTP can halt HCMV DNA replication by inhibiting the viral DNA polymerase and that LBV's phosphorylation can occur in the absence of viral factors including the UL97 protein kinase. In addition, LBV may be effective in the treatment of GCV-resistant HCMV. LBV has undergone clinical trials as an anti-HBV agent.

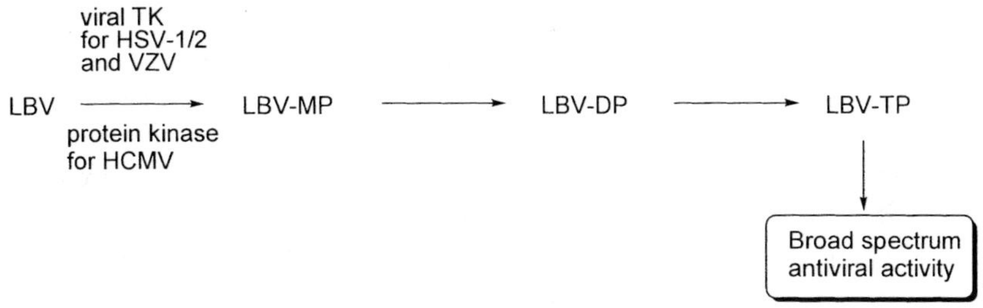

Figure 57. Metabolism of lobucavir.

3'-Fluorocarbocyclic oxetanocin A (**142**, Figure 58) exhibits a broad spectrum of antiviral activity especially against HCMV with an EC_{50} of 0.18 μg/mL, which is 4-fold more potent than that of ganciclovir.[248] However, this compound is slightly cytotoxic at higher concentrations (100 μg/mL) in HEL or MT-4 cells, although this toxicity is minor compared with that of carbocyclic OXT-A (CC_{50} for HEL; 8 μg/mL, CC_{50} for MT-4; 12 μg/mL).[248]

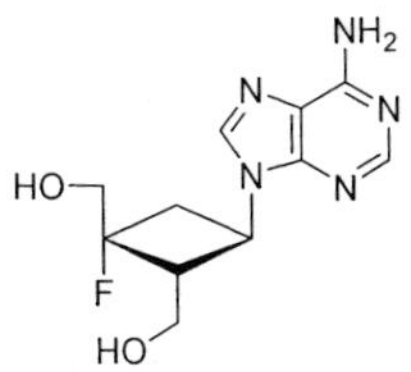

142

Figure 58. Carbocyclic 3'-fluoro-oxetanocin A.

(±)-9-{[(*Z*)-2-(Hydroxymethyl)cyclopropyl]methyl}guanine (**143**) (Figure 59) displays significant antiherpetic activity *in vitro*, and the (*E*)-adenine analog has a modest antiviral activity despite an apparent inability to be enzymatically phosphorylated.[249]

Enzymatic phosphorylation studies indicate that in the (cyclopropyl)methyl derivatives, both the *cis*- and *trans*-(hydroxymethyl) derivatives are reasonably good substrates for the HSV-1 TK in comparison with ACV and its carba analog. The *cis*-isomer is converted efficiently to the triphosphate, but it inhibits the HSV-1 DNA polymerase poorly. The *trans*-isomer accumulates as the diphosphate with little triphosphate detected, but the triphosphate appears to be a better inhibitor of HSV-1 DNA polymerase.[249]

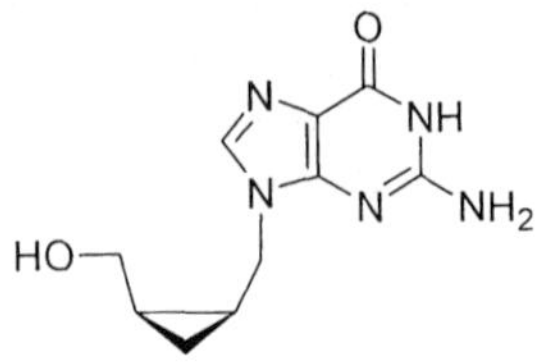

143

Figure 59. (±)-9-{[(Z)-2-Hydroxymethyl)-cyclopropyl]methyl}guanine.

Recently, Chu and co-workers also reported the synthesis of enantiomeric 1-[2-(hydroxymethyl)cyclopropyl]methyl nucleosides, among which the adenosine and guanosine analogs show moderate antiviral activity against HIV-1 and HBV in PBM and 2.2.15 cells, respectively, without significant cytotoxicity.[250]

9-{[*cis*-1',2'-Bis(hydroxymethyl)cycloprop-1'-yl]methyl}guanine (A-5021, **144**, Figure 60) is an extremely potent anti-HSV-1 agent without significant cytotoxicity.[251] Both enantiomers were prepared from chiral epichlorohydrins, and only one A-5021 (1'*S*,2'*R*-configuration) exhibits strong antiherpetic activity (EC_{50} of 0.020 μg/mL against HSV-1 Tomioka vs 0.81 μg/mL for ACV). A-5021 is more inhibitory than ACV against HSV-2 and VZV but ineffective against HIV. Its enantiomer (**145**) has modest anti-HSV-1 activity. A-5021 is monophosphorylated by viral TKs.[252] A-5021 triphosphate accumulates more than ACVTP but less than penciclovir (PCV) triphosphate in MRC-5 cells infected with HSV-1 or VZV, whereas HSV-2 infected MRC-5 cells show comparable levels of A-5021 and ACV triphosphates. A-5021TP competitively inhibits HSV DNA polymerases with respect to dGTP (ACVTP > A-5021TP > PCVTP) and is incorporated into DNA instead of dGTP terminating elongation, although limited chain extension has been observed. Thus, the stronger antiviral activity of A-5021 appears to depend on a more rapid and stable accumulation of its triphosphate in infected cells than that of ACV as well as on stronger inhibition of viral DNA polymerase by its triphosphate than that of PCV.

A number of 5-substituted uracil nucleoside derivatives with a 1-(1'*S*,2'*R*)-[1',2'-bis(hydroxymethyl)cyclopropyl]methyl group have also been reported to exhibit antiherpetic activity.[253] Among them, the BVU analog (**146**, Figure 60) is the most potent (EC_{50} 0.027 μg/mL), 40 to 60-fold more than ACV (EC_{50} 3.4 μg/mL) against clinical isolates of VZV. Catabolism of **146** does not produce BVU, responsible for toxic drug interactions (*vide supra*).[253]

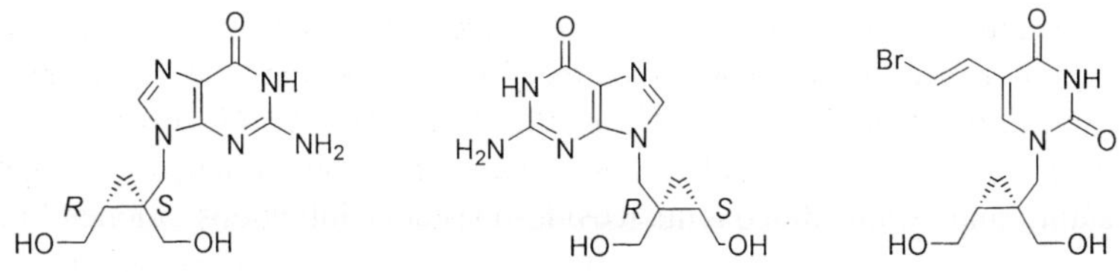

144 (A-5021) **145** **146**

Figure 60. *Cis*-1',2'-bis(hydroxymethyl)-cyclopropyl]methyl nucleosides.

Other types of cyclopropyl-containing nucleosides (**147**, **148** and **149**), where the base is directly linked to the cyclopropyl ring, have been prepared, but none of them showed significant antiviral activity (Figure 61).[254]

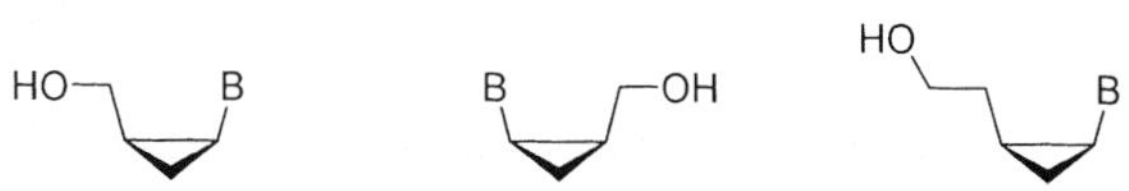

147 **148** **149**

Figure 61. 2-(Hydroxymethyl)cyclopropyl nucleosides.

Nucleoside analogs based on a methylenecyclopropane structure show broad-spectrum antiviral activity against HCMV, EBV, human herpes virus type 6 (HHV-6), VZV and HBV (Figure 62).[255] (*Z*)-2-{[(Hydroxymethyl)cyclopropylidene]-methyl}adenine (**150**, synadenol), -2-amino-6-chloropurine (**151**) and -guanine (**152**, synguanol) are the most effective agents against HCMV (EC_{50} 1-2.1, 0.04-2.1 and 0.8-5.6 μM, respectively) and EBV in H-1 cells (EC_{50} 0.2, 0.3 and 0.7 μM, respectively). Synadenol is moderately active against HIV, HBV, VZV and HHV-6. It is a substrate for ADA from calf intestine and is also deaminated by AMP deaminase from *Aspergillus* sp.

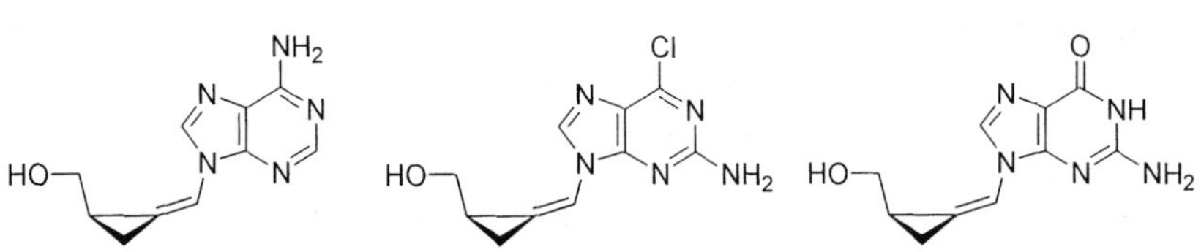

150 (synadenol) **151** **152** (synguanol)

Figure 62. Methylenecyclopropyl nucleosides.

Conversion of the modestly active analogs to their methyl phenyl phosphoro-L-alaninate esters results in potentiation of their anti-HIV-1 activity.[256] Among these prodrugs, the 2,6-diaminopurine (**153**) and adenine derivatives (**154**, Figure 63) are the most potent against HIV-1 *in vitro* with EC_{50} of 0.034 and 0.0026 μM, respectively in MT-2 cell-based assays. Both compounds are interestingly active against AZT-resistant, ddI-resistant and multi-dideoxynucleoside-resistant infectious clones *in vitro*. Analogously, synguanol phosphoralaninate prodrug QYL-678 (**155**) inhibits EBV with EC_{50} (in the viral capsid antigen expression assay) of 0.05 μM *vs* 5.6 μM for synguanol and 6.3 μM for ACV.[257]

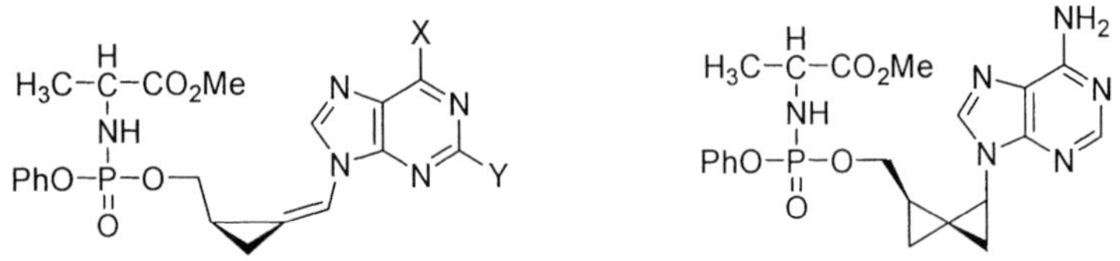

153 (X = NH$_2$, Y = H)
154 (X = NH$_2$, Y = NH$_2$)
155 (X = OH, Y = NH$_2$)

156

Figure 63. Phosphoralaninate prodrugs of methylenecyclopropyl and spiropentane nucleosides.

Spirocyclic analogs of 2'-deoxynucleosides have been synthesized and evaluated by Zemlicka and co-workers. Among these novel derivatives, the phosphoralaninate **156** (Figure 63) is an effective inhibitor of HCMV (EC_{50} 0.38 μM *vs* 2.9 μM for GCV in HFF cells). It also shows interesting activity against HSV-1 (EC_{50} 7.0 μM in BSC-1 and 20 μM in Vero cells), HSV-2 (EC_{50} 31 μM in Vero cells), VZV (EC_{50} 1.4 μM in the cytopathic effect inhibition effect assay), EBV (EC_{50} 8.4 μM in Daudi cells), HIV-1 (EC_{50} 3.5 μM in CEM-SS cells) and HBV (EC_{50} 3.1 μM in HepG 2.2.15 cells). Unfortunately, this compound also shows varying degrees of toxicity in cell lines (e.g. EC_{50} 27 μM in Vero cells).[258]

1.8. Acyclonucleosides

Acyclonucleosides are characterized by the absence of a cyclic sugar moiety and, thus, a higher conformational flexibility than other nucleoside analogs. A consequence of this flexibility is that acyclonucleosides possess biological properties despite their lack of chirality. Whenever chirality or prochirality (as in ganciclovir and penciclovir) is present, quite surprisingly the *S* configuration at the 4'-equivalent position gives the only active enantiomer.[5g]

Acyclovir (ACV, **157**, Figure 64) was one of the first antiviral agents that showed potent and selective viral inhibition and is still one of the most effective anti-herpetic drugs,[259,260] although its use is limited by poor oral bioavailability (20-25%).[261] This is much improved in its prodrug valaciclovir (*vide infra*).

Ganciclovir (GCV, **158**) is the drug of choice for the treatment of HCMV retinitis in AIDS patients.[262,263,264] Like acyclovir, its systemic use is limited by poor oral bioavailability (2-7%).[265]

Penciclovir (PCV, **159**), the *carba*-analog of GCV, has a broad-spectrum antiviral activity, being active against HSV-1, HSV-2, VZV, EBV and HBV.[266,267,268] It is currently approved for the treatment of herpes zoster infections.[269] Like its congeners, it has a low oral bioavailability.

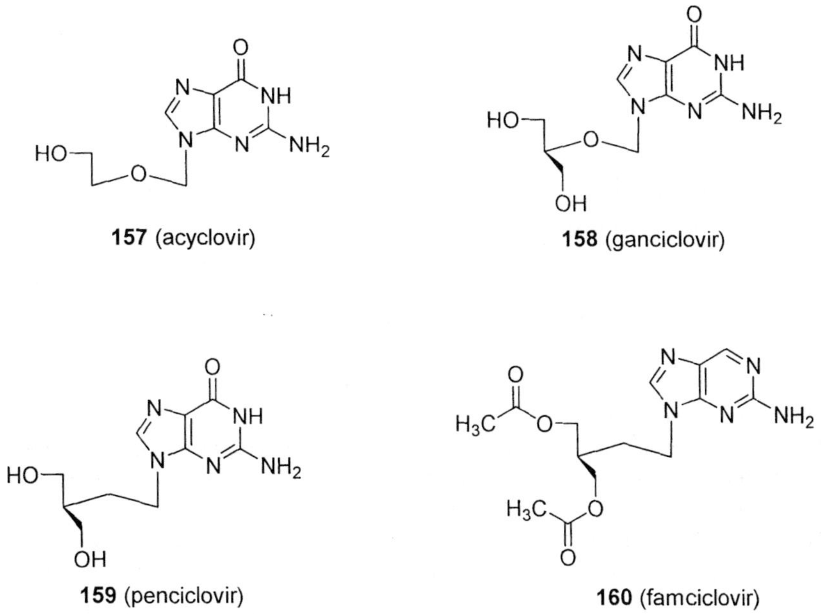

Figure 64. Acyclovir and related compounds.

Famciclovir (FCV, **160**) is a prodrug of PCV which is converted to its parent drug by three metabolic steps, two hydrolytic and one oxidative.[270] Bioavailability of PCV is 77% upon oral administration of FCV.[271] PCV is used for the treatment of VZV retinitis.

As mentioned above, a common drawback of most acyclonucleosides is their low oral bioavailability. This is mainly due to the low solubility of these compounds.[272] In order to overcome this problem, many modifications of the parent structures have been tried to give prodrugs with favorable absorption/distribution properties.

The conversion of a nucleoside into its phosphonate offers several advantages. In particular, a phosphonate can mimic a phosphate group and can be converted to the triphosphate-like analog, thus bypassing the first phosphorylation necessary for the activation of the drug. As a result, inactive drugs can be converted into active phosphonates if their lack of activity is due to a non-efficient initial phosphorylation step. Furthermore, unlike a phosphate, a phosphonic group is not cleaved by chemical or enzymatical hydrolysis, therefore it is metabolically stable.

Adefovir, 9-(2-phosphonylmethoxyethyl)adenine (PMEA, **161**, Figure 65) is a broad-spectrum antiviral agent, active against retroviruses, hepadnaviruses, and herpesviruses.[273,274,275] However, the possibility of PMEA of becoming an orally administered drug is limited by its poor bioavailability,[276] due to the negative charge of the phosphonate functionality at physiological pH, which limits its gastroenteric absorption. In order to increase the low bioavailability and decrease the toxicity of PMEA, a number of prodrugs have been prepared,[277] and the bis-pivaloyloxymethyl derivative (bis-POM-PMEA, adefovir dipivoxil, **163**)[278] has recently been approved as an anti-HBV agent. *In vitro* studies show that adefovir dipivoxil is able to increase the intracellular concentration of PMEA by 2 logs.[273] It has comparable antiviral activity in HIV-1 infected CEM cells and HCMV-infected MRC-5 cells, and it is even more potent than the parent compound on HSV-1 and HSV-2-infected Vero cells.[279] In clinical trials, oral bioavailability was found to be greater than 40%.[279] However, the bis-POM functionality has been found to confer cytostatic effects, probably due to the liberation of formaldehyde and pivalic acid.[279] In clinical trials, adefovir has shown modest anti-HIV activity with nephrotoxicity at dose levels of 60-120 mg once daily.[280] However, adefovir does not show cross-resistance with lamivudine,[281] and a combination of the two drugs has proved effective in the therapy of HBV-HIV-coinfected patients.[280] In this combination, the use of adefovir dipivoxil at a suboptimal concentration for HIV activity (10 mg once daily) prevents the occurrence of mutations at codons 65 and 70 of HIV RT, responsible for HIV resistance to adefovir.[280]

9-[2-(R)-(phosphonomethoxy)propyl]adenine (PMPA, tenofovir **162**)[282] is an effective anti-HIV agent. In a phase I/II clinical study, it showed a 1.1 log reduction in HIV RNA levels after administration of only eight doses.[283] Tenofovir is less toxic towards erythroid progenitor cells than AZT, 3TC and d4T.[284] Its resistance to phosphorolysis and nucleotide-dependent chain-terminator removal is greater than AZT or 3TC.[285] As in the case of adefovir, also tenofovir displays low bioavailability in animals. This problem has been overcome with the bis-isopropyloxycarbonyloxymethyl derivative (bis-POC-PMPA, tenofovir disoproxil, **164**),[87,88] the first nucleotide analog approved by the FDA for the treatment of AIDS.[89,90] Bis-POC derivatives were designed in order to eliminate the side effects of the POM group. Tenofovir disoproxil retains the antiviral activity of the parent drug, showing an oral bioavailability of 30% with minimal toxicity.[87,88,89,90,286] Its efficacy on antiretroviral-experienced patients makes it one of the best choices in salvage therapy.[90]

Cidofovir (HPMPC, **165**) exhibits potent *in vitro* and *in vivo* activity against a broad spectrum of herpes viruses, including HCMV,[287] and has been approved for the treatment of HCMV retinitis in AIDS patients.[288] Its adenine congener HPMPA (**166**) has anti-HBV activity in both duck hepatocytes and 2.2.15 cells with an EC_{50} of 1.2 μM.[289] The cyclic prodrug of cidofovir, cHPMPC (**167**) has antiviral activity similar to the parent compound with reduced toxicity.[290]

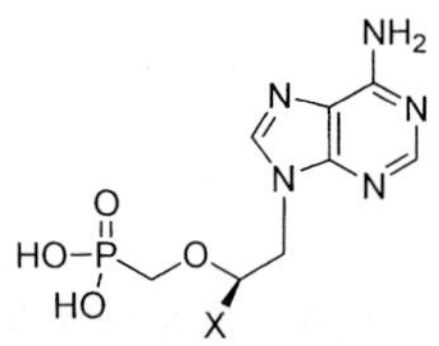

161 (adefovir, X = H)
162 (tenofovir, X = CH_3)

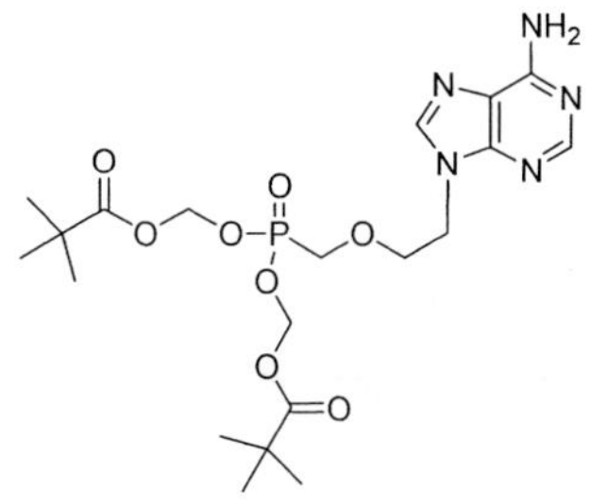

163 (adefovir dipivoxil)

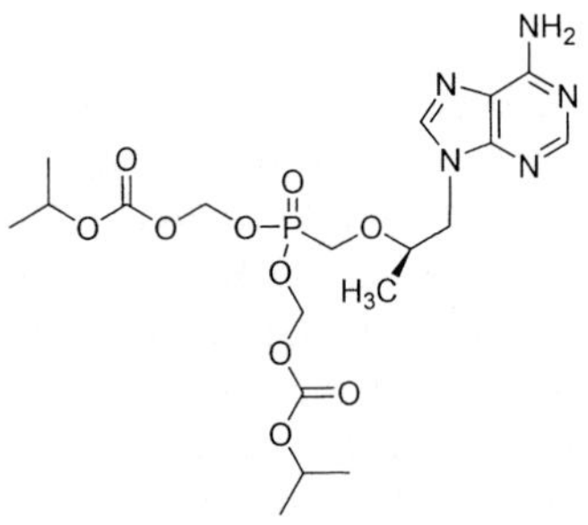

164 (tenofovir disoproxil)

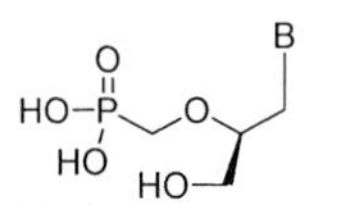

165 (cidofovir, B = Cytosine)
166 (HPMPA, B = Adenine)

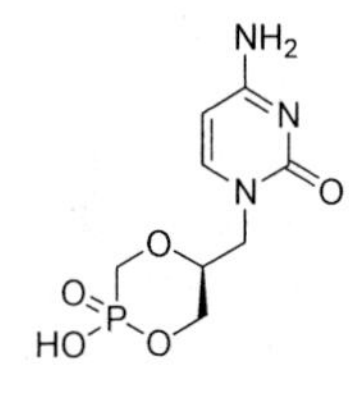

167 (cHPMPC)

Figure 65. Acyclonucleoside phosphonates and their “bis-POM” and “bis-POC” prodrugs.

As discussed above, the major drawback of acyclovir and its phosphonates is their poor oral bioavailability, due to low water solubility. The valine conjugate valacyclovir (VCV, **168**, Figure 66) has more than twice-greater bioavailability compared to that of ACV. This is due to the increased water solubility as well as the presence of an amino-acid moiety, which probably allows VCV to be absorbed in the intestin *via* the saturable dipeptide transporter system.[291]

The lipophilic prodrug 1-*O*-hexadecylpropanediol-3-phosphoacyclovir (HDP-P-ACV, **169**) also has improved bioavailability compared to the parent drug. Besides, unlike ACV, it is active against HBV in 2.2.15 cells.[292] This is due to the fact that, unlike herpes viruses, HBV does not encode for a TK, which catalyses the conversion of ACV to ACVMP. HDP-P-ACV delivers ACVMP, which can then be further phosphorylated to the active triphosphate form.

The ganciclovir prodrug HDP-P-GCV (**170**) has given promising results in the therapy of HSV-1 or HCMV retinitis. In the rabbit model, intravitreal injections with resultant 0.2 μM intravitreal concentration of prodrugs allowed a 4 to 6 weeks complete protection of the retina against HSV-1 with an IC_{50} of 0.6 μM.[293,294] HDP-P-GCV has also been evaluated in HCMV-infected human lung fibroblasts. Its IC_{50} was 0.6 μM.

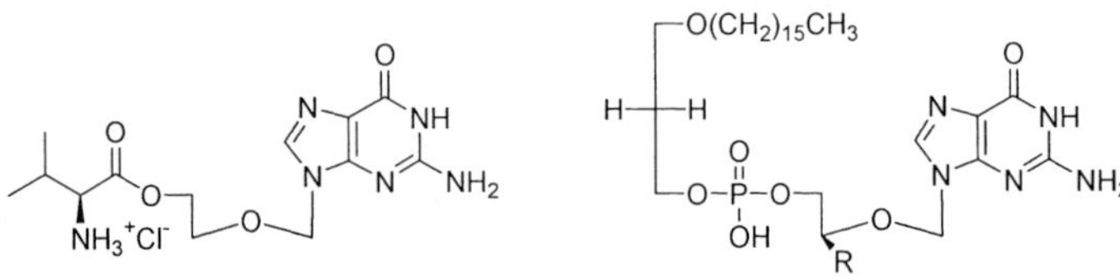

168 (VCV)

169 (HDP-P-ACV), R = H
170 (HDP-P-GCV), R = CH_2OH

Figure 66. Prodrugs of acyclovir and ganciclovir.

Recently, the same liphophilic prodrug approach has been applied to the synthesis of prodrugs of cidofovir and cyclic cidofovir.[295] The alkoxyalkyl esters 1-*O*-hexadecyloxypropyl-cidofovir (HDP-CDV, **171**, Figure 67) and 1-*O*-octadecyloxyethyl-cidofovir (ODE-CDV, **172**) and their cyclic analogs HDP-cCDV (**173**) and ODE-cCDV (**174**) were more active against HSV-1, HSV-2, HCMV, VZV, EBV and human herpes virus type 6 (HHV-6) and 8 (HHV-8) than the parents compounds, with the cidofovir analogs generally more active than the cyclic ones. [295b] All the prodrugs were active against several ganciclovir- and cidofovir-resistant HCMV strains, particularly.[295a] An *in vitro* study using ^{14}C-labeled HDP-CDV and cidofovir showed that the cellular drug content of HDP-CDV increased progressively to 24 hours, whereas the content of cidofovir reached a peak after 1-4 hours, remaining stable or slightly declining at 24 hours. The cellular content of cidofovir after 24 hours was 73-fold higher after incubation with the prodrug, which may explain the increased activity.[295c] Cidofovir is an effective inhibitor of variola, monkeypox, cowpox and vaccina viruses. In a study on the use of cidofovir for the treatment of smallpox, a single intravenous dose fully protected mice against a lethal cowpox virus aerosol.[295d] In the search for orally active anti-smallpox agents, alkoxyalkyl esters of cidofovir and cCDV proved able to deliver, upon oral administration in mice, plasma levels of drugs in the low micromolar range, that is 10-fold the EC_{50} for smallpox.[295d] Analogous pharmacokinetic studies on HDP-CDV for the treatment of CMV retinitis showed plasma levels of drug that should allow antiviral activity. [295e]

Recently, 5-(1-azidovinyl)-substituted acyclic pyrimidine nucleosides (**175**, **176** and **177**, Figure 68) have shown potent and selective anti-HBV activity, with EC_{50} values of ranging from 0.01 to 0.1 μM in duck hepatitis B virus-infected primary duck hepatocytes, without significant toxicity.[296]

1.9. Ribofuranosyl nucleosides

Because of their close resemblance with natural nucleosides, the class of ribofuranosyl analogs has not produced many useful antiviral agents. In recent years, however, a number of carbocyclic analogs have shown promising antiviral activity, particularly against DNA virus.

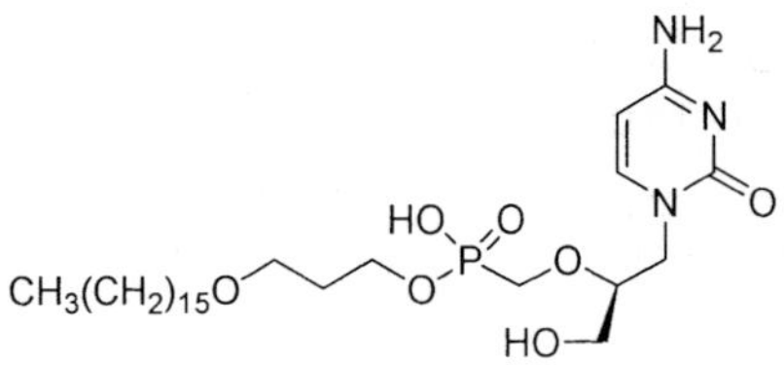

171 (HDP-CDV)

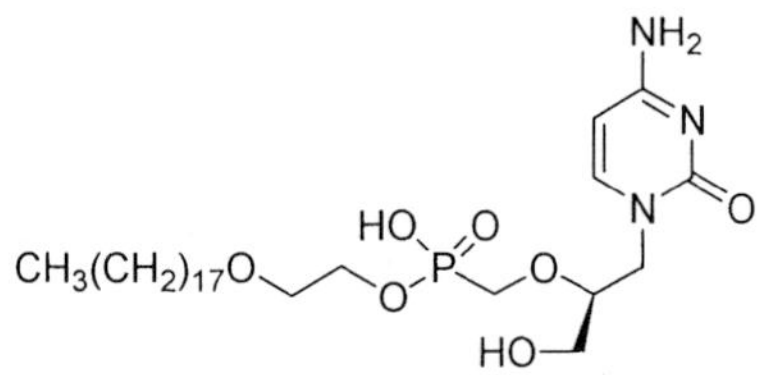

172 (ODE-CDV)

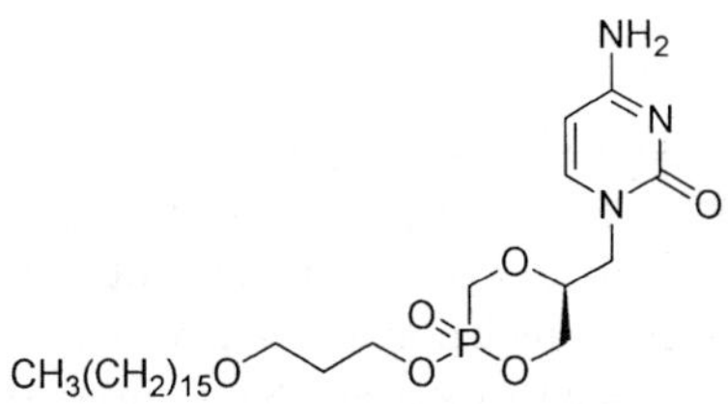

173 (HDP-cCDV)

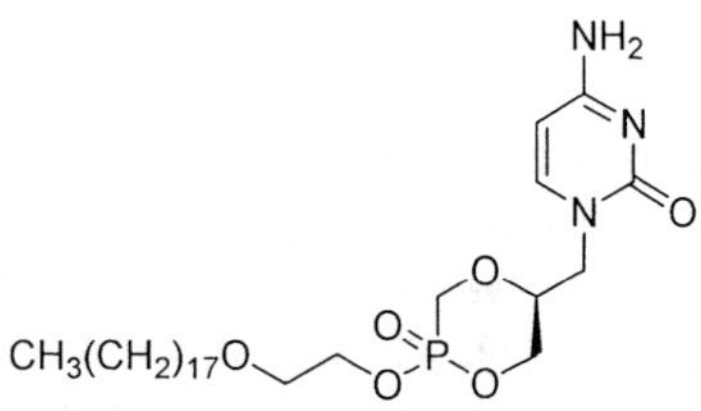

174 (ODE-cCDV)

Figure 67. Liphophilic prodrugs of cidofovir and cyclic cidofovir.

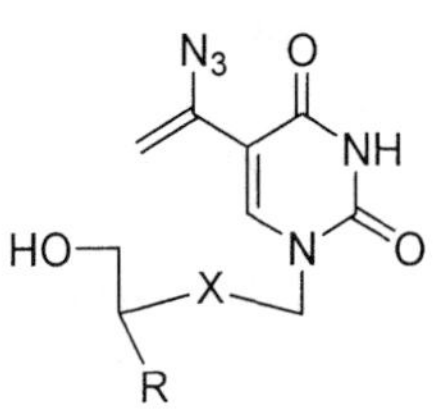

175 X = O, R = H
176 X = O, R = CH_2OH
177 X = C, R = CH_2OH

Figure 68. 5-(1-Azidovinyl)-substituted acyclic pyrimidine nucleosides.

The most important analog bearing a ribose sugar moiety is ribavirin (**178**, Figure 69), one of the first discovered compounds endowed with anti-respiratory syncytial virus, which has been shown to be effective as an anti-HBV and anti-HCV agent. It is currently approved, in combination with interferon-α, for the treatment of chronic hepatitis C.[297]

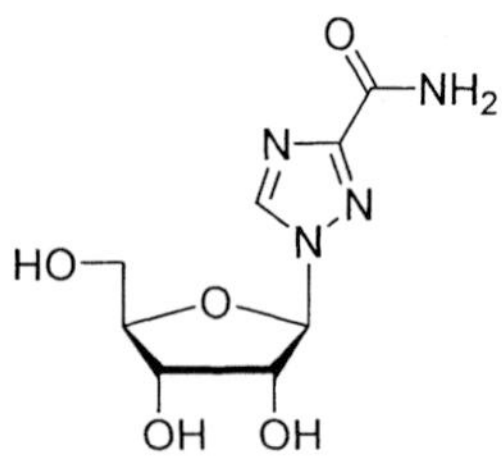

178 (ribavirin)

Figure 69. Ribofuranosyl nucleoside analogs

The unusual ring-expanded ("fat") ribonucleosides **179-182** (Figure 70) are endowed with anti-HBV activity in 2.2.15 cells, with EC_{50} in the micromolar range.[298] Analog **180** also displayed potent and selective antitumor activity against a number of leukemia, non-small cell lung cancer, colon cancer, CNS cancer, melanoma, ovarian cancer, renal cancer, prostate cancer and breast cancer.[299] Its tribenzoyl ester **181** showed similar activity profile as **180**, but was considerably more active probably because of better cell permeation.[299]

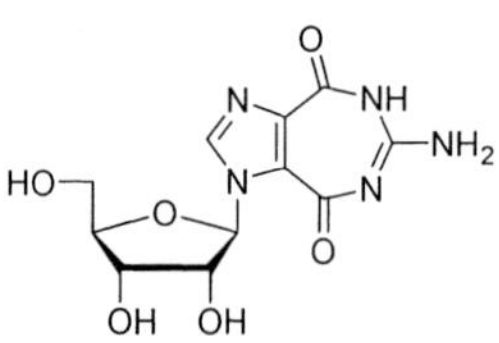

179

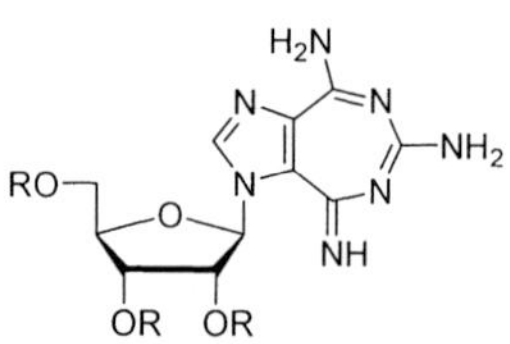

180 R = H
181 R = Bz

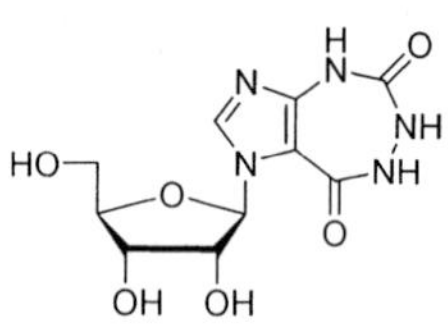

182

Figure 70. "Fat" nucleosides.

Neplanocin A (**183**, Figure 71) is a natural unsaturated nucleoside endowed with antiviral and antitumor properties.[300] Structure-activity relationships of neplanocin A analogs revealed interesting antiviral activity. Particularly, the D-cytosine and 5-F-cytosine derivatives **184** and **185** show anti-HIV activity (EC_{50} 0.06 and 5.34 μM, respectively) and are also the first nucleosides active against West Nile virus (EC_{50} 0.2 and 15 μM, respectively). However, their severe toxicity will prevent their development as antiviral agents.[301]

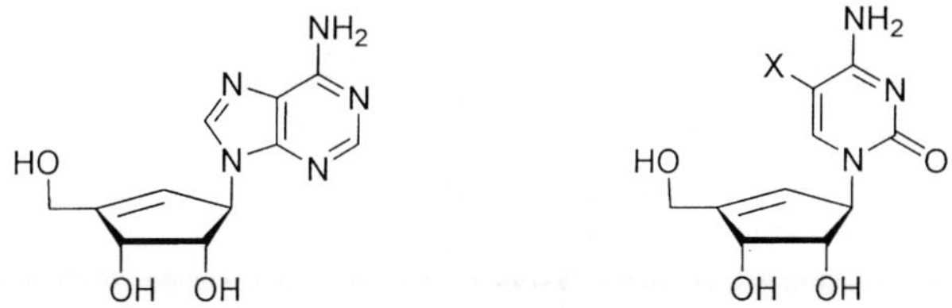

183 (neplanocin A)

184 X = H
185 X = F

Figure 71. Neplanocin A and its D-cytidine analogs.

Aristeromycin (**186**, Figure 72) is a natural carbocyclic ribonucleoside endowed with antitumor properties.[302] Its 5'-nor analog **187** is active against HCMV,[303,304] vaccinia virus and measles, while its (+)-enantiomer **188** is active against HBV.[303,305] The guanine analog **189** has anti-EBV activity.[303]

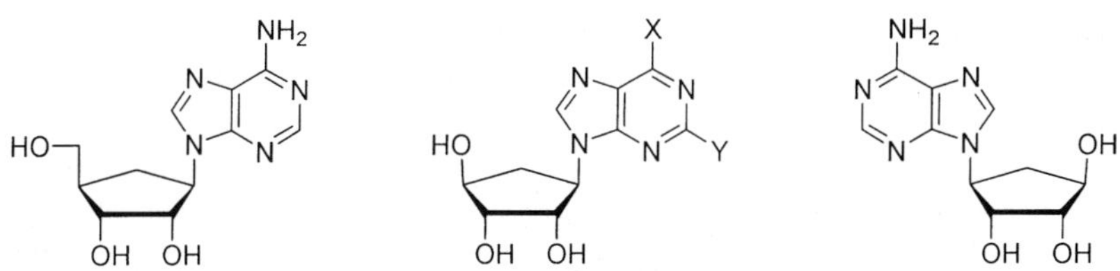

186 (aristeromycin)

187 X = NH_2, Y = H
189 X = OH, Y = NH_2

188

Figure 72. Aristeromycin and its 5'-nor analogs.

1.10. References

1. Kinchington, D. Recent advances in antiviral therapy. *J. Clin. Pathol.* **1999**, *52*, 89-94.
2. (a) Agrofoglio, L.; Suhas, E.; Fares, A.; Condom, R.; Challand, S. R.; Earl, R. A. Synthesis of carbocyclic nucleosides. *Tetrahedron* **1994**, *50*, 10611-10670. (b) Nair, V.; Jahnke, T. S. Antiviral activities of isomeric dideoxynucleosides of D- and L-related stereochemistry. *Antimicrob. Agents Chemother.* **1995**, *39*, 1017-1029. (c) Marquez, V. E. Carbocyclic nucleosides. *Adv. Antiviral Drug Design* **1996**, *2*, 89-145. (d) Freeman, S.; Gardiner, J. M. Acyclic nucleosides as antiviral compounds. *Mol. Biotechnol.* **1996**, *5*, 125-137. (e) Naesens, L.; Snoeck, R.; Andrei, G.; Balzarini, J.; Neyts, J.; De Clercq, E. HPMPC (cidofovir), PMEA (adefovir) and related acyclic nucleoside phosphonate analogs: a review of their

pharmacology and clinical potential in the treatment of viral infections. *Antiviral Chem. Chemther.* **1997**, *8*, 1-23. (f) Wang, P.; Hong, J. H.; Cooperwood, J. S.; Chu, C. K. Recent advances in L-nucleosides: chemistry and biology. *Antiviral Res.* **1998**, *40*, 19-44. (g) Crimmins, M. New development in the enantioselective synthesis of cyclopentyl carbocyclic nucleosides. *Tetrahedron* **1998**, *54*, 9229-9272. (h) Zhu, X.-F. The latest progress in the synthesis of carbocyclic nucleosides. *Nucleosides Nucleotides* **2000**, *19*, 651-690. (i) Gumina, G.; Song, G. -Y.; Chu, C. K. L-Nucleosides as chemotherapeutic agents. **2001**, *202*, 9-15. (j) Cooperwood, J. S.; Gumina, G.; Boudinot, F. D.; Chu, C. K. Nucleoside and nucleotide prodrugs. In *Recent Advances in Nucleosides: Chemistry and Chemotherapy*. C. K. Chu Ed. **2002**, 91-147.

3. (a) Périgaud, C.; Gosselin, G.; Imbach, J. -L. Nucleoside analogs as chemotherapeutic agents: a review. *Nucleosides Nucleotides* **1992**, *11*, 903-945. (b) Mansour, T. S.; Storer, R. Antiviral nucleosides. *Curr. Pharm. Design* **1997**, *3*, 227-264.
4. Huryn, D. M.; Okabe, M. AIDS-driven nucleoside chemistry. *Chem. Rev.* **1992**, *92*, 1745-1768.
5. (a) De Clercq, E. HIV inhibitors targeted at the reverse transcriptase. *AIDS Res. Hum. Retroviruses* **1992**, *8*, 119-134. (b) Schinazi, R. F.; Mead, J. R.; Feorino, P. M. Insight into HIV chemotherapy. *AIDS Res. Hum. Retroviruses* **1992**, *8*, 963-990. (c) De Clercq, E. Antiviral activity spectrum and target of action of different classes of nucleoside analogs. *Nucleosides Nucleotides* **1994**, *13*, 1271-1295. (d) De Clercq, E. In search of a selective antiviral chemotherapy. *Clin. Microbiol. Rev.* **1997**, *10*, 674-693. (e) Hong, J. H.; Choi, Y.; Chun, B. K.; Lee, K.; Chu, C. K. Current status of anti-HBV chemotherapy. *Arch. Pharm. Res.* **1998**, *21*, 89-105. (f) Tan, X.; Chu, C. K.; Boudinot, F. D. Development and optimization of anti-HIV nucleoside analogs and prodrugs: a review of their cellular pharmacology, structure-activity relationships and pharmacokinetics. *Adv. Drug Delivery Reviews* **1999**, *39*, 117-151. (g) Zemlicka, J. Enantioselectivity of the antiviral effects of nucleoside analogs. *Pharmacol. Ther.* **2000**, *85*, 251-266. (h) Gumina, G.; Song, G. -Y.; Chu, C. K. Advances in antiviral agents for hepatitis B virus. *Antiviral Chem. Chemother.* **2001**, *12*, 89-112.
6. Challand, R.; Young, R. J. *Antiviral Chemotherapy* Mann, J. Ed.; Biochemical & Medicinal Chemistry Series; Spektrum: UK, **1997**.
7. (a) Crick, F. *Life Itself*, Simon and Shuster: New York, **1981**. (b) Spadari, S.; Maga, G.; Verri, A.; Bendiscioli, A.; Tondelli, L.; Capobianco, M.; Colonna, F.; Garbesi, A.; Focher, F. Lack of stereospecificity of some cellular and viral enzymes involved in the synthesis of deoxyribonucleotides and DNA: molecular basis for the antiviral activity of unnatural L-β-nucleosides. *Biochimie* **1995**, *77*, 861-867.
8. Focher, F.; Maga, G.; Bendiscioli, A.; Capobianco, M.; Colonna, F.; Garbesi, A.; Spadari, S. Stereospecificity of human DNA polymerases α, γ, δ and ε, HIV-reverse transcriptase, HSV-1 DNA polymerase, calf thymus transferase and *Escherichia coli* DNA polymerase I in recognizing D- and L-thymidine 5'-triphosphate as substrate. *Nucleic Acids Research* **1995**, *23*, 2840-2847.
9. Spadari, S.; Ciarrocchi, G.; Focher, F.; Verri, A.; Maga, G.; Arcamone, F.; Iafrate, E.; Manzini, S.; Garbesi, A.; Tondelli, L. 5-Iodo-2'-deoxy-L-uridine and (*E*)-5-(2-bromovinyl)-2'-deoxy-L-uridine: selective phosphorylation by herpes simplex virus type 1 thymidine kinase, antiherpetic activity, and cytotoxicity studies. *Mol. Pharmacol.* **1995**, *47*, 1231-1238.
10. Faraj, A.; Loi, A. G.; Xie, M. Y.; Bryant, M.; Bridges, E. G.; Juodawlkis, A.; Pierra, C.; Dukhan, D.; Gosselin, G.; Imbach, J.-L.; Schinazi, R. F.; Pai, S. B.; Korba, B.; Marion, P.; Sommadossi, J.-P. β-L-Thymidine and β-L-2'-deoxycytidine are potent, selective and specific anti-hepatitis B virus agents. *Abstracts of Papers*, 3rd International Conference on Therapies for Viral Hepatitis, Maui, USA; International Society for Antiviral Therapy: **1999**; Abstract 119.

11. De Clercq, E. Vaccinia virus inhibitors as a paradigm for the chemotherapy of poxvirus infections. *Clin. Microb. Rev.* **2001**, *14*, 382-397.
12. Chu, C. K.; Ma, T.; Shanmuganathan, K.; Wang, C.; Xiang, Y.; Pai, S. B.; Tao, G.-Q.; Sommadossi, J. -P.; Cheng, Y. -C. Use of 2'-fluoro-5-methyl-β-L-arabinofuranosyluracil as a novel antiviral agent for hepatitis B virus and Epstein-Barr virus. *Antimicrob. Agents Chemother.* **1995**, *39*, 979-981.
13. (a) Prusoff, W. H. Synthesis and biological activities of iododeoxyuridine, an analog of thymidine. *Biochim. Biophys. Acta* **1959**, *32*, 295-296. (b) Kaufman, H. E.; Martola, E.; Heidelberger, C. Therapeutic antiviral action of 5-trifluoromethyl-2-deoxyuridine in herpes simplex keratitis. *Science* **1964**, *145*, 585-586. (c) De Clercq, E. Synthetic pyrimidine nucleoside analogs. In *Approaches to Antiviral Agents*, Harnden, M. R., Ed.; Macmillan: London, **1985**, pp. 57-99. (d) De Clercq, E. In *Advances in Drug Research*, Testa, B., Ed.; Academic Press Ltd.: London, **1988**, Vol. 17, pp. 1-59.
14. De Clercq, E.; Descamps, J.; De Somer, P.; Barr, P. J.; Jones, A. S.; Walker, R. T. (*E*)-5-(2-Bromovinyl)-2'-deoxyuridine: a potent and selective anti-herpes agent. *Proc. Natl. Acad. Sci. USA* **1979**, *76*, 2947-2951.
15. Desgranges, C.; Razaka, G.; Rabaud, M.; Bricaud, H.; Balzarini, J.; De Clercq, E. Phosphorolysis of (*E*)-5-(2-bromovinyl)-2'-deoxyuridine (BVDU) and other 5-substituted-2'-deoxyuridines by purified human thymidine phosphorylase and intact blood-platelets. *Biochem. Pharmacol.* **1983**, *32*, 3583-3590.
16. Harris, S. A.; McGuigan, C.; Andrei, G.; Snoeck, R.; De Clercq, E.; Balzarini, J. Synthesis and antiviral evaluation of phsphoramidate derivatives of (*E*)-5-(2-bromovinyl)-2'-deoxyuridine. *Antiviral Chem. Chemother.* **2001**, *12*, 293-300.
17. De Clercq, E.; Descamps, J.; Balzarini, J.; Giziewicz, J.; Barr, P. J.; Robins, M. J. Nucleic acid related compounds. 40. Synthesis and biological activities of 5-alkynyluracil nucleosides. *J. Med. Chem.* **1983**, *26*, 661-666.
18. Wigerinck, P.; Snoeck, R.; Claes, P.; De Clercq, E.; Herdewijn, P. Synthesis and antiviral activity of 5-heteroaryl-substituted 2'-deoxyuridines. *J. Med. Chem.* **1991**, *34*, 1767-1772.
19. (a) Wigerinck, P.; Pannecouque, C.; Snoeck, R.; Claes, P.; De Clercq, E.; Herdewijn, P. 5-(5-Bromothien-2-yl)-2'-deoxyuridine and 5-(5-chlorothien-2-yl)-2'-deoxyuridine are equipotent to (*E*)-5-(2-bromovinyl)-2'-deoxyuridine in the inhibition of herpes simplex virus type I replication. *J. Med. Chem.* **1991**, *34*, 2383-2389. (b) Wigerinck, P.; Kerremans, L.; Claes, P.; Snoeck, R.; Maudgal, P.; De Clercq, E.; Herdewijn, P. Synthesis and antiviral activity of 5-thien-2-yl-2'-deoxyuridines. *J. Med. Chem.* **1993**, *36*, 538-543.
20. Choi, Y.; Li, L.; Grill, S.; Gullen, E.; Lee, C. S.; Gumina, G.; Cheng, Y.-C.; Chu, C. K. Structure-activity relationships of (*E*)-5-(2-bromovinyl)uracil nucleosides and related analogs as anti-herpesvirus agents. *J. Med. Chem.* **2000**, *43*, 2538-2546.
21. Bryant, M. L.; Bridges, E. G.; Placidi, L.; Faraj. A.; Loi, A. G.; Pierra, C.; Dukhan, D.; Gosselin, G.; Imbach, J. -L.; Hernandez, B.; Juodawlkis, A.; Tennant, B.; Korba, B, Cote, P.; Marion, P.; Cretton-Scott, E.; Schinazi, R. F.; Sommadossi, J. -P. Antiviral L-nucleosides specific for hepatitis B virus infection. *Antimicrob. Agents Chemother.* **2001**, *45*, 229-235.
22. Pharmacology of β-L-thymidine and β-L-2'-deoxycytidine in HepG2 cells and primary human hepatocytes: relevance to chemotherapeutic efficacy against hepatitis B virus. *Antimicrob. Agents Chemother.* **2002**, *46*, 1728-1733.
23. McGuigan, C.; Yarnold, C. J.; Jones, G.; Velazquez, S.; Barucki, H.; Brancale, A.; Andrei, G.; Snoeck, R.; De Clercq, E.; Balzarini, J. Potent and selective inhibition of varicella-zoster virus (VZV) by nucleoside analogs with an unusual bicyclic base. *J. Med. Chem.* **1999**, *42*, 4479-4484.
24. Dyson, M. R.; Coe, P. L.; Walker, R. T. The synthesis and antiviral properties of *E*-5-(2-bromovinyl)-4'-thio-2'-deoxyuridine. *J. Chem. Soc., Chem. Commun.* **1991**, 741-742.

25. Rahim, S. G.; Trivedi, N.; Bogunovic-Batchelor, M. V.; Hardy, G. W.; Mills, G.; Selway, J. W. T.; Snowden, W.; Littler, E.; Coe, P. L.; Basnak, I.; Whale, R. F. Walker, R. T. Synthesis and anti-herpes virus activity of 2'-deoxy-4'-thiopyrimidine nucleosides. *J. Med. Chem.* **1996**, *39*, 789-795.
26. Basnak, I.; Sun, M.; Coe, P. L.; Walker, R. T. The synthesis of some 5-alkyl (cycloalkyl) substituted 2'-deoxy-4'-thiouridines. *Nucleosides Nucleotides* **1996**, *15*, 121-134.
27. Draanen, N. A.; Freeman, G. A.; Short, S. A.; Harvey, R.; Jansen, R.; Szczech, G.; Koszalka, G. W. Synthesis and antiviral activity of 2'-deoxy-4'-thio purine nucleosides. *J. Med. Chem.* **1996**, *39*, 538-542.
28. Uenishi, J.; Takahashi, K.; Motoyama, M.; Akashi, H.; Sasaki, T. Syntheses and antitumor activities of D- and L-2'-deoxy-4'-thio pyrimidine nucleosides. *Nucleosides Nucleotides* **1994**, *13*, 1347-1361.
29. Herdewijn, P.; De Clercq, E.; Balzarini, J.; Vanderhaeghe, H. Synthesis and antiviral activity of the carbocyclic analogs of (*E*)-5-(2-halovinyl)-2'-deoxyuridine and (*E*)-5-(2-halovinyl)-2'-deoxycytidines. *J. Med. Chem.* **1985**, *28*, 550-555.
30. Balzarini, J.; Baumgartner, H.; Bodenteich, M.; De Clercq, E.; Griengl, H. Synthesis and antiviral activity of the enantiomeric forms of carba-5-iodo-2'-deoxyuridine and carba-(*E*)-5-(2-bromovinyl)-2'-deoxyuridine. *J. Med. Chem.* **1989**, *32*, 1861-1865.
31. Shealy, Y. F.; O'Dell, C. A.; Shannon, W. M.; Arnett, G. Synthesis and antiviral activity of carbocyclic analogs of 2'-deoxyribofuranosides of 2-amino-6-substituted-purines and of 2-amino-6-substituted-8-azapurines. *J. Med. Chem.* **1984**, *27*, 1416-1421.
32. Secrist III, J. A.; Montgomery, J. A.; Shealy, Y. F.; O'Dell, C. A.; Clayton, S. J. Resolution of racemic carbocyclic analogs of purine nucleosides through the action of adenosine deaminase. Antiviral activity of the carbocyclic 2'-deoxyguanosine enantiomers. *J. Med. Chem.* **1987**, *30*, 746-749.
33. (a) Bennett Jr., L. L.; Shealy, Y. F.; Allan, P. W.; Rose, L. M.; Shannon, W. M.; Arnett, G. Phosphorylation of the carbocyclic analog of 2'-deoxyguanosine in cells infected with herpes viruses. *Biochem. Pharmacol.* **1990**, *40*, 1515-1522. (b) Bennett Jr., L. L.; Parker, W. B.; Allan, P. W.; Rose, L. M.; Shealy, Y. F.; Secrist III, J. A.; Montgomery, J. A.; Arnett, G.; Kirkman, R. L.; Shannon, W. M. Phosphorylation of the enantiomers of the carbocyclic analog of 2'-deoxyguanosine in cells infected with herpes simplex virus type 1 and in uninfected cells. Lack of enantiomeric selectivity with viral thymidine kinase. *Mol. Pharmacol.* **1993**, *44*, 1258-1266.
34. Bennett Jr., L. L.; Allan, P. W.; Arnett, G.; Shealy, Y. F.; Shewach, D. S.; Mason, W. S.; Fourel, I.; Parker, W. B. Metabolism in human cells of the D and L enantiomers of the carbocyclic analog of 2'-deoxyguanosine: substrate activity with deoxycytidine kinase, mitochondrial deoxyguanosine kinase, and 5'-nucleotidase. *Antimicrob. Agents Chemother.* **1998**, *42*, 1045-1051.
35. Price, P. M.; Banjit, R.; Acs, G. Inhibition of the replication of hepatitis B virus by the carbocyclic analog of 2'-deoxyguanosine. *Proc. Natl. Acad. Sci. USA* **1989**, *86*, 8541-8544.
36. Price, P. M.; Banerjee, R.; Jeffrey, A. M.; Acs, G. The mechanism of inhibition of hepatitis B virus replication by the carbocyclic analog of 2'-deoxyguanine. *Hepatology* **1992**, *16*, 8-12.
37. Jansen, R. W.; Johnson, L. C.; Averett, D. R. High capacity *in vitro* assessment of anti-hepatitis B virus compound selectivity by a virion-specific polymerase chain reaction assay. *Antimicrob. Agents Chemother.* **1993**, *37*, 441-447.
38. (a) Slusarchyk, W. A.; Field, A. K.; Greytok, J. A.; Taunk, P.; Tuomari, A. V.; Young, M.G.; Zahler, R. 4-Hydroxy-3-(hydroxymethyl)-2-methylenecyclopentyl purines and pyrimidines, a new class of anti-herpesvirus agents. *Antiviral Res.* **1992**, *17*, suppl.1, 98. (b) Bisacchi, G. S.; Chao, S. T.; Bachard, C.; Daris, J. P.; Innaimo, S.; Jacobs, G. A.; Kocy, O.; Lapointe, P.; Martel, A.; Merchant, Z.; Slusarchyk, W. A.; Sundeen, J. E.; Young, M. G.; Colonno, R.; Zahler, R. BMS-200475, A novel carbocyclic 2'-deoxyguanosine analog with potent and selective anti-hepatitis B virus activity in vitro. *Bioorg. Med.*

Chem. Lett. **1997**, *7*, 127-132. (c) Innaimo, S. F.; Seifer, M.; Bisacchi, G. S.; Standring, D. N.; Zahler, R.; Colonno, R. J. Identification of BMS-200475 as a potent and selective inhibitor of hepatitis B virus. *Antimicrob. Agents Chemother.* **1997**, *41*, 1444-1448.

39. Genovesi, E. V.; Lamb, L.; Medina, I.; Taylor, D.; Seifer, M.; Innaimo, S.; Colonno, R. J.; Standring, D. N.; Clark, J. M. Efficacy of the carbocyclic 2'-deoxyguanosine nucleoside BMS-200475 in the woodchuck model of hepatitis B virus infection. *Antimicrob. Agents Chemother.* **1998**, *42*, 3209-3217.
40. Yamanaka, G.; Wilson, T.; Innaimo, S.; Bisacchi, G. S.; Egli, P.; Rinehart, J. K.; Zahler, R.; Colonno, R. J. Metabolic studies on BMS-200475, a new antiviral compound active against hepatitis B virus. *Antimicrob. Agents Chemother.* **1999**, *43*, 190-193.
41. (a) Shannon, W. M. In *Antiviral Agents and Viral diseases of Man*; Galasso, G. J., Ed.; Raven Press: New York, **1984**; Chapter 3, pp. 55-121. (b) Robins, R. K.; Revenkar, G. R. In *Antiviral Drug Development. A Multidisciplinary Approach*; De Clercq, E; Walker, R. T., Eds.; Nato Advanced Institutes Series, Series A: Life Sciences; Plenum Press: New York, **1987**; Vol. 143, pp. 11-36.
42. Jacyna, M. R.; Thomas, H. C. Antiviral therapy: hepatitis B virus. *Br. Med. Bull.* **1990**, *46*, 368-382.
43. Machida, H.; Sakata, S.; Kumnaka, A.; Yoshino, H. Anti-herpesviral and anti-cellular effects of 1-β-D-arabinofuranosyl-*E*-5-(2-halogenovinyl)uracils. *Antimicrob. Agents Chemother.* **1981**, *20*, 47-52.
44. (a) Miyauchi, S.; Imaoka, T.; Okada, T.; Motoyama, M.; Kawaguchi, T.; Akiyama, H.; Odomi, M. Leukopenia-inducing effect of a combination of a new 5-fluorouracil (5-FU)-derived drug, BOF-A2 (emitefur), with other 5-FU-derived drugs or BV-araU (sorivudine) in rats. *Jpn. J. Pharmacol.* **1996**, *70*, 139-148. (b) Watabe, T.; Okuda, H.; Ogura, K. Lethal drug interaction of the new antiviral, sorivudine, with anticancer prodrugs of 5-fluorouracil. *Yakugaku Zasshi* **1997**, *117*, 910-921.
45. Lin, T.-S.; Luo, M.-Z.; Liu, M.-C. Synthesis of several pyrimidine L-nucleoside analogs as potential antiviral agents. *Tetrahedron* **1995**, *51*, 1055-1068.
46. Machida, H.; Sakata, S. 5-Substituted arabinofuranosyluracil nucleosides as antiherpesvirus agents: from araT to BV-araU. In *Nucleosides and Nucleotides as Antitumor and Antiviral Agents*; Chu, C. K.; Baker, D. C., Eds.; Plenum Press: New York, **1993**; pp. 245-264.
47. (a) Rahim, S. G.; Trevidi, N.; Selway, J.; Darby, G.; Collins, P.; Powell, K. L.; Purifoy, D. J. M. 5-Alkynyl pyrimidine nucleosides as potent selective inhibitors of Varicella-Zoster Virus. *Antiviral Chem. Chemother.* **1992**, *3*, 293-297. (b) Purifoy, D. J. M.; Beauchamp, L. M.; de Miranda, P.; Ertl, P.; Lacey, S.; Roberts, G.; Rahim, S. G.; Darby, G.; Krenitsky, T. A.; Powell, K. L. Review of research leading to new anti-herpesvirus agents in clinical and 882C, a specific agent for varicella zoster virus. *J. Med. Virol.* **1993**, suppl. 1, 139-145.
48. (a) Machida, H.; Ashida, N.; Miura, S.; Endo, M.; Yamada, K.; Kitano, K.; Yoshimura, Y.; Sakata, S.; Ijichi, O.; Eizuru, Y. Anti-herpesvirus activity profile of 4'-thioarabinofuranosyl purine and uracil nucleosides and activity of 1-β-D-2'-fluoro-4'-thioarabinofuranosyl guanine and 2,6-diaminopurine against clinical isolates of human cytomegalovirus. *Antiviral Res.* **1998**, *39*, 129-137. (b) Satoh, H.; Yoshimura, Y.; Sakata, S.; Miura, S.; Machida, H. Synthesis of L-enantiomers of 4'-thioarabinofuranosyl pyrimidine nucleosides. *Bioorg. Med. Chem. Lett.* **1998**, *8*, 989-992.
49. (a) Watanabe, K. A.; Reichman, U.; Hirota, K.; Lopez, C.; Fox, J. J. Nucleosides. 110. Synthesis and anti-herpes virus activity of some 2'-fluoro-2'-deoxy-arabinofuranosyl pyrimidine nucleosides. *J. Med. Chem.* **1979**, *22*, 21-24. (b) Watanabe, K. A.; Su, T.-L.; Klein, R. S.; Chu, C. K.; Matsuda, A.; Chun, M. W.; Lopez, C.; Fox, J. J. Nucleosides. 123. Synthesis of antiviral nucleosides, 5-substituted 1-(2-deoxy-2-halogeno-β-D-arabinofuranosyl) cytosines and uracils, some structure-activity relationships. *J. Med. Chem.* **1983**, *26*, 152-156. (c) Watanabe, K. A.; Su, T. -L.; Reichman, U.; Greenberg, N.; Lopez, C.; Fox, J. J. Nucleosides. 129. Synthesis of antiviral nucleosides, 5-alkenyl-1-(2-deoxy-2-fluoro-β-D-

arabinofuranosyl)uracils. *J. Med. Chem.* **1984**, *27*, 91-94.

50. Kreis, W.; Damin, L.; Colacino, J.; Lopez, C. *In vitro* metabolism of 1-β-D-arabinofuranosylcytosine and 1-β-D-2'-fluoroarabino-5-iodocytosine in normal and herpes simplex type 1 infected cells. *Biochem. Pharmacol.* **1982**, *31*, 767-773.
51. Abbruzzese, J. L.; Schmidt, S.; Raber, M. N.; Levy, J. K.; Castellanos, A. M.; Legha, S. S.; Krakoff, I. H. Phase I trial of 1-(2-deoxy-2-fluoro-β-D-arabinofuranosyl)-5-methyluracil (FMAU) terminated by severe neurologic toxicity. *Invest. New Drugs* **1989**, *7*, 195-201.
52. (a) Macilwain, C. NIH, FDA seek lessens from the hepatitis B drug trial deaths. *Nature* **1993**, *364*, 275-275. (b) Touchette, N. HBV-drug deaths prompt restudy of similar antivirals. *J. NIH Res.* **1993**, *5*, 33-36.
53. (a) Parker, W. B.; Cheng, Y. -C. Mitochondrial toxicity of antiviral nucleoside analogs. *J. NIH Res.* **1994**, *6*, 57-61. (b) Cui, L.; Yoon, S.; Schinazi, R. F.; Sommadossi, J.-P. Cellular and molecular events leading to mitochondrial toxicity of 1-(2-deoxy-2-fluoro-1-β-D-arabinofuranosyl)-5-iodouracil in human liver cells. *J. Clin. Invest.* **1995**, *95*, 555-563.
54. Chu, C. K.; Boudinot, F. D.; Peek, S. F.; Hong, J. H.; Choi, Y. S.; Korba, B. E.; Gerin, J. L.; Cote, P. J.; Tennant, B. C.; Cheng, Y. -C. Preclinical investigation of L-FMAU as an anti-hepatitis B virus agent. In *Therapies for viral Hepatitis*; Schinazi, R. F.; Sommadossi, J.-P.; Thomas, H. C. Eds.; International Medical Press: UK, **1999**; Chapter 32, pp 303-312.
55. Jeong, L. S.; Moon, H. R.; Yoo, S. J.; Lee, S. N.; Chun, M. W.; Lim, Y. -H. Ring contraction and rearrangement of 4-thiofuranose deravatives observed during DAST fluorination in the synthesis of L-2'-"up"-fluoro-4'-thiothymidine. *Tetrahedron Lett.* **1998**, *39*, 5201-5204.
56. Borthwick, A. D.; Evans, D. N.; Kirk, B. E.; Biggadike, K.; Exall, A. M.; Youds, P.; Roberts, S. M.; Knight, D. J.; Coates, J. A. V. Fluoro carbocyclic nucleosides: synthesis and antiviral activity of 2'- and 6'-fluoro carbocyclic pyrimidine nucleosides including carbocyclic 1-(2-deoxy-2-fluoro-β-D-arabinofuranosyl)-5-methyluracil and carbocyclic 1-(2-deoxy-2-fluoro-β-D-arabinofuranosyl)-5-iodouracil. *J. Med. Chem.* **1990**, *33*, 179-186.
57. Borthwick, A. D.; Butt, S.; Biggadike, K.; Exall, A. M.; Roberts, A. M.; Youds, P. M.; Kirk, B. E.; Booth, B. R.; Cameron, J. M.; Cox, S. W.; Marr, C. L. P.; Shill, M. D. Synthesis and enzymatic resolution of carbocyclic 2'-*ara*-fluoro-guanosine: a potent new anti-herpetic agent. *J. Chem. Soc., Chem. Commun.* **1988**, 657-658.
58. Biggadike, K.; Borthwick, A. D.; Exall, A. M.; Kirk, B. E.; Ward, R. A. 2'-Deoxy-2'-fluoro-*ara*-aristeromycin, a new anti-herpes agent: the first direct introduction of a 2'-fluoro substituent into a carbocyclic nucleoside. *J. Chem. Soc., Chem. Commun.* **1988**, 898-900.
59. Hertel, L. W.; Kroin, J. S.; Misner, J. W.; Tustin, J. M. Synthesis of 2-deoxy-2,2-difluoro-D-ribose and 2-deoxy-2,2-difluoro-D-ribofuranosyl nucleosides. *J. Org. Chem.* **1988**, *53*, 2406-2409.
60. Plunkett, W.; Huang, P.; Ganghi, V. Gemcitabine: actions and interactions. *Nucleosides Nucleotides* **1997**, *16*, 1261-1270.
61. Gandhi, V.; Mineishi, S.; Huang, P.; Chapman, A. J.; Young, Y.; Chen, F.; Nowak, B.; Chubb, S.; Hertel, L. W.; Plunkett, W. Cytotoxicity, metabolism, and mechanism of action of 2',2'-difluorodeoxyguanosine in Chinese hamster ovary cells. *Cancer Res.* **1995**, *55*, 1517-1524.
62. Kotra, L. P.; Xiang, Y.; Newton, M. G.; Schinazi, R. F.; Cheng, Y. -C.; Chu, C. K. Structure-activity relationships of 2'-deoxy-2',2'-difluoro-L-*erythro*pentofuranosyl nucleosides. *J. Med. Chem.* **1997**, *40*, 3635-3644.
63. Yoshimura, Y.; Kitano, K.; Yamada, K.; Satoh, H.; Watanabe, M.; Miura, S.; Sakata, S.; Sasaki, T.; Matsuda, A. A novel synthesis of 2'-modified 2'-deoxy-4'-thiocytidines from D-glucose. *J. Org. Chem.* **1997**, *62*, 3140-3152.

64. Jeong, L. S.; Moon, H. R.; Choi, Y. J.; Chun, M. W.; Kim, H. O. A short and efficient synthesis of L-thioarabinose derivative: a versatile synthon for the synthesis of L-2'-deoxy-2',2'-disubstituted-4'-thionucleosides. *J. Org. Chem.* **1998**, *63*, 4821-4825.
65. Takenuki, K.; Matsuda, A.; Ueda, T.; Sasaki, T.; Fujii, A.; Yamagami, K. Design, synthesis, and antineoplastic activity of 2'-deoxy-2'-methylidenecytidine (DMDC). *J. Med. Chem.* **1988**, *31*, 1063-1064.
66. Yamagami, K.; Fujii, A.; Arita, M.; Okumoto, T.; Sakata, S.; Matsuda, A.; Ueda, T.; Sasaki, T. Antitumor activity of 2'-deoxy-2'-methylidenecytidine, a new 2'-deoxycytidine derivative. *Cancer Res.* **1991**, *51*, 2319-2323.
67. Baker, C. H.; Banzon, J.; Bollinger, J. M.; Stubbe, J.; Samano, V.; Robins, M. J.; Lippert, B.; Jarvi, E.; Resvick, R. 2'-Deoxy-2'-methylidenecytidine and 2'-deoxy-2',2'-difluorocytidine 5'-diphosphates: potent mechanism-based inhibitors of ribonucleotide reductase. *J. Med. Chem.* **1991**, *34*, 1879-1884.
68. Machida, H.; Sakata, S.; Ashida, N.; Takenuki, K.; Matsuda, A. *In vitro* anti-herpesvirus activities of 5-substituted 2'-deoxy-2'-methylidene pyrimidine nucleosides. *Antiviral Chem. Chemother.* **1993**, *4*, 11-17.
69. Satoh, H.; Yoshimura, Y.; Watanabe, M.; Ashida, N.; Ijichi, K.; Sakata, S.; Machida, H. Synthesis and antiviral activities of 5-substituted 1-(2-deoxy-2-*C*-methylene-4-thio-β-D-*erythro*-pentofuranosyl)uracils. *Nucleosides Nucleotides* **1998**, *17*, 65-79.
70. Maag, H.; Rydzewski, R. M.; McRoberts, M. J.; Crawford-Ruth, D.; Verheyden, J. P. H.; Prisbe, E. J. Synthesis and anti-HIV activity of 4'-azido- and 4'-methoxynucleosides. *J. Med. Chem.* **1992**, *35*, 1440-1451.
71. Chen, M. S.; Suttmann, R. T.; Wu, J. C.; Prisbe, E. J. Metabolism of 4'-azidothymidine. *J. Biol. Chem.* **1992**, *267*, 257-260.
72. Chen, M. S.; Suttmann, R. T.; Papp, E.; Cannon, P. D.; McRoberts, M. J.; Bach, C.; Copeland, W. C.; Wang, T. S.-F. Selective action of 4'-azidothymidine triphosphate on reverse transcriptase of human immunodeficiency virus type 1 and human DNA polymerases α and β. *Biochemistry* **1993**, *32*, 6002-6010.
73. Nomura, M.; Shuto, S.; Tanaka, M.; Sasaki, T.; Mori, S.; Shigeta, S.; Matsuda, A. Nucleosides and nucleotides. 185. Synthesis and biological activities of 4'α-*C*-branched-chain sugar pyrimidine nucleosides. *J. Med. Chem.* **1999**, *42*, 2901-2908.
74. Kitano, K.; Machida, H.; Miura, S.; Ohrui, H. Synthesis of novel 4'-*C*-methyl-pyrimidine nucleosides and their biological activities. *Bioorg. Med. Chem. Lett.* **1999**, *9*, 827-830.
75. Borthwick, A. D.; Biggadike, K. Eur. Pat. Appl. 0345 016, 1989.
76. Ohrui, H.; Kohgo, S.; Kitano, K.; Sakata, S.; Kodama, E.; Yoshimura, K.; Matsuoka, M.; Shigeta, S.; Mitsuya, H. Syntheses of 4'-*C*-ethynyl-β-D-*arabino*- and 4'-*C*-ethynyl-2'-deoxy-β-D-*ribo*-pentofuranosy lpyrimidines and –purines and evaluation of their anti-HIV activity. *J. Med. Chem.* **2000**, *43*, 4516-4525.
77. Marquez, V. E.; Siddiqui, M. A.; Ezzitouni, A.; Russ, P.; Wang, J.; Wagner, R. W.; Matteucci, M. D. Nucleosides with a twist. Can fixed forms of sugar ring pucker influence biological activity in nucleosides and oligonucleotides? *J. Med. Chem.* **1996**, *39*, 3739-3747.
78. Huheihel, M.; Salah, L.; Konson, A.; Manor, E.; Ford Jr., H.; Marquez, V. E.; Johns, D. G.; Agbaria, R. Intracellular phosphorylation of *N*-methanocarbathymidine, a novel antiherpetic agent. *Antiviral Res.* **2000**, *53*, A55-A55.
79. Chun, B. K.; Olgen, S.; Hong, J. H.; Newton, M. G.; Chu, C. K. Enantiomeric syntheses of conformationally restricted D- and L-2',3'-dideoxy-2',3'-*endo*-methylene nucleosides from carbohydrate chiral templates. *J. Org. Chem.* **2000**, *65*, 685-693.
80. Herdewijn, P.; De Clercq, E. The cyclohexene ring as bioisostere of a furanose ring: synthesis and antiviral activity of cyclohexenyl nucleosides. *Bioorg. Med. Chem. Lett.* **2001**, *11*, 1591-1597.
81. Wang, J.; Froeyen, M.; Hendrix, C.; Andrei, C.; Snoeck, R.; Lescrinier, E.; De Clercq, E.; Herdewijn, P. (D)- And (L)-cyclohexenyl-G, a new class of antiviral agents: synthesis, conformational analysis, molecular modeling, and biological activity. *Nucleosides Nucleotides* **2001**, *20*, 727-730.

82. Mitsuya, H.; Broder, S. Inhibition of the in vitro infectivity and cytopathic effect of human T-lymphotropic virus type III/lymphadenopathy-associated virus (HTLV-III/LAV) by 2',3'-dideoxynucleosides. *Proc. Natl. Acad. Sci. USA* **1986**, *83*, 1911-1915.
83. (a) Mitsuya, H.; Weinhold, K. J.; Furman, P. A.; St Clair, M. H.; Lehman, S. N.; Gallo, R. C.; Bolognesi, D.; Barry, D. W.; Broder, S. 3'-Azido-3'-deoxythymidine (BW A509U): an antiviral agent that inhibits the infectivity and cytopathic effect of human T-lymphotropic virus type III/lymphadenopathy-associated virus in vitro. *Proc. Natl. Acad. Sci. USA* **1985**, *82*, 7096-7100. (b) Perno, C.-F.; Yarchoan, R.; Cooney, D. A.; Hartman, N. R.; Gartner, S.; Popovic, M.; Hao, Z.; Gerrard, T. L.; Wilson, Y. A. Johns, D. G.; Broder, S. Inhibition of human immunodeficiency virus (HIV-1/HTLV-III_{Ba-L}) replication in fresh and cultured human peripheral blood monocytes/macrophages by azidothymidine and related 2',3'-dideoxynucleosides. *J. Exp. Med.* **1988**, *168*, 1111-1125.
84. (a) Lin, T.-S.; Schinazi, R. F.; Prusoff, W. H. Potent and selective in vitro activity of 3'-deoxythymidine-2'-ene (3'-deoxy-2',3'-didehydrothymine) against human immunodeficiency virus. *Biochem. Pharmacol.* **1987**, *36*, 2713-2718. (b) Baba, M.; Pauwels, R.; Herdewijn, P.; De Clercq, E.; Desmyter, J.; Vandeputte. Both 2',3'-dideoxythymidine and its 2',3'-unsaturated derivative (2',3'-dideoxythymidinene) are potent and selective inhibitors of human immunodeficiency virus replication *in vitro*. *Biochem. Biophys. Res. Commun.* **1987**, *142*, 128-134. (c) Hamamoto, Y.; Nakashima, H.; Matsuda, A.; Ueda, T.; Yamamoto, N. Inhibitory effect of 2',3'-didehydro-2',3'-dideoxynucleosides on infectivity, cytopathic effects, and replication of human immunodeficiency virus. *Antimicrob. Agents Chemother.* **1987**, *31*, 907-910.
85. Soudeyns, H.; Yao, X. J.; Gao, Q. Anti-human immunodeficiency virus type 1 activity and in vitro toxicity of 2'-deoxy-3'-thiacytidine (BCH-189), a novel heterocyclic nucleoside analog. *Antimicrob. Agents Chemother.* **1991**, *35*, 1386-1390.
86. Daluge, S. M.; Good, S. S.; Faletto, M. B.; Miller, W. H.; St Clair, M. H.; Boone, L. R.; Tisdale, M.; Parry, N. R.; Reardon, J. E.; Dornsife, R. E.; Averett, D. R.; Krenitsky, T. A. 1592U89, a novel carbocyclic nucleoside analog with potent, selective anti-human immunodeficiency virus activity. *Antimicrob. Agents Chemother.* **1997**, *41*, 1082-1093.
87. Balzarini, J.; Aquaro, S.; Perno, C. -F.; Witvrouw, M.; Holy, A.; De Clercq, E. Activity of the (*R*)-enantiomers of 9-(2-phosphonylmethoxypropyl)adenine and 9-(2-phosphonylmethoxypropyl)-2,6-diaminopurine against human immunodeficiency virus in different human cell systems. *Biochem. Biophys. Res. Comm.* **1996**, *219*, 337-341.
88. Balzarini, J.; Holy, A.; Jindrich, J.; Naesens, L.; Snoeck, R.; Schols, D.; De Clercq E. Differential antiherpesvirus and antiretrovirus effects of the (*S*) and (*R*) enantiomers of acyclic nucleoside phosphonates: potent and selective *in vitro* and *in vivo* antiretrovirus activities of (*R*)-9-(2-phosphonylmethoxypropyl)-2,6-diaminopurine. *Antimicrob. Agents and Chemother.* **1993**, *37*, 332-338.
89. First nucleotide analog for treatment of HIV-1 infection; dosed once daily. *Formulary* **2001**, *36*, 826-827.
90. Coleman, C.; Ross, J.; Reddy, P. Tenofovir disoproxil fumarate. The first nucleotide reverse transcriptase inhibitor for treatment of patients with HIV-1 infection. *Formulary* **2002**, *37*, 15-22.
91. Norbeck, D. W.; Spanton, S.; Broder, S.; Mitsuya, H. (±)-Dioxolane-T ((±)-1-[(2β,4β)-2-(hydroxymethyl-4-dioxolanyl]thymine). A new 2',3'-dideoxy nucleoside prototype with in vitro activity against HIV. *Tetrahedron Lett.* **1989**, *30*, 6263-6266. (b) Belleau, B., Dixit, D.M., Ba, N.N., Brown, W.L., Gulini, U., Lafleur, D., Cameron, J.M. Synthesis and anti-HIV activity of (±)-1,3-dioxolane nucleoside analogs and their enzymatic resolution. *Abstr. Papers Am. Chem. Soc.* **1992**, MEDI 138.
92. Storer, R.; Clemens, I. R.; Lamont, B.; Noble, S. A.; Williamson, C.; Belleau, B. The resolution and absolute stereochemistry of enantiomers of *cis*-1,2-hydroxymethyl-1,3-oxathiolan-5-yl-cytosine BCH-189 equipotent anti-HIV agents. *Nucleosides Nucleotides* **1993**, *12*, 225-236.

93. (a) Johnson, M. A.; Fridland, A. Phosphorylation of 2',3'-dideoxyinosine by cytosolic 5'-nucleotidase of human lymphoid cells. *Mol. Pharmacol.* **1989**, *36*, 291-295. (b) Ahluwalia, G.; Cooney, D. A.; Mitsuya, H.; Fridland, A.; Flora, K. P.; Hao, Z.; Dalal, M.; Broder, S.; Johns, D. G. Initial studies on the cellular pharmacology of 2',3'-dideoxyinosine, an inhibitor or HIV infectivity. *Biochem. Pharmacol.* **1987**, *36*, 3797-3800. (c) Johnson, M. A.; Ahluwalia, G.; Connelly, M. C.; Cooney, D. A.; Broder, S.; Johns, D. G.; Fridland, A. Metabolic pathways for the activation of the antiretroviral agent 2',3'-dideoxyadenosine in human lymphoid cells. *J. Biol. Chem.* **1988**, *263*, 15354-15357.

94. Ahluwalia, G.; Johnson, M. A.; Fridland, A.; Cooney, D. A.; Broder, S.; Johns, D. G. Cellular pharmacology of the anti-HIV agent 2',3'-dideoxyadenosine. *Proc. Am. Acad. Cancer Res.* **1988**, *29*, 349.

95. (a) Starnes, M. C.; Cheng, Y. -C. Cellular metabolism of 2',3'-dideoxycytidine, a compound against human immunodeficiency virus. *J. Biol. Chem.* **1987**, *262*, 988-991. (b) Chen, C. -H.; Cheng, Y. -C. Delayed cytotoxicity and selective loss of mitochondrial DNA in cells treated with the anti-human immunodeficiency virus compound 2',3'-dideoxycytidine. *J. Biol. Chem.* **1989**, *264*, 11934-11937.

96. Lewis, L. D.; Hamzeh, F. M.; Lietman, P. S. Ultrastructure changes associated with reduced mitochonrial DNA and impaired mitochondrial function in the presence of 2',3'-dideoxycytidine. *Antimicrob. Agents Chemother.* **1992**, *36*, 2061-2065.

97. Lin, T. -S.; Luo, M. -Z.; Liu, M. -C.; Pai, S. B.; Dutschman, G. E.; Cheng, Y. -C. Synthesis and biological evaluation of 2',3'-dideoxy-L-pyrimidine nucleosides as potential antiviral agents against human immunodeficiency virus (HIV) and hepatitis B virus (HBV). *J. Med. Chem.* **1994**, *37*, 798-803.

98. Pélicano, H.; Pierra, C.; Eriksson, S.; Gosselin, G.; Imbach, J. -L.; Maury, G. Enzymatic properties of the unnatural β-L-enantiomers of 2',3'-dideoxyadenosine and 2',3'-didehydro-2',3'-dideoxyadenosine. *J. Med. Chem.* **1997**, *40*, 3969-3973.

99. (a) Secrist III, J. A.; Riggs, R. M.; Tiwari, K. N.; Montgomery, J. A. Synthesis and anti-HIV activity of 4'-thio-2',3'-dideoxynucleosides. *J. Med. Chem.* **1992**, *35*, 533-538. (b) Young, R. J.; Shaw-Ponter, S.; Thomson, J. B.; Miller, J. A.; Cumming, J. G.; Pugh, A. W.; Rider, P. Synthesis and antiviral evaluation of enantiomeric 2',3'-dideoxy- and 2',3'-didehydro-2',3'-dideoxy-4'-thionucleosides. *Bioorg. Med. Chem. Lett.* **1995**, *5*, 2599-2604. (c) Beres, J.; Sagi, G.; Tomoskozi, I.; Gruber, L.; Gulacsi, E.; Otvos, L. Stereospecific synthesis of (-)-carbocyclic 2',3'-dideoxythymidine, a potential anti-AIDS agent. *Tetrahedron Lett.* **1988**, *29*, 2681-2684. (d) Wolffkugel, D.; Halazy, S. Studies towards the synthesis of the saturated and unsaturated carbocyclic methylene phosphonate analogs of dideoxyadenosine. *Nucleosides Nucleotides* **1993**, *12*, 279-294. (e) Wang, P.; Gullen, B.; Newton, M. G.; Cheng, Y. -C.; Schinazi, R. F.; Chu, C. K. Asymmetric synthesis and antiviral activities of L-carbocyclic 2',3'-didehydro-2',3'-dideoxy and 2',3'-dideoxy nucleosides. *J. Med. Chem.* **1999**, *42*, 3390-3399. (d) Rassu, G.; Zanardi, F.; Battistini, L.; Gaetani, E.; Casiraghi, G. Expeditious syntheses of sugar-modified nucleosides and collections thereof exploiting furan-, pyrrole-, and thiophene-based siloxy dienes. *J. Med. Chem.* **1997**, *40*, 168-180.

100. Lin, T. -S.; Guo, J. -Y.; Schinazi, R. F.; Chu, C. K.; Xiang, J. -N.; Prusoff, W. H. Synthesis and antiviral activity of various 3'-azido analogs of pyrimidine deoxyribonucleosides against human immunodeficiency virus (HIV-1, HTLV/LAV). *J. Med. Chem.* **1988**, *31*, 336-340.

101. Baba, M.; Pauwels, R.; Balzarini, J.; Herdewijn, P.; De Clercq, E. Selective inhibition of human immunodeficiency virus (HIV) by 3'-azido-2',3'-dideoxyguanosine *in vitro*. *Biochem. Biophys. Res. Commun.* **1987**, *145*, 1080-1086.

102. Balzarini, J.; Baba, M.; Pauwels, R.; Herdewijn, P.; Wood, S. G.; Robins, M. J.; De Clercq, E. Potent and selective activity of 3'-azido-2,6-diaminopurine-2',3'-dideoxyriboside, 3'-fluoro-2,6-diaminopurine-2',3'-dideoxyriboside, and 3'-fluoro-2',3'dideoxyguanine against human immunodeficiency virus. *Mol.*

Pharmacol. **1988**, *33*, 243-249.

103. Furman, P. A.; Fyfe, J. A.; St Clair, M. H.; Weihold, K.; Rideut, J. L.; Freeman, G. A.; Lehrman, S. N.; Bolognesi, D. P.; Broder, S.; Mitsuya, H.; Barry, D. W. Phosphorylation of 3'-azido-3'-deoxythymidine and selective interaction of the 5'-triphosphate with human immunodeficiency virus reverse transcriptase. *Proc. Natl. Acad. Sci. USA* **1986**, *83*, 8333-8337.
104. St Clair, M. H.; Richard, C. A.; Spector, T.; Weinhold, K.; Miller, W. H.; Langloir, A. J.; Furman, P. A. 3'-Azido-3'-deoxythymidine triphosphate as an inhibitor and substrate of purified human immunodeficiency virus reverse transcriptase. *Antimicrob. Agents Chemother.* **1987**, *31*, 1972-1977.
105. Ono, K.; Nakane, H.; Herdewijn, P.; Balzarini, J.; De Clercq. Differential inhibitory effects of several pyrimidine 2',3'-dideoxynucleoside 5'-triphosphates on the activities of reverse transcriptase and various cellular DNA polymerases. *Mol. Pharmacol.* **1989**, *35*, 578-583.
106. Sommadossi, J.-P.; Carlisle, R.; Zhou, Z. Cellular pharmacology of 3'-azido-3'-deoxythymidine with evidence of incorporation into DNA of human bone marrow cells. *Mol. Pharmacol.* **1989**, *36*, 9-14.
107. Larder, B. A.; Darby, G.; Richman, D. D. HIV with reduced sensitivity to zidovudine (AZT) isolates during prolonged therapy. *Science* **1989**, *243*, 1731-1734.
108. Richman, D. D. Subsceptibility to nucleoside analogs of zidovudine-resistant isolates of human-immunodeficiency-virus. *Am. J. Med.* **1990**, *88*, 8-10.
109. Wengel, J.; Lau, J.; Pederson, E. B.; Nielson, C. M. Synthesis of L-3'-azido-3'-deoxythymidine and its stereoisomers. *J. Org. Chem.* **1991**, *56*, 3591-3594.
110. Gosselin, G.; Boudou, V.; Griffon, J.-F.; Pavia, G.; Pierra, C.; Imbach, J.-L.; Aubertin, A.-M.; Schinazi, R. F.; Faraj, A.; Sommadossi, J.-P. New unnatural L-nucleoside enantiomers: from their stereospecific synthesis to their biological activities. *Nucleosides Nucleotides* **1997**, *16*, 1389-1398.
111. Eriksson, B. F. H.; Chu, C. K.; Schinazi, R. F. Phosphorylation of 3'-azido-2',3'-dideoxyuridine and preferential inhibition of human and simian immunodeficiency virus reverse transcriptases by its 5'-triphosphate. *Antimicrob. Agents Chemother.* **1989**, *33*, 1729-1734.
112. Chu, C. K.; Schinazi, R. F.; Ahn, M. K.; Ullas, G. V.; Gu, Z. P. Structure-activity relationships of pyrimidine nucleosides as antiviral agents for human immunodeficiency virus type 1 in peripheral blood mononuclear cells. *J. Med. Chem.* **1989**, *32*, 612-617.
113. Schinazi, R. F.; Chu, C. K.; Ahn, M. K.; Sommadossi, J. -P. Selective *in vitro* inhibition of human immunodeficiency virus (HIV) replication by 3'-azido-2',3'-dideoxyuridine (CS-87). *J. Cell Biochem.* **1987**, *11* (suppl. D), 405.
114. Zhu, Z.; Schinazi, R. F.; Chu, C. K.; Williams, G. J.; Colby, C. B.; Sommadossi, J. -P. Cellular metabolism of 3'-azido-2',3'-dideoxyuridine with formation of 5'-*O*-diphosphohexose derivative by previously unrecognized metabolic pathways for 2'-deoxyuridine analogs. *Mol. Pharmacol.* **1990**, *38*, 929-938.
115. Schinazi, R. F.; Chu, C. K.; Eriksson, B. F.; Sommadossi, J. -P.; Doshi, K. J.; Boudinot, F. D.; Oswald, B.; McClure, H. M. Antiretroviral activity, biochemistry, and pharmacokinetics of 3'-azido-2',3'-dideoxy-5'-methylcytidine. *Ann. New York Acad. Sci.* **1990**, *616*, 385-397.
116. Watanabe, K. A.; Harada, K.; Zeidler, J.; Matulic-Adamic, J.; Takahashi, K.; Ren, W. -Y.; Cheng, L. -C.; Fox, J. J.; Chou, T. -C.; Zhu, Q. -Y.; Polsky, B.; Gold, J. W. M.; Armstrong, D. Synthesis and anti-HIV activity of 2'-"up"-fluoro and analogs of active anti-AIDS nucleosides 3'-azido-3'-deoxythymidine (AZT) and 2',3'-dideoxycytidine (DDC). *J. Med. Chem.* **1990**, *33*, 2145-2150.
117. Tber, B.; Fahmi, N.-E.; Ronco, G.; Villa, P.; Ewing, D. F.; Mackenzie, G. An alternative strategy for the synthesis of 3'-azido-2',3'-dideoxy-4'-thionucleosides starting from D-xylose. *Carbohydr. Res.* **1995**, *267*, 203-215.
118. Bodenteich, M.; Griengl, H. Synthesis of enantiomerically pure carbocyclic 3'-azido-2',3'-

dideoxythymidine, a potential anti-AIDS drug. *Tetrahedron Lett.* **1978**, *28*, 5311-5312.

119. Marquez, V. E.; Ezzitouni, A.; Russ, P.; Siddiqui, M. A.; Ford, Jr., H.; Feldman, R. J.; Mitsuya, H.; George, C.; Barchi, Jr., J. J. HIV-1 reverse transcriptase can discriminate between two conformationally locked carbocyclic AZT triphosphate analogs. *J. Am. Chem. Soc.* **1998**, *120*, 2780-2789.
120. Herdewijn, P.; Balzarini, J.; De Clercq, E.; Pauwels, R.; Baba, M.; Broder, S.; Vanderhaeghe, H. 3'-Substituted 2',3'-dideoxynucleoside analogs as potential anti-HIV (HTLV-III/LAV) agents. *J. Med. Chem.* **1987**, *30*, 1270-1278.
121. Balzarini, J.; Baba, M. Pauwels, R.; Herdewijn, J.; De Clercq, E. Anti-retrovirus activity of 3'-fluoro- and 3'-azido-substituted pyrimidine 2',3'-dideoxynucleoside analogs. *Biochem. Pharmacol.* **1988**, *37*, 2847-2856
122. Hartmann, H.; Vogt, M. W.; Durno, A. G.; Hirsch, M. S.; Hunsmann, G.; Eckstein, F. Enhanced *in vitro* inhibition of HIV-1 replication by 3'-fluoro-3'-deoxythymidine compared to several other nucleoside analogs. *AIDS Res. Human Retroviruses* **1988**, *4*, 457-466.
123. Matthes, E.; Lehmann, Ch.; Scholz, D.; Rosenthal, H. A.; Langen, P. Phosphorylation, anti-HIV activity and cytotoxicity of 3'-fluorothymidine. *Biochem. Biophys. Res. Commun.* **1988**, *153*, 825-831.
124. Matthes, E.; Lehmann, C. H.; Scholz, D.; von Janta-Lipinski, M.; Gaertner, K.; Rosenthal, H. A.; Langen, P. Inhibition of HIV-associated reverse transcriptase by sugar-modified derivatives of thymidine 5'-triphosphate in comparison to cellular DNA α and β. *Biochem. Biophys. Res. Commun.* **1987**, *148*, 78-85.
125. Mansuri, M. M.; Hitchcock, M. J. M.; Buroker, R. A.; Bregman, C. L.; Ghazzouli, I.; Desiderio, J. V.; Starrett, J. E.; Sterzycki, R. Z.; Martin, J. C. Comparison of *in vitro* biological properties and mouse toxicities of three thymidine analogs active against human immunodeficiency virus. *Antimicrob. Agents Chemother.* **1990**, *34*, 637-641.
126. Lietamn, P. S. Lessons learned from pharmacodynamic studies of anti-HIV drugs. *Antiviral Res.* **1993**, *20* (Suppl. 1), 22.
127. Balzarini, J.; Aerschot, A.; Pauwels, R.; Baba, M.; Scholz, D.; Herdewijn, P.; De Clercq, E. 5-Halogeno-3'-fluoro-2',3'-dideoxyuridines as inhibitors of human immunodeficiency virus (HIV): potent and selctive anti-HIV activity of 3'-fluoro-2',3'-dideoxy-5-chlorouridine. *Mol. Phamacol.* **1989**, *35*, 571-577.
128. Daluge, S.; Purifoy, D. J. M.; Savina, P. M.; St Clair, M. H.; Parry, N. R.; Dev, I. K.; Novak, P.; Ayers, K. M.; Reardon, J. E.; Roberts, G. B.; Fyfe, J. A.; Blum, M. B.; Averett, D. R.; Dornsife, R.E.; Domin, B. A.; Ferone, R.; Lewis, D. A.; Krenitsky, T. A. 5-Chloro-2',3'-dideoxy-3'-fluorouridine (93U83), a selective anti-human immunodeficiency virus agent with an improved metabolic and toxicological profile. *Antimicrob. Agents Chemother.* **1994**, *38*, 1590-1603.
129. Chun, B. K.; Schinazi, R. F.; Cheng, Y. -C.; Chu, C. K. Synthesis of L-2',3'-dideoxy-3'-fluoro-L-ribonucleosides as potential antiviral agents from D-sorbitol. *Carbohydr. Res.* **2000**, *328*, 49-59.
130. Baumgarten, H.; Bodenteich, M.; Griengl, H. Chemo-enzymatic synthesis of (-)-carba-2',3'-dideoxythymidine and (-)-carba-2',3'-dideoxy-3'-fluorothymidine. *Tetrahedron Lett.* **1988**, *29*, 5745-5746.
131. Nakayama, T.; Matsumura, Y.; Morizawa, Y.; Yasuda, A.; Uchida, K.; Takase, H.; Murakami, Y.; Atarashi, S.; Ikeuchi, T.; Osada, Y. Regio- and stereoselective synthesis of carbocyclic 2',3'-dideoxy-3'-fluoro nucleosides as potential antiviral agents. *Chem. Pharm. Bull.* **1989**, *42*, 183-187.
132. Marquez, V. E.; Tseng, C. K. -H.; Mitsuya, H.; Aoki, S.; Kelly, J. A.; Ford Jr., H.; Roth, J. S.; Broder, S.; Johns, D. G.; Driscoll, J. S. Acid-stable 2'-fluoro purine dideoxynucleosides as active agents against HIV. *J. Med. Chem.* **1990**, *33*, 978-985.
133. Masood, R.; Ahluwalia, G. S.; Cooney, D. A.; Fridland, A.; Marquez, V. E.; Driscoll, J. S.; Hao, Z.; Mitsuya, H.; Perno, C.-F.; Broder, S. Johns, D. G. 2'-Fluoro-2',3'-dideoxyarabinosyladenine: a metabolic

stable analog of the antiretroviral agent 2',3'dideoxyadenosine. *Mol. Pharmacol.* **1990**, *37*, 590-596.

134. Ahluwalia, G. S.; Cooney, D. A.; Shirasaka, T.; Mitsuya, H.; Driscoll, J. S.; Johns, D. G. Enhancement by 2'-deoxycoformycin of the 5'-phosphorylation and anti-human immunodeficiency virus activity of 2',3'-dideoxyadenosine and 2'-β-fluoro-2',3'-dideoxyadenosine. *Mol. Pharmacol.* **1994**, *46*, 1002-1008.

135. Sterzycki, R. Z.; Ghazzouli, I.; Brankovan, V.; Martin, J. C.; Mansuri, M. M. Synthesis and anti-HIV activity of several 2'-fluoro-containing pyrimidine nucleosides. *J. Med. Chem.* **1990**, *33*, 2150-2157.

136. Van Aerschot, A.; Herdewijn, P.; Balzarini, J.; Pauwels, R.; De Clercq, E. 3'-Fluoro-2',3'-dideoxy-5-chlorouridine: most selective anti-HIV-1 agent among a series of new 2'- and 3'-fluorinated 2',3'-dideoxynucleoside analogs. *J. Med. Chem.* **1989**, *32*, 1743-1749.

137. (a) Huang, J.-T.; Chen, L.-C.; Wang, L.; Kim, M.-H.; Warshaw, J. A.; Armstrong, D.; Zhu, Q.-Y.; Chou, T.-C.; Watanabe, K. A.; Matulic-Adamic, J.; Su, T.-L.; Fox, J. J.; Polsky, B.; Baron, P. A.; Gold, J. W. M.; Hardy, W. D.; Zuckerman, E. Fluorinated sugar analogs of potential anti-HIV nucleosides. *J. Med. Chem.* **1991**, *34*, 1640-1646. (b) McAtee, J. J.; Schinazi, R. F.; Liotta, D. C. A completely diastereoselective electrophilic fluorination of a chiral, noncarbohydrate sugar ring precusor: application to the synthesis of several novel 2'-fluoronucleosides. *J. Org. Chem.* **1998**, *63*, 2161-2167.

138. Cavalcanti, S. C. H.; Xiang, Y.; Newton, M. G.; Schinazi, R. F.; Cheng, Y. -C.; Chu, C. K. Synthesis of 2',3'-dideoxy-2'-fluoro-L-*threo*-pentofuranosyl nucleosides as potential antiviral agents. *Nucleosides Nucleotides* **1999**, *18*, 2233-2252.

139. Kotra, L. P.; Newton, M. G.; Chu, C. K. Synthesis of 2,3-dideoxy-2,2-difluoro-L-*glycero*-pentofuranosyl nucleosides. *Carbohydr. Res.* **1998**, *306*, 69-80.

140. Marquez, V. E.; Jeong, L. S.; Nicklaus, M. C.; George, C. Synthesis and biological activity of sugar-fluorinated 2',3'-dideoxy-4'-thioribofuranosyl nucleosides. *Nucleosides Nucleotides* **1995**, *14*, 555-558.

141. Acton, E. M.; Goener, R. N.; Uh, H. S.; Ryan, K. J.; Henry, D. W. Improved antitumor effects in 3'-branched homologues of 2'-deoxythioguanosine. Synthesis and evaluation of thioguanosine nucleosides of 2,3-dideoxy-3-(hydroxymethyl)-D-*erythro*-pentofuranose. *J. Med. Chem.* **1979**, *22*, 518-525.

142. (a) Svansson, L.; Kvarnström, I.; Classon, B.; Samuelsson, B. Synthesis of 2',3'-dideoxy-3'-*C*-hydroxymethyl nucleosides as potential inhibitors of HIV. *J. Org. Chem.* **1991**, *56*, 2993-2997. (b) Lee-Ruff, E.; Ostrowski, M.; Ladha, A.; Stynes, D. V.; Vernik, I.; Jiang, J.-L.; Wan, W.-Q.; Ding, S.-F.; Joshi, S. Synthesis and HIV inhibition activity of 2',3'-dideoxy-3'-*C*-hydroxymethyl nucleosides. *J. Med. Chem.* **1996**, *39*, 5276-5280.

143. (a) Branalt, J.; Kvarnström, I.; Niklasson, G.; Svensson, S. C. T. Classon, B.; Samuelsson, B. Synthesis of 2',3'-dideoxy-3'-*C*-(hydroxymethyl)-4'-thionucleosides as potential inhibitors of HIV. *J. Org. Chem.* **1994**, *59*, 1783-1788. (b) Janson, M.; Svansson, L.; Svensson, S. C. T.; Kvarnström, I.; Classon, B.; Samuelsson, B. Synthesis of some purine carbocyclic isosters of 2',3'-dideoxy-3'-*C*-hydroxymethyl nucleosides as potential inhibitors of HIV. *Nucleosides Nucleotides* **1992**, *11*, 1739-1747.

144. Lin, T. -S.; Zhu, J. -L.; Dutschman, G. F.; Cheng, Y. -C.; Prusoff, W. H. Syntheses and biological evaluations of 3'-deoxy-3'-*C*-branched-substituted nucleosides. *J. Med. Chem.* **1993**, *36*, 353-362.

145. Chun, M. W.; Lee, K.; Choi, Y.; Lee, J.; Kim, J. H.; Lee, C. K.; Choi, B. G.; Xu, Y. C. Synthesis and antiviral activity of fluoro sugar nucleosides 2: synthesis and biological evaluations of 2',3'-dideoxy-2'-fluoro-3'-*C*-hydroxymethyl-β-D-arabinofuranosyl nucleosides. *Arch. Pharm. Res.* **1996**, *19*, 243-245.

146. Wachtmeister, J.; Classon, B.; Samuelsson, B. Synthesis of carbocyclic 2',3'-dideoxy-2'-fluoro-3'-*C*-hydroxymethyl nucleoside analogs as potential inhibitors of HIV and HSV. *Tetrahedron* **1997**, *53*, 1861-1872.

147. (a) Ho, H. -T.; Hitchcock, M. J. M. Cellular pharmacology of 2',3'-dideoxy-2',3'-didehydrothymidine, a nucleoside analog active against human immunodeficiency. *Antimicrob. Agents Chemother.* **1987**, *33*, 844-849. (b) Balzarini, J.; Herdewijn, P.; De Clercq, E. Differential patterns of intracellular metabolism

of 2',3'-didehydro-2',3'-dideoxythymidine, two potent anti-human immunodeficiency virus compounds. *J. Biol. Chem.* **1989**, *264*, 6127-6133.

148. Huang, P.; Farquhar, D.; Plunkett, W. Selective action of 2',3'-didehydro-2',3'-dideoxythymidine triphosphate on human immunodeficiency virus reverse transcriptase and human DNA polymerases. *J. Biol. Chem.* **1992**, *267*, 2817-2822.
149. Mansuri, M. M.; Strarrett Jr., J. E.; Ghazzouli, I.; Hitchcock, M. J. M.; Sterzycki, R. Z.; Brankovan, V.; Lin, T.-S.; August, E. M.; Prusoff, W. H.; Sommadossi, J.-P.; Martin, J. C. (1-(2,3-Dideoxy-β-D-*glycero*-pent-2-enofuranosyl)thymine. A potent and selective anti-HIV agent. *J. Med. Chem.* **1989**, *32*, 461-466.
150. Zhu, Z.; Hitchcock, M. J. M.; Sommadossi, J. -P. Metabolism and DNA interaction of 2',3'-didehydro-2',3'-dideoxythymidine in human bone marrow cells. *Mol. Pharmacol.* **1991**, *40*, 838-845.
151. Chen, C. -H.; Vazquez-Padua, M.; Cheng, Y. -C. Effect of anti-human immunodeficiency virus nucleoside analogs on mitochondrial DNA and its implication for delayed toxicity. *Mol. Pharmacol.* **1991**, *39*, 625-628.
152. Lin, T. -S.; Luo, M. -Z.; Liu, M. -C.; Zhu, Y. -L.; Gullen, E.; Dutschman, G. E.; Cheng, Y. -C. Design and synthesis of 2',3'-dideoxy-2',3'-didehydro-β-L-cytidine (β-L-d4C) and of 2',3'-dideoxy-2',3'-didehydro-β-L-5-fluorocytidine (β-L-Fd4C), two exceptionally potent inhibitors of human hepatitis B virus (HBV) and potent inhibitors of human immunodeficiency virus (HIV) *in vitro*. *J. Med. Chem.* **1996**, *39*, 1757-1759.
153. (a) Faraj, A.; Schinazi, R. F.; Joudawlkis, A.; Lesnikowski, Z.; McMillan, A.; Morrow, C. D.; Sommadossi, J. -P. Effects of β-D-2',3'-didehydro-2',3'-dideoxy-5-fluorocytidine 5'-triphosphate (β-D-D4FCTP) and its β-L-enantiomer 5'-triphosphate (β-L-D4FCTP) on viral DNA polymerases. *Antiviral Res.* **1997**, *34*, A66. (b) Ma, L.; Hurwitz, S. J.; Shi, J.; McAtee, J. J.; Liotta, D. C.; McClure, H. M.; Schinazi, R. F. Pharmacokinetics of the antiviral agent β-D-2',3'-didehydro-2',3'-dideoxy-5-fluorocytidine in rhesus monkeys. *Antimicrob. Agents Chemother.* **1999**, *43*, 381-384.
154. Shi, J.; McAtee, J. J.; Schlueter Wirts, S.; Tharnish, P.; Juodawlkis, A.; Liotta, D. C.; Schinazi, R. F. Synthesis and biological evaluation of 2',3'-didehydro-2',3'-dideoxy-5-fluorocytidine (D4FC) analogs: discovery of carbocyclic nucleoside triphosphates with potent inhibitory activity against HIV-1 reverse transcriptase. *J. Med. Chem.* **1999**, *42*, 859-867.
155. Chen, S. -H.; Lin, S.; King, I.; Spinka, T.; Dutschman, G. F.; Gullen, E. A.; Cheng, Y. -C.; Doyle, T. W. Synthesis and comparative evaluation of two antiviral agents: β-L-Fd4C and β-D-Fd4C. *Bioorg. Med. Chem. Lett.* **1998**, *8*, 3245-3250.
156. Schinazi, R. F.; Mellors, J.; Bazmi, H.; Diamond, S.; Garber, S.; Gallagher, K.; Geleziunas, R.; Klabe, R.; Pierce, M.; Rayner, M.; Wu, J. –T.; Zhang, H.; Hammond, J.; Bacheler, L.; Manion, D. J.; Otto, M. J.; Stuyver, L.; Trainor, G.; Liotta, D. C.; Erickson-Viitanen, S. DPC 817: a cytidine nucleoside analog with activity against zidovudine- and lamivudine-resistant viral variants. *Antimicrob. Agents Chemother.* **2002**, *46*, 1394-1401.
157. Bolon, P. J.; Wang, P.; Chu, C. K.; Gosselin, G.; Boudou, V.; Pierra, C.; Mathé, C.; Imbach, J. -L.; Faraj, A.; Alaoui, M. A.; Sommadossi, J. -P.; Pai, S. B.; Zhu, Y. -L.; Lin, J. -S.; Cheng, Y. -C.; Schinazi, R. F. Anti-human immunodeficiency and anti-hepatitis B virus activities of β-L-2',3'-dideoxy purine nucleosides. *Bioorg. Med. Chem. Lett.* **1996**, *6*, 1657-1662.
158. Chu, C. K.; Schinazi, R. F.; Arnold, B. H.; Cannon, D. L.; Doboszewski, B.; Bhadti, V. B.; Gu, Z. Comparative activity of 2',3'-saturated and unsaturated pyrimidine and purine nucleosides against human immunodeficiency virus type 1 in peripheral blood mononuclear cells. *Biochem. Pharmacol.* **1988**, *37*, 3543-3548.
159. Ray, A. S.; Yang, Z.; Chu, C. K.; Anderson, K. S. Novel use of a guanosine prodrug approach to convert 2',3'-didehydro-2',3'-dideoxyguanosine into a viable antiviral agent. *Antimicrob. Agents Chemother.*

2002, *46*, 887-891.

160. Lee, K.; Choi, Y.; Schinazi, R. F.; Cheng, Y. -C.; Chu, C. K. Structure-activity relationships of 2'-fluoro-2',3'-unsaturated D-nucleosides as anti-HIV agents. *Abstracts of Papers*, 218th American Chemical Society National Meeting, **1999**, Part 1, MEDI 9.
161. Lee, K.; Choi, Y.; Gumina, G.;Zhou, W.; Schinazi, R. F.; Chu, C. K. Structure-activity relationships of 2'-fluoro-2',3'-unsaturated D-nucleosides as anti-HIV agents. *J. Med.Chem.* **2002**, *45*, 1313-1320.
162. (a) Choi, Y.; Lee, K.; Hong, J. H.; Schinazi, R. F.; Chu, C. K. Synthesis and anti-HIV activity of L-2'-fluoro-2',3'-unsaturated purine nucleosides. *Tetrahedron Lett.* **1998**, *39*, 4437-4440. (b) Lee, K.; Choi, Y.; Gullen, E.; Schlueter-Wirtz, S.; Schinazi, R. F.; Cheng, Y.-C.; Chu, C. K. Synthesis and anti-HIV and anti-HBV activities of 2'-fluoro-2',3'-unsaturated L-nucleosides. *J. Med. Chem.* **1999**, *42*, 1320-1328.
163. Pai, S. B.; Liotta, D. C.; Chu, C. K.; Schinazi, R. F. Inhibition of hepatitis B virus by novel 2'-fluoro-2',3'-unsaturated D- and L-nucleosides. *Abstrats of Papers*, 3rd International Conference on Therapies for Viral Hepatitis, Maui, USA; International Society for Antiviral Therapy: **1999**; Abstract 125.
164. Koshida, R.; Cox, S.; Harmenberg, J.; Gilljam, G.; Wahren, B. Structure-activity relationships of fluorinated nucleoside analogs and their synergistic effect in combination with phosphonoformate against human immunodeficiency virus type 1. *Antimicrob. Agents Chemother.* **1989**, *33*, 2083-2088.
165. Chu, C. K.; Gumina, G.; Zhou, W.; Schinazi, R. F. Synthesis and anti-HIV activity of D- and L-3'-fluoro-2',3'-unsaturated cytidines. *Antiviral Res.* **2002**, *53*, A41-A41.
166. Gumina, G.; Schinazi, R. F.; Chu, C. K. Synthesis and potent anti-HIV activity of L-3'-fluoro unsaturated cytidine. *Org. Lett.* **2001**, *3*, 4177-4180.
167. Choi, Y.; Choo, H.; Chong, Y.; Lee, S.; Olgen, S.; Schinazi, R. F.; Chu, C. K. Synthesis and potent anti-HIV activity of L-2',3'-didehydro-2',3'-dideoxy-2'-fluoro-4'-thiocytidine. *Org. Lett.* **2002**, *4*, 305-307.
168. Choi, Y.; Choo, H.; Chong, Y.; Lee, S.; Schinazi, R. F.; Chu, C. K. Synthesis and antiviral activities of L-2',3'-didehydro-2',3'-dideoxy-2'-fluoro-4'-thionucleosides. *Abstr. Papers Am. Chem. Soc.* **2001**, CARB 029.
169. Chu, C. K.; Choi, Y.; Choo, H.; Chong, Y.; Lee, S.; Zhu, W.; Schinazi, R. F. Synthesis and anti-HIV activity of D- and L-2',3'didehydro-2',3'-dideoxy-2'-fluoro-4'-thiocytidines. *Antiviral Res.* **2002**, *53*, A36-A36.
170. Vince, R.; Hua, M.; Brownell, J.; Daluge, S.; Lee, F.; Shannon, W. M.; Lavelle, G. C.; Qualls, J.; Weislow, O. S.; Kiser, R.; Canonico, P. G.; Schitz, R. H.; Narayanan, V. L.; Mayo, J. G.; Shoemaker, R. H.; Boyd, M. R. Potent and selective activity of a new carbocyclic nucleoside analog (carbovir: NSC 614846) against human immunodeficiency virus in vitro. *Biochem. Biophys. Res. Commun.* **1988**, *156*, 1046-1053.
171. White, E. L.; Parker, W. B.; Macy, L. J.; Shaddis, S. C.; McCaleb, G.; Secrist III, J. A.; Vince, R.; Shannon, W. M. Comparison of the effect of carbovir, AZT, and dideoxynucleoside triphosphates on the activity of human immunodeficiency virus reverse transcriptase and selected human polymerases. *Biochem. Biophys. Res. Commun.* **1989**, *161*, 393-398.
172. Bondoc Jr., L. L.; Shannon, W. M.; Secrist III, J. A.; Vince, R.; Fridland, A. Metabolism of the carbocyclic nucleoside analog carbovir, an inhibitor of human immunodeficiency virus, in human lymphoid cells. *Biochemistry* **1990**, *29*, 9839-9843.
173. Miller, W. A.; Daluge, S.; Garvey, E. P.; Hopkins, S.; Reardon, J. E.; Boyd, F. L.; Miller, R. L. Phosphorylation of carbovir enantiomers by cellular enzymes determines the stereoselectivity of antiviral activity. *J. Biol. Chem.* **1992**, *267*, 21220-21224.
174. Parker, W. B.; White, E. L.; Shaddix, S. C.; Rose, L. J.; Buckheit Jr., R. W.; Germay, J. M.; Secrist III, J. A.; Vince, R.; Shannon, W. M. Mechanism of inhibition of human immunodeficiency virus type 1 reverse transcriptase α, β, and γ by the 5'-triphosphates of carbovir, 3'-azido-3'-deoxythymidine, 2',3'-

dideoxyguanosine, and 3'-deoxythymidine. *J. Biol. Chem.* **1991**, *266*, 1754-1762.

175. Orr, D. C.; Figueiredo, H. T.; Mo, C. -L.; Penn, C. R.; Cameron, J. M. DNA chain termination activity and inhibition of human immunodeficiency virus reverse transcriptase by carbocyclic 2',3'-didehydro-2',3'-dideoxyguanosine triphosphate. *J. Biol. Chem.* **1992**, *267*, 4177-4182.
176. Parker, W. B.; Shaddix, S.; Bowdon, B. J.; Rose, L. M.; Vince, R.; Shannon, W. M. Bennett Jr., L. L. Metabolism of carbovir, a potent inhibitor of human immunodeficiency virus type 1, and its effects on cellular metabolism. *Antimicrob. Agents Chemother.* **1993**, *37*, 1004-1009.
177. Parker, W. B.; Shaddix, S.; Vince, R.; Bennett Jr., L. L. Lack of mitochondrial toxicity in CEM cells treated with carbovir. *Antiviral Res.* **1997**, *34*, 131-136.
178. Faletto, M. B.; Miller, W. H.; Garvey, E. P.; St Clair, M. H.; Daluge, S. M.; Good, S. S. Unique intracellular activation of the potent anti-human immunodeficiency virus agent 1592U89. *Antimicrob. Agents Chemother.* **1997**, *41*, 1099-1107.
179. Dobkin, J. F. Two new agents for HIV: Kaletra and Trizivir. *Infect. Med.* **2000**, *17*, 777-777.
180. Trizivir - Abacavir sulfate/lamivudine/zidovudine tablets - Glaxo Wellcome - Fixed-dose combination of 3 nucleoside analogs for twice-daily Tx of HIV. *Formulary* **2001**, *36*, 13-13.
181. Katagiri, N.; Nomura, M.; Sato, H.; Kaneko, C.; Tsuruo, T. Synthesis of nucleosides and related compounds - Synthesis and anti-HIV activity of 9-(c-4,t-5-bis[hydroxymethyl]cyclopent-2-en-R-1-yl)-9*H*-adenine. *J. Med. Chem.* **1992**, *35*, 1882-1886.
182. Katagiri, N.; Shiraishi, T.; Sato, H.; Toyota, A.; Kaneko, C.; Yusa, K.; Ohhara, T.; Tsuruo, T. Resolution of 9-(c-4,t-5-Bis[hydroxymethyl]cyclopent-2-en-r-1-yl)-9*H*-adenine and selective inhibition of human immunodeficiency virus by the (-)-enantiomer. *Biochem. Biophys. Res. Commun.* **1992**, *184*, 154-159.
183. (a) Valette, G.; Pompon, A.; Giradet, J. -L.; Gosselin, G.; Périgaud, C.; Gosselin, G.; Imbach, J. -L. Decomposition pathways and *in vitro* HIV inhibitory effects of isoddA pronucleotides: toward a rational approach for intracellular delivery of nucleoside 5'-monophosphates. *J. Med. Chem.* **1996**, *39*, 1981-1990. (b) Siddiqui, A. Q.; Ballatore, C.; McGuigan, C.; De Clercq, E.; Balzarini, J. The presence of substituents on the aryl moiety of the aryl phosphoramidate derivative of d4T enhances anti-HIV efficacy in cell culture: a structure-activity relationship. *J. Med. Chem.* **1999**, *42*, 393-399. (c) McGuigan, C.; Cahard, D.; Sheeka, H. M.; De Clercq, E.; Balzarini, J. Aryl phosphoramidate derivatives of d4T have improved anti-HIV efficacy in tissue culture and may act by the generation of a novel intracellular metabolite. *J. Med. Chem.* **1996**, *39*, 1748-1753. (d) Balzarini, J.; Karlsson, A.; Aquaro, S.; Perno, C. –F. Cahard, D.; Naesens, L.; De Clercq, E.; McGuigan, C. Mechanism of anti-HIV action of masked alaninyl d4T-MP derivatives. *Proc, Natl. Acad. Sci. USA* **1996**, *93*, 7295-7299. (e) Balzarini, J.; Egberink, H.; Hartmann, K.; Cahard, D.; Vahlenkamp, T.; Thormar, H.; De Clercq, E.; McGuigan, C. Antiretrovirus specificity and intracellular metabolism of 2',3'-didehydro-2',3'-dideoxythymidine (stavudine) and its 5'-monophosphate triester prodrug So324. *Mol. Pharmacol.* **1996**, *50*, 1207-1213. (f) Beltran, T.; Egron, D.; Lefebvre, I.; Perigaud, C.; Pompon, A.; Gosselin, G.; Aubertin, A. –M.; Imbach, J. –L. Anti-HIV pronucleotides: SATE *versus* phenyl as a protecting group of AZT phosphoramidate derivatives. *Nucleosides Nucleotides* **1999**, *4-5*, 973-975. (g) McGuigan, C.; Bidois, L.; Hiouni, A.; Ballatore, C.; De Clercq, E.; Balzarini, J. Phosphoramidate derivatives of stavudine as inhibitors of HIV: unnatural amino acids may substitute for alanine. *Antiviral Chem. Chemother.* **2000**, *11*, 111-116.
184. Balzarini, J.; Wedgwood, O.; Kruining, J.; Pelemans, H.; Heijtink, R.; De Clercq, E.; McGuigan, C. Anti-HIV and anti-HBV activity and resistance profile of 2', 3'-dideoxy-3'-thiacytidine (3TC) and its arylphosphoramidate derivative CF 1109. *Biochem. Biophys. Res. Comm.* **1996**, *225*, 363-369.
185. McGuigan, C.; Pathirana, R. N.; Balzarini, J.; De Clercq, E. Intracellular delivery of bioactive AZT nucleotides by aryl phosphate derivatives of AZT. *J. Med. Chem.* **1993**, *36*, 1048-1052.

186. Balzarini, J.; Kruining, J.; Wedgwood, O.; Pannecouque, C.; Aquaro, S.; Perno, C. –F.; Naessens, L.; Witvrouw, M.; Heijtink, R.; De Clercq, E.; McGuigan, C. Conversion of 2', 3'-dideoxyadenosine (ddA) and 2', 3'-didehydro-2', 3'-dideoxyadenosine (d4A) to their corresponding aryloxyphosphoramidate derivatives markedly potentiates their activity against human immunodeficiency virus and hepatitis B virus. *FEBS Letters* **1997**, *410*, 324-328.

187. McGuigan, C.; Wedgwood, O. M.; De Clercq, E.; Balzarini, J. Phosphoramidate derivatives of 2',3'-didehydro-2',3'-dideoxyadenosine [d4A] have markedly improved anti-HIV potency and selectivity. *Bioorg. Med. Chem. Lett.* **1996**, *6*, 2359-2362.

188. Meier, C. 4H-1.3.2.-Benzodioxaphosphorin-2-nucleosyl-2-oxide: a new concept for lipophilic, potential prodrugs of biologically active nucleoside monophosphates. *Angew. Chem.* **1996**, *108*, 77-79.

189. Meier, C.; Knispel, T.; Lorey, M.; Balzarini, J. *Cyclo*Sal-pronucleotides: the design and biological evaluation of a new class of lipophilic nucleotide prodrugs. *Int. Antiviral News* **1997**, *5*, 183-186.

190. Meier, C. Pro-nucleotides – recent advances in the design of efficient tools for the delivery of biologically active nucleoside monophosphate. *Synlett* **1997**, 233-242.

191. Meier, C.; Lorey, M.; De Clercq, E.; Balzarini, J. Cyclic saligenyl phosphotriesters of 2',3'-dideoxy-2',3'-didehydrothymidine (d4T) - a new pro-nucleotide approach. *Bioorg. Med. Chem. Lett.* **1997**, *7*, 99-104.

192. Meier, C.; Lorey, M.; De Clercq, E.; Balzarini, J. *Cyclo*Sal-2',3'-dideoxy-2',3'-didehydrothymidine monophosphate (*cyclo*Sal-d4TMP): synthesis and antiviral evaluation of a new d4TMP delivery System. *J. Med. Chem.* **1998**, *41*, 1417-1427.

193. Balzarini, J.; Aquaro, S.; Knispel, T.; Rampazzo, C.; Bianchi, V.; Perno, C. -F.; De Clercq, E. Meier, C. Cyclosaligenyl-2',3'-didehydro-2',3'-dideoxythymidine monophosphate: efficient intracellular delivery of d4TMP. *Mol. Pharm.* **2000**, *58*, 928-935.

194. Lorey, M.; Meier, C.; De Clercq, E.; Balzarini, J. New synthesis and antitumor activity of *cyclo* Sal-derivatives of 5-fluoro-2'-deoxyuridine monophosphate. *Nucleosides Nucleotides* **1997**, *16*, 789-792.

195. Meier, C.; De Clercq, E.; Balzarini, J. Cyclosaligenyl-3'-azido-2',3'-dideoxythymidine monophosphate (cycloSal-AZTMP): a new pro-nucleotide approach. *Nucleosides Nucleotides* **1997**, *16*, 793-796.

196. Meier, C.; De Clercq, E.; Balzarini, J. Nucleotide delivery from cyclosaligenyl-3'-azido-3'-deoxythymidine monophosphates (*cyclo*Sal-AZTMP). *Eur. J. Org. Chem.* **1998**, 837-846.

197. Balzarini, J.; Naesens, L.; Aquaro, S.; Knispel, T.; Perno, C. -F.; De Clercq, E.; Meier, C. Intracellular metabolism of cyclosaligenyl 3'-azido-2',3'-dideoxythimidine monophosphate, a prodrug of 3'-azido-2',3'-dideoxythymidine (zidovudine). *Mol. Pharmacol.* **1999**, *56*, 1354-1361.

198. Meier, C.; Knispel, T.; De Clercq, E.; Balzarini, J. ADA-bypass by lipophilic *cyclo*Sal-ddAMP pronucleotides: a second example of the efficiency of the *cyclo*Sal concept. *Bioorg. Med. Chem. Lett.* **1997**, *7*, 1577-1582.

199. Meier, C.; Knispel, T.; De Clercq, E.; Balzarini, J. *Cyclo*Sal pronucleotides of 2',3'-dideoxyadenosine and 2',3'-dideoxy-2',3'-didehydroadenosine: synthesis and antiviral evaluation of a highly efficient nucleotide delivery system. *J. Med. Chem.* **1999**, *42*, 1604-1614.

200. Meier, C.; Knispel, T.; Marquez, V. E.; Siddiqui, M. A.; De Clercq, E.; Balzarini, J. *Cyclo*Sal pronucleotides of 2'-fluoro-*ara*- and 2'-fluoro-*ribo*-2',3'-dideoxyadenosine as a strategy to bypass a metabolic blockade. *J. Med. Chem.* **1999**, *42*, 1615-1624.

201. Schinazi, R. F.; McMillan, A.; Cannon, D.; Mathis, R.; Lloyd, R. M.; Peck, A.; Sommadossi, J. -P.; St Clair, M.; Wilson, J.; Furman, P. A.; Painter, G.; Choi, W. B.; Liotta, D. C. Selective inhibition of human immunodeficiency viruses by racemates and enantiomers of *cis*-5-fluoro-1-[2-(hydroxymethyl)-1,3-oxathiolan-5-yl]cytosine. *Antimicrob. Agents Chemother.* **1992**, *36*, 2423-2431.

202. (a) Chu, C. K.; Ahn, S. K.; Kim, H. O.; Beach, J. W.; Alves, A. J.; Jeong, L. S.; Islam, Q.; Van Roey, P.;

Schinazi, R. F. Asymmetric synthesis of enantiomerically pure (-)-(1'R,4'R)-dioxolane-thymine and its anti-HIV activity. *Tetrahedron Lett.* **1991**, *32*, 3791-3794. (b) Kim, H. O.; Shanmuganathan, K. Alves, A. J.; Shanmuganathan, K.; Cannon, D. L.; Alves, A. J.; Jeong, L. S.; Beach, J. W.; Chu, C. K. Potent anti-HIV and anti-HBV activities of (-)-L-β-dioxolane-C and (+)-L-β-dioxolane-T and their asymmetric synthesis. *Tetrahedron Lett.* **1992**, *33*, 6899-6902. (c) Kim, H. O.; Schinazi, R. F.; Nampalli, S.; Shanmuganathan, K.; Cannon, D. L.; Alves, A. J.; Jeong, L. S.; Beach, J. W.; Chu, C. K. 1,3-Dioxolanylpurine nucleosides (2R,4R) and (2R,4S) with selective anti-HIV-1 activity in human lymphocytes. *J. Med. Chem.* **1993**, *36*, 30-37. (d) Kim, H. O.; Schinazi, R. F.; Shanmuganathan, K.; Jeong, L. S.; Beach, J. W.; Nampalli, S.; Cannon, D. L.; Chu, C. K. L-β-(2S,4S)- and L-α-(2S,4R)-Dioxolanyl nucleosides as potential anti-HIV agents: asymmetric synthesis and structure-activity relationships. *J. Med. Chem.* **1993**, *36*, 519-528.

203. Doong, S. -L.; Tsai, C. H.; Schinazi, R. F.; Liotta, D. C.; Cheng, Y. -C. Inhibition of the replication of hepatitis B virus in vitro by 2'-deoxy-3'-thiacytidine and related analogs. *Proc. Natl. Acad. Sci. USA* **1991**, *88*, 8495-8499.
204. (a) Mahmoudian, M.; Bains, B. S.; Drake, C. S.; Hale, R. S.; Jones, P.; Piercery, J. E.; Montgomery, D. S.; Purvis, I. J.; Storer, R.; Dawson, M. J.; Lawrence, G. C. Enzymatic production of optically pure (2'*R-cis*)-2'-deoxy-3'-thiacytidine (3TC, lamivudine): a potent anti-HIV agent. *Enzym. Microb. Technol.* **1993**, *15*, 749-755. (b) Choi, W. B.; Wilson, L. J.; Yeola, S.; Liotta, D. C. *In situ* complexation directs the stereochemistry of *N*-glycosylation in the synthesis of oxathiolanyl and dioxolanyl nucleoside analogs. *J. Am. Chem. Soc.* **1991**, *113*, 9377-9379.
205. (a) Coates, J. A. V.; Cammack, N.; Jenkinson, H. J.; Mutton, I. M.; Pearson, B. A.; Storer, R.; Cameron, J. M.; Penn, C. R. The separated enantiomers of 2'-deoxy-3'-thiacytidine (BCH-189) both inhibit human immunodeficiency virus *in vitro*. *Antimicrob. Agents Chemother.* **1992**, *36*, 202-205. (b) Schinazi, R. F.; Chu, C. K.; Peck, A.; McMillan, A.; Mathis, S.; Cannon, D.; Jeong, L. S.; Beach, J. W.; Choi, W. B.; Yeola, S.; Liotta, D. C. Activities of the four optical isomers of 2',3'-dideoxy-3'-thiacytidine (BCH-189) against human immunodeficiency virus type I in human lymphocytes. *Antimicrob. Agents Chemother.* **1992**, *36*, 672-676.
206. (a) Chu, C. K.; Beach, J. W.; Jeong, L. S.; Choi, B. G.; Comer, F. I.; Alves, A. J.; Schinazi, R. F. Enantiomeric synthesis of (+)-BCH-189 [(+)-(2S,5R)-1-[2-(hydroxymethyl)-1,3-oxathiolan-5-yl]cytosine] from D-mannose and its anti-HIV activity. *J. Org. Chem.* **1991**, *56*, 6503-6505. (b) Beach, J. W.; Jeong, L. S.; Alves, A. J.; Pohl, D.; Kim, H. O.; Chang, C. -N.; Doong, S. -L.; Schinazi, R. F.; Cheng, Y. -C.; Chu, C. K. Synthesis of enantiomerically pure 2'*R*,5'*S*-(-)-[2-(hydroxymethyl)-oxathiolan-5-yl]cytosine as a potent antiviral agent against hepatitis B virus (HBV) and human immunodeficiency virus (HIV). *J. Org. Chem.* **1992**, *57*, 2217-2219. (c) Jeong, L. S.; Schinazi, R. F.; Beach, J. W.; Kim, H. O.; Shanmuganathan, K.; Nampalli, S.; Chun, M. W.; Chung, W. K.; Choi, B. G.; Chu, C. K. Structure-activity relationships of β-D-(2*S*,5*R*)- and α-D-(2*S*,5*S*)-1,3-oxathiolanyl nucleosides as potential anti-HIV agents. *J. Med. Chem.* **1993**, *36*, 2627-2638. (d) Jeong, L. S.; Schinazi, R. F.; Beach, J. W.; Kim, H. O.; Nampalli, S.; Shanmuganathan, K.; Alves, A. J.; McMillan, A.; Chu, C. K.; Mathis, R. Asymmetric synthesis and biological evaluation of β-L-(2*R*,5*S*)- and α-L-(2*R*,5*R*)-1,3 oxathiolane-pyrimidine and -purine nucleosides as potential anti-HIV agents. *J. Med. Chem.* **1993**, *36*, 181-195.
207. Chang, C. -N.; Doong, S. -L.; Zhou, J. H.; Beach, J. W.; Jeong, L. S.; Chu, C. K.; Tsai, C. H.; Cheng, Y. -C. Deoxycytidine deamination-resistant stereoisomer is the active form of (-)-2',3'-dideoxy-3'-thiacytidine in the inhibition of hepatitis B virus replication. *J. Biol. Chem.* **1992**, *267*, 13938-13942.
208. Severini, A.; Liu, X.-Y.; Wilson, J. S.; Tyrrell, D. L. J. Mechanism of inhibition of duck hepatitis B virus polymerase by (-)-β-L-2',3'-dideoxy-3'-thiacytidine. *Antimicrob. Agents Chemother.* **1995**, *39*, 1430-1435.
209. Shewach, D. S.; Liotta, D. C.; Schinazi, R. F. Affinity of the antiviral enantiomers of oxathiolane cytosine

nucleosides for human 2'-deoxycytidine kinase. *Biochem. Pharmacol.* **1993**, *45*, 1540-1543.

210. Hart, G. J.; Orr, D. C.; Penn, C. R.; Figueiredo, H. T.; Gray, N. M.; Boehme, R. E.; Cameron, J. M. Effects of (-)-2'-deoxy-3'-thiacytidine (3TC) 5'-triphosphate on human immunodeficiency virus reverse transcriptase and mammalian DNA polymerase α, β, and γ. *Antimicrob. Agents Chemother.* **1992**, *36*, 1688-1694.

211. Gish, R. G.; Leung, N. W. Y.; Wright, T. L.; Trinh, H.; Lang, W.; Kessler, H. A.; Fang, L.; Wang, L. H.; Delehanty, J.; Rigney, A.; Mondou, E.; Snow, A.; Rousseau, F. Dose range study of pharmacokinetics, safety, and preliminary antiviral activity of emtricitabine in adults with hepatitis B virus infection. *Antimicrob. Agents Chemother.* **2002**, *46*, 1734-1740.

212. Furman, P. A.; Davis, M.; Liotta, D. C.; Paff, M.; Frick, L. W.; Nelson, D. J.; Dornsife, R. E.; Wyrster, J. A.; Wilson, L. J.; Fyfe, J. A.; Tuttle, J. V.; Miller, W. H.; Condreay, L.; Averett, D. R.; Schinazi, R. F.; Painter, G. R. The anti-hepatitis B virus activities, cytotoxicities, and anabolic profiles of the (-) and (+) enantiomers of *cis*-5-fluoro-1-[2-(hydroxymethyl)-1,3-oxathiolan-5-yl]cytosine. *Antimicrob. Agents Chemother.* **1992**, *36*, 2686-2692.

213. (a) Schinazi, R. F.; Boudinot, F. D.; Ibrahim, S. S.; Manning, C.; McClure, H. M.; Liotta, D. C. Pharmacokinetics and metabolism of racemic 2',3'-dideoxy-5-fluoro-3'-thiacytidine in rhesus monkeys. *Antimicrob. Agents Chemother.* **1992**, *36*, 2432-2438. (b) Schinazi, R. F.; Lloyd, R. M.; Nguyen, M. H.; Cannon, D. L.; McMillan, Ilksoy, N.; Chu, C. K.; Liotta, D. C.; Bazmi, H. Z.; Mellors, J. W. Characterization of human immunodeficiency viruses resistant to oxathiolane-cytosine nucleosides. *Antimicrob. Agents Chemother.* **1993**, *38*, 875-881.

214. (a) Grove, K. L.; Guo, X.; Liu, S.-H.; Gao, Z.; Chu, C. K.; Cheng, Y. -C. Anticancer activity of β-L-dioxolane-cytidine, a novel nucleoside analog with the unnatural L-configuration. *Cancer Res.* **1995**, *55*, 3008-3011. (b) Grove, K. L.; Cheng, Y. -C. Uptake and metabolism of the new anticancer compound β-L-(-)-dioxolane-cytidine in human prostate carcinoma DU-145 cells. *Cancer Res.* **1996**, *56*, 4187-4191.

215. Kukhanova, M.; Liu, S.-H.; Mozzherin, D.; Lin, T. -S.; Chu, C. K.; Cheng, Y. -C. L- and D-enantiomers of 2',3'-dideoxycytidine 5'-triphosphate analogs as substrates for human DNA polymerases. *J. Biol. Chem.* **1995**, *270*, 23055-23059.

216. (a) Rajagopalan, P.; Boudinot, F. D.; Chu, C. K.; Tennant, B. C.; Baldwin, B. H.; Schinazi, R. F. Pharmacokinetics of (-)-β-D-2,6-diaminopurine dioxolane and its metabolite, dioxolane guanosine, in woodchucks (*Marmota monax*). *Antiviral Chem. Chemother.* **1996**, *7*, 65-70. (b) Chen, H.; Boudinot, F. D.; Chu, C. K.; McClure, H. M.; Schinazi, R. F. Pharmacokinetics of (-)-β-D-2-aminopurine dioxolane and (-)-β-D-2-amino-6-chloropurine dioxolane and their antiviral metabolite (-)-β-D-dioxolane guanine in rhesus monkeys. *Antimicrob. Agents Chemother.* **1996**, *40*, 2332-2336.

217. Bednarski, K.; Dixit, D. M.; Wang, W.; Evans, C. A.; Jin, H.; Yuen, L.; Mansour, T. S. Inhibitory activities of herpes simplex viruses type 1 and 2 and human cytomegalovirus by stereoisomers of 2'-deoxy-3'-oxa-5-(*E*)-(2-bromovinyl)uridines and their 4'-thioanalogs. *Bioorg. Med. Chem. Lett.* **1994**, *4*, 2667-2672.

218. (a) Li, L.; Dutschman, G. E.; Gullen, E. A.; Grill, S. P.; Choi, Y.; Chu, C. K.; Cheng, Y.-C. Inhibition of the growth of varicella-zoster virus by L-β-5-bromovinyl-(2-hydroxymehtyl)-(1,3-dioxolanyl)uracil. *Abstracts of Papers*, 91th Annual Meeting of the American Assciation for Cancer Research, San Francisco, CA; American Association for Cancer Research: **2000**; A22. (b) Li, L.; Dutschman, G. E.; Gullen, E. A.; Tsujii, E.; Grill, S. P.; Choi, Y.; Chu, C. K.; Cheng, Y. -C. Metabolism and mode of inhibition of varicella-zoster virus by L-5-bromovinyl-(2-hydroxymethyl)-(1,3-dioxolanyl)uracil is dependent on viral thymidine kinase. *Mol. Pharmacol.* **2000**, *58*, 1109-1114.

219. Lin, J. -S.; Kira, T.; Gullen, E.; Choi, Y.; Qu, F.; Chu, C. K.; Cheng, Y. -C. Structure-activity relationships

of L-dioxolane uracil nucleosides as anti-Epstein Barr virus agents. *J. Med. Chem.* **1999**, *42*, 2212-2217.

220. Kira, T.; Grill, S. P.; Dutschman, G. E.; Lin J. -S.; Qu, F.; Choi, Y.; Chu, C. K.; Cheng, Y. -C. Anti-Epstein-Barr virus (EBV) activity of -L-5-iododioxolane uracil is dependent on EBV thymidine kinase. *Antimicrob. Agents Chemother.* **2000**, *44*, 3278-3284.
221. Mansour, T. S.; Jin, H.; Wang, W.; Hooker, E. U.; Ashman, C.; Cammack, N.; Salomon, H.; Belmonte, A. R.; Wainberg, M. A. Anti-human immunodeficiency virus and anti-hepatitis B virus activities and toxicities of the enantiomers of 2'-deoxy-3'-oxa-4'-thiocytidine and their 5-fluoro analogs *in vitro*. *J. Med. Chem.* **1995**, *38*, 1-4.
222. De Muys, J. -M.; Gourdeau, H.; Nguyen-Ba, N.; Taylor, D. L.; Ahmed, P. S.; Mansour, T.; Locas, C.; Richard, N.; Wainberg, M. A.; Rando, R. F. Anti-human immunodeficiency virus type 1 activity, intracellular metabolism, and pharmacokinetic evaluation of 2'-deoxy-3'-oxa-4'-thiocytidine. *Antimicrob. Agents Chemother.* **1999**, *43*, 1835-1844.
223. Du, J.; Surzhykov, S.; Lin, J. S.; Newton, M. G.; Cheng, Y. -C.; Schinazi, R. F.; Chu, C. K. Synthesis, anti-human immunodeficiency virus and anti-hepatitis B virus activities of novel oxaselenolane nucleosides. *J. Med.Chem.* **1997**, *40*, 2991-2993.
224. Chu, C. K.; Pierra, C.; Ma, L.; Olgen, S.; Du, J.; Gumina, G.; Gullen, E.; Cheng, Y. -C.; Schinazi, R. F. Synthesis and antiviral activities of oxaselenolane nucleosides. *J. Med.Chem.* **2000**, *43*, 3906-3912.
225. Du, J.; Chu, C. K. Asymmetric synthesis of oxazolidine nucleosides and related chemistry. *Nucleosides Nucleotides* **1998**, *17*, 1-13.
226. Nair, V.; Jahnke, T. Antiviral activities of isomeric dideoxnucleosidesof D- and L-related stereochemistry. *Antimicrob. Agents Chemother.* **1995**, *39*, 1017-1029.
227. Huryn, D. M.; Sluboski, B. C.; Tam, S. Y. Weigele, M.; Sim, I.; Anderson, B. D.; Mitsuya, H.; Broder, S. Synthesis and anti-HIV activity of isonucleosides. *J. Med. Chem.* **1992**, *35*, 2347-2354.
228. Nair, V.; St Clair, M. H.; Reardon, J. E.; Krasny, H. C.; Hazen, R. J.; Paff, M. T.; Boone, L. R.; Tisdale, M.; Najera, I.; Dornsife, R. E.; Averett, D. R.; Borroto-Esoda, K.; Yale, J. L.; Zimmerman, T. P.; Rideout, J. L. Antiviral, metabolic, and pharmacokinetic properties of the isomeric dideoxynucleoside 4(*S*)-(6-amino-9H-purin-9-yl)tetrahydro-2(*S*)-furanmethanol. *Antimicrob. Agents Chemother.* **1995**, *39*, 1993-1999.
229. Frank, K. B.; Connell, E. V.; Holman, M. J.; Huryn, D. M.; Sluboski, B. C.; Tam, S. Y.; Todaro, L. J.; Weigele, M.; Richman, D. D.; Mitsuya, H.; Broder, S.; Sim, I. S. Anabolism and mechnism of action of Ro24-5098, an isomer of 2',3'-dideoxyadenosine (ddA) with anti-HIV activity. *Ann. N. Y. Acad. Sci.* **1990**, *616*, 408-414.
230. (a) Ng, K. M. E.; Orgel, L. E. Replacement of the 3'-CH group by nitrogen in the carbocyclic analog of thymidine. *J. Med. Chem.* **1989**, *32*, 1754-1757. (b) Peterson, M. L.; Vince, R. Synthesis and biological evaluation of 4-purinylpyrrolidine nucleosides. *J. Med. Chem.* **1991**, *34*, 2787-2797.
231. (a) Jones, M. F.; Nobel, S. A.; Robertson, C. A.; Storer, R. Tetrahydrothiophenenucleosides as potential anti-HIV agents. *Tetrahedron Lett.* **1991**, *32*, 247-250. (b) Jones, M. F.; Nobel, S. A.; Robertson, C. A.; Storer, R.; Highcock, R. M.; Lamont, R. B. Enantiospecific synthesis of 3'-hetero-dideoxy nucleoside analogs as potential anti-HIV agents. *J. Chem. Soc. Perkin Trans. I* **1992**, 1427-1436.
232. Pal, S.; Nair, V. Phosphorylation of the anti-HIV compound (*S,S*)-isodideoxyadenosine by human recombinant deoxycytidine kinase. *Biochem. Pharmacol.* **2000**, *60*, 1505-1508.
233. Sells, T. B.; Nair, V. Synthetic approaches novel cis and trans dideoxynucleosides of the apiose family. *Tetrahedron* **1994**, *50*, 117-138.
234. (a) Hammerschmidt, F.; Öhler, E.; Polsterer, J.-P.; Zbiral, E.; Balzarini, J.; De Clercq, E. A convenient route to D-apio-β-D-furanosyl- and 2'-deoxyapio-β-D-furanosyl nucleosides. *Liebigs Ann.* **1995**, 551-558.

(b) Hammerschmidt, F.; Öhler, E.; Polsterer, J.-P.; Zbiral, E.; Balzarini, J.; De Clercq, E. A convenient route to 3'-deoxy-α-L- and 3'-deoxy-β-D-apionucleosides. *Liebigs Ann.* **1995**, 559-565.

235. Hong, J. H.; Lee, K.; Choi, Y.; Chu, C. K. Enantiomeric synthesis of 3'-fluoro-apionucleosides using Claisen rearrangement. *Tetrahedron Lett.* **1998**, *39*, 3443-3446.

236. Jeong, L. S.; Lee, Y. A.; Moon, H. R.; Chun, M. W. Syntheses and antiviral activity of apio dideoxy nucleosides with azido or amino substituent. *Nucleosides Nucleotides* **1998**, *17*, 1473-1487.

237. Tino, J. A.; Clark, J. M.; Kirk Field, A.; Jacobs, G. A.; Lis, K. A.; Michalik, T. L.; McGeever-Rubin, B.; Slusarchyk, W. A.; Spergel, S. H.; Sundeen, J. E.; Tuomari, V.; Weaver, E. R.; Young, M. G.; Zahler, R. Synthesis and antiviral activity of novel isonucleoside analogs. *J. Med. Chem.* **1993**, *36*, 1221-1229.

238. (a) Nair, V.; Purdy, D. F. Synthetic approaches to new regioisomers of AZT and AZU. *Heterocycles* **1993**, *36*, 421-425. (b) Purdy, D. F.; Zintek, L. B.; Nair, V. Synthesis of isonucleosides related to AZT and AZU. *Nucleosides Nucleotides* **1994**, *13*, 109-126. (c) Zintek, L. B.; Jeon, G. S.; Nair, V. The synthesis of (*S*) and (*R*) enantiomers of novel hydroxymethylated isodideoxynucleosides. *Heterocycles* **1994**, *37*, 1853-1864.

239. Shimada, N.; Hasegawa, N.; Harada, T.; Tomisawa, T.; Fujii, A.; Takita, T. Oxetanocin, a novel nucleoside from bacteria. *J. Antibiot.* **1986**, *39*, 1623-1625.

240. Seki, J. -I.; Shimada, N.; Takahashi, K.; Takita, T.; Takeuchi, T.; Hoshino, H. Inhibition of infectivity of human immunodeficiency virus by a novel nucleoside, oxetanocin, and related compounds. *Antimicrob. Agents Chemother.* **1989**, *33*, 773-775.

241. Nishiyama, Y.; Yamamoto, N.; Takahashi, K.; Shimada, N. Selective inhibition of human cytomegalovirus replication by a novel nucleoside, oxetanocin G. *Antimicrob. Agents Chemother.* **1988**, *32*, 1053-1056.

242. Alder, J.; Mitten, M.; Norbeck, D.; Marsh, K.; Kern, E. R.; Clement, J. Efficacy of A-73209, a potent orally active agent against VZV and HSV infections. *Antiviral Res.* **1994**, *23*, 93-105.

243. Gumina, G.; Chu, C. K. Synthesis of L-Oxetanocin. *Org. Lett.* **2002**, *4*, 1147-1149.

244. Hayashi, S.; Norbeck, D. W.; Rosenbrook, W.; Fine, R. L.; Matskura, M.; Plattner, J. J.; Broder, S.; Mitsuya, H. Cyclobut-A and cyclobut-G, carbocyclic oxetanocin analogs that inhibit the replication of human immunodeficiency virus in T cells and monocytes and macrophages *in vitro. Antimicrob. Agents Chemother.* **1990**, *34*, 287-294.

245. Field, A. K.; Tuomari, A. V.; McGreever-Rubin, B.; Terry, B. J.; Mazina, K. E.; Haffey, M. L.; Hagen, M. E.; Clark, J. M.; Braitman, A.; Slusatchyk, W. A.; Young, M. G.; Zahler, R. (±)-(1α,2β,3α)-9-[2,3-*Bis*(hydroxymethyl)-cyclobutyl]guanine [(±)-BHCG or SQ 33054]: a potent and selective inhibitor of herpesvirus. *Antiviral Res.* **1990**, *13*, 41-52.

246. Norbeck, D. W.; Kern, E.; Hayashi, S.; Rosenbrook, W.; Sham, H.; Herrin, T.; Plattner, J. J.; Erickson, J.; Clement, J.; Swanson, R.; Shipkowitz, N.; Hardy, D.; Marsh, K.; Arnett, G.; Shannon, W.; Broder, S.; Mitsuya, H. Cyclobut-A and cyclobut-G: broad-spectrum antiviral agents with potential utility for the therapy of AIDS. *J. Med. Chem.* **1990**, *33*, 1281-1285.

247. Tenney, D. J.; Yamanaka, G.; Voss, S. M.; Cianci, C. W.; Tuomari, A. V.; Sheaffer, A. K.; Alam, M.; Colonno, R. J. Lobucavir is phosphorylated in human cytomegalovirus infected and -uninfected cells and inhibits the viral DNA polymerase. *Antimicrob. Agents Chemother.* **1997**, *41*, 2680-2685.

248. Sato, Y.; Maruyama, T. Synthesis and antiviral activity of 3'-fluorocarbocyclic oxetanocin A. *Chem. Pharm. Bull.* **1995**, *43*, 91-95.

249. Ashton, W. T.; Meurer, L. C.; Cantone. C. L.; Field, K.; Hannah, J.; Karkas, J. D.; Liou, R.; Patel, G. F.; Perry, H. C.; Wagner, A. F.; Walton, E.; Tolman, R. L. Synthesis and antiherpetic activity of (±)-9-[[(*Z*)-2-(hydroxymethyl)cyclopropyl]methyl]guanine and related compounds. *J. Med. Chem.* **1988**, *31*, 2304-2315.

250. Pierra, C.; Olgen, S.; Cavalcanti, S. C. H.; Cheng, Y. -C.; Schinazi, R. F.; Chu, C. K. Synthesis and antiviral activities of enantiomeric 1-[2-(hydroxymethyl)cyclopropyl]methyl nucleosides. *Nucleosides*

Nucleotides **2000**, *19*, 253-268.

251. Sekiyama, T.; Hatsuya, S.; Tanaka, Y.; Uchiyama, M.; Ono, N.; Iwayama, S.; Oikawa, M.; Suzuki, K.; Okunishi, M.; Tsuji, T. Synthesis and antiviral activity of novel acyclic nucleosides: discovery of cyclopropyl nucleoside with potent inhibitory activity against herpesviruses. *J. Med. Chem.* **1998**, *41*, 1284-1298.
252. Iwayama, S.; Ono, N.; Ohmura, Y.; Suzuki, K.; Aoki, M.; Nakazawa, H.; Oikawa, M.; Kato, T.; Okunishi, M.; Mishiyama, Y.; Yamanishi, K. Antiherpesvirus activities of (1'S,2'R)-9-{[1',2'-bis(hydroxymethyl)cyclo prop-1'-yl]methyl}guanine (A-5021) in cell culture. *Antimicrob. Agents Chemother.* **1998**, *42*, 1666-1670.
253. Onoshi, T.; Mukai, C.; Nakagawa, R.; Sekiyama, T.; Aoki, M.; Suzuki, K.; Nakazawa, H.; Ono, N.; Ohmura, Y.; Iwayama, S.; Okunishi, M.; Tsuji, T. Synthesis and antiviral activity of novel anti-VZV 5-substituted uracil nucleosides with a cyclopropane sugar moiety. *J. Med. Chem.* **2000**, *43*, 278-282.
254. (a) Csuk, R.; Scholz, Y. Synthesis of cyclopropyl carbocyclic nucleosides. *Tetrahedron* **1994**, *50*, 10431-10442. (b) Zhao, Y. F.; Yang, T. -F.; Lee, M. G.; Lee, D. W.; Newton, M. G.; Chu, C. K. Asymmetric synthesis of (1'*S*,2'*R*)-cyclopropyl carbocyclic nucleosides. *J. Org. Chem.* **1995**, *60*, 5236-5242. (c) Lee, M. G.; Du, J. F.; Chun, M. W.; Chu, C. K. Enantiomeric synthesis of L- (or 1'R,2'S)-carbocyclic cyclopropyl nucleosides. *J. Org. Chem.* **1997**, *62*, 1991-1995. (d) Yang, T. -F.; Kim, H.; Kotra, L. P.; Chu, C. K. Design and synthesis of 2'-hydroxyethylcyclopropyl carbocyclic nucleosides. *Tetrahedron Lett.* **1996**, *37*, 8849-8852.
255. (a) Qiu, Y. -L.; Ksebati, M. B.; Ptak, R. G.; Fan, B. Y.; Breitenbach, J. M.; Lin, J. -S.; Cheng, Y. -C.; Kern, E. R.; Drach, J. C.; Zemlicka, J. (*Z*)- and (*E*)-2-((Hydroxymethyl)cyclopropylidene)methyladenine and -guanine. New nucleoside analogs with a broad-spectrum antiviral activity. *J. Med. Chem.* **1998**, *41*, 10-23. (b) Qiu, Y. -L.; Hempel, A.; Camerman, N.; Camerman, A.; Geiser, F.; Ptak, R. G.; Breitenbach, J. M.; Kira, T.; Li, L.; Gullen, E.; Cheng, Y. -C.; Drach, J. C.; Zemlicka, J. (*R*)-(-)- And (*S*)-(+)-synadenol: synthesis, absolute configuration, and enantioselectivity of antiviral effect. *J. Med. Chem.* **1998**, *41*, 5257-5264. (c) Rybak, R. J.; Zemlicka, J.; Qiu, Y. -L.; Hartline, C. B.; Kern, E. R. Effective treatment of murine cytomegalovirus infections with methylenecyclopropane analogs of nucleosides. *Antiviral Res.* **1999**, *43*, 175-188.
256. Uchida, H.; Komada, E. N.; Yoshimura, K.; Maeda, Y.; Kosalaraksa, P.; Maroun, V.; Qiu, Y. -L.; Zemlicka, J.; Mitsuya, H. *In vitro* anti-human immunodeficiency virus activities of *Z*- and *E*-methylenecyclopropane nucleoside analogs and their phosphoro-L-alaninate diesters. *Antimicrob. Agents Chemother.* **1999**, *43*, 1487-1490.
257. Rybak, R. J.; Hartline, C. B.; Qiu, Y. -L.; Zemlicka, J.; Harden, E.; Marshall, G.; Sommadossi, J. -P.; Kern, E. R. *In vitro* activities of methylenecyclopropane analogs of nucleosides and their phosphoralaninate prodrugs against cytomegalovirus and other herpesvirus infections. *Antimicrob. Agents Chemother.* **2000**, *44*, 1506-1511.
258. Guan, H. -P.; Ksebati, M. B.; Cheng, Y. -C.; Drach, J. C.; Kern, E. R.; Zemlicka, J. Spiropentane mimics of nucleosides: analogs of 2'-deoxyadenosine and 2'-deoxyguanosine. Synthesis of all stereoisomers, isomeric assignment, and biological activity. *J. Org. Chem.* **2000**, *65*, 1280-1290.
259. Whitley, R. J.; Gnann, J. W. Acyclovir: a Decade Later. *N. Engl. J. Med.* **1992**, *327*, 782-788.
260. Schaeffer, H. J.; Beauchamp, L.; De Miranda, P.; Elion, G. B.; Bauer, D. J.; Collins, P. 9-(2-Hydroxyeth oxymethyl)guanine activity against viruses of the herpes group. *Nature* **1978**, *272*, 583-585.
261. Lewis, L. D.; Fowle, A. S. E.; Bittiner, S. B. Human gastrointestinal absorption of acyclovir from tablet, duodenal infusion and sipped solution. *Br. J. Clin. Pharmacol.* **1986**, *21*, 459-462.
262. Laskin, O. L.; Cederberg, D. M.; Mills, J.; Eron, L. J.; Mildvan, D.; Spector, S. S. Ganciclovir for the treatment and suppression of serious infections caused by cytomegaloviruses. *Am. J. Med.* **1987**, *83*, 201-207.

263. Buhles, W. C.; Mastre, B. J. Jr.; Tinker, A. J. Ganciclovir treatment of life or sight-threatening cytomegalovirus infection: experience in 314 immunocompromised patients. *Rev. Infect. Dis.* **1988**, *10* (Suppl. 3), S495-506.
264. Drew, W. L; Ives, D.; Lalezari, J. P.; Crumpacker, C.; Follansbee, S. E.; Spector, S. A.; Benson, C. A.; Friedberg, D. N.; Hubbard, L. Oral ganciclovir as maintenance treatment for cytomegalovirus retinitis in patients with AIDS. *N. Engl. J. Med.* **1995**, *333*, 615-620.
265. Spector, S. A.; Busch, D. F.; Follansbee, S.; Squires, K.; Lalezari, J. P.; Jacobson, M. A.; Connor, J. D.; Jung, D.; Shadman, A.; Mastre, B.; Buhles, W.; Drew, W. L. Pharmacokinetics, safety, and antiviral profiles of oral ganciclovir in persons infected with human immunodeficiency virus: a phase II study. *J. Infect. Dis.* **1995**, *171*, 1431-1437.
266. Harden, M. R.; Jarvest, R. L.; Boyd, M. R.; Sutton, D.; Vere Hodge, R. A. Prodrugs of selective antiherpesvirus agent 9-[4-hydroxy-3-(hydroxymethyl)but-1-yl]guanine (BRL 39123) with improved gastrointestinal absorption properties. *J. Med. Chem.* **1989**, *32*, 1738-1743.
267. Boyd, M. R.; Bacon, T. H.; Sutton, D. Comparative activity of penciclovir and acyclovir in mice infected intraperitoneally with herpes simplex virus type 1 SC 16. *Antimicrob. Agents Chemother.* **1993**, *37*, 642-645.
268. Boyd, M. R.; Bacon, T. H.; Sutton, D. Antiherpesvirus activity of 9-(4-hydroxy-3-hydroxymethylbut-1-yl)guanine (BRL 39123) in animals. *Antimicrob. Agents Chemother.* **1988**, *32*, 358-363.
269. Vere Hodge, R. A.; Cheng, Y. -C. The mode of action of penciclovir. *Antiviral Chem. Chemother.* **1993**, *4*, 13-24.
270. Rolan, P. Pharmacokinetics of new antiherpetic agents. *Clin. Pharmacokinet.* **1995**, *29*, 333-340.
271. Pue, M. A.; Benet, L. Z. Pharmacokinetics of famciclovir in man. *Antiviral Chem. Chemother.* **1993**, *4* (Suppl. 1), 47-55.
272. Parang, K.; Knaus, E. E.; Wiebe, L. I.; Sardari, S.; Daneshtalab, M.; Csizmadia, F. Synthesis and antifungal activities of myristic acid analogs. *Arch. Pharm.-Pharm. Med. Chem.* **1996**, *329*, 475-482.
273. Naesens, L.; Balzarini, J.; De Clercq, E. Therapeutic potential of PMEA as an antiviral drug. *Med. Virol.* **1994**, *4*, 147-159.
274. De Clercq, E. Broad-spectrum anti-DNA virus and anti-retrovirus activity of phosphonylmethoxyalkylpu rine and pyrimidines. *Biochem. Pharmacol.* **1991**, *42*, 963-972.
275. Walker, R. E.; Vogel, S. E.; Jaffe, H.S.; Polis, M. A.; Kovacs, J.; Markowitz, N.; Masur, H.; Lane, H. C. A phase I/II study of PMEA in HIV infected patients. Abstracts of Papers; 1st National Conference on Human Retroviruses and Related Infections, **1993**, Washington DC.
276. Starrett, J. E., Jr.; Tortolani, D. R.; Russel, J.; Hitchcock, M. J. M.; Whiterock, V.; Martin, J. C.; Mansuri, M. M. Synthesis, oral bioavailability determination, and *in vitro* evaluation of prodrugs of the antiviral agent 9-[2-(phosphonomethoxy)-ethyl]adenine (PMEA). *J. Med. Chem.* **1994**, *37*, 1857-1864.
277. De Clercq, E. Acyclic nucleoside phosphonates in the chemotherapy of DNA virus and retrovirus infections. *Intervirology* **1997**, *40*, 295-303.
278. Srinivas, R. V.; Robbins, B. L.; Connelly, M. C.; Gong, Y. -F., Bischofberger, N.; Fridland, A. Metabolism and *in vitro* antiretroviral activities of bis(pivaloyloxymethyl) prodrugs of acyclic nucleoside phosphonates. *Antimicrob. Agents Chemother.* **1993**, *37*, 2247-2250.
279. Starrett, J. E. Jr.; Tortolani, D. R.; Hitchcock, M. J. M.; Martin, J. C.; Mansuri, M. M. Synthesis and *in vitro* evaluation of a phosphonate prodrug: bis(pivaloyloxymethyl) 9-(2-phosphonylmethoxyethyl)ade nine. *Antiviral Res.* **1992**, *19*, 267-273.
280. Delaugerre, C.; Marcelin, A. -G.; Thibalut, V.; Peytavin, G.; Bombled, T.; Bochet, M. -V.; Katlama, C.; Benhamou, Y.; Calvez, V. Human immunodeficiency virus (HIV) type 1 reverse transcriptase resistance mutations in hepatitis B virus (HBV)-HIV-coinfected patients treated for HBV chronic infecion once

daily with 10 milligrams of adefovir dipivoxil combined with lamivudine. *Antimicrob. Agents Chemother.* **2002**, *46*, 1586-1588.

281. Ying, C.; De Clercq, E.; Nicholson, W.; Furman, P.; Neyts, J. Inhibition of the replication of the DNA polymerase M550V mutation variant of human hepatitis B virus by adefovir, tenofovir, L-FMAU, DAPD, penciclovir and lobucavir. *J. Viral Hepat.* **2000**, 7, 161-165.
282. (a) Srinivas, R. V.; Fridland, A. Antiviral activities of 9-R-2-phosphonomethoxypropyl adenine (PMPA) and bis(isopropyloxymethylcarbonyl)PMPA against various drug-resistant human immunodeficiency virus strains. *Antimicrob. Agents Chemother.* **1998**, *42*, 1484-1487. (b) Palmer, S.; Buckheit, R. W.; Gilbert, H.; Shaw, N.; Miller, M. D. Anti-HIV-1 activity of PMEA, PMPA (9-(2-phosphonylmethoxypropyl)adenine) and AZT against HIV-1 Subtypes A, B, C, D, E, F, G and O. *Antiviral Res.* **2000**, *46*, A42-A42.
283. Arimilli, M. N.; Kim, C. U.; Dougherty, J.; Mulato, A.; Oliyai, R.; Shaw, J. P.; Cundy, K. C.; Bischofberger, N. Synthesis, *in vitro* biological evaluation and oral bioavailability of 9-[2-(phosphonome thoxy)propyl]adenine (PMPA) prodrugs. *Antiviral Chem. Chemother.* **1997**, *8*, 557-564.
284. Cihlar, T.; Birkus, G.; Greenwalt, D. E.; Hitchcock, M. J. M. Tenofovir exhibits low cytotoxicity in various human cell types: comparison with other nucleoside reverse transcriptase inhibitors. *Antiviral Res.* **2002**, *54*, 37-45.
285. Naeger, L. K.; Margot, N. A.; Miller, M. D. Tenofovir (PMPA) is less susceptible to pyrophosphorolysis and nucleotide-dependent chain-terminator removal than zidovudine or stavudine. *Nucleosides Nucleotides* **2001**, *20*, 635-639.
286. Fridland, A.; Robbins, B. L.; Srinivas, R. V. Antiretroviral activity and metabolism of bis(POC)PMPA, an oral bioavailable prodrug of PMPA. *Antiviral Res.* **1997**, *34*, A49.
287. De Clercq, E.; Sakuma, T.; Baba, M.; Pauwels, R.; Balzarini, J.; Rosenberg, I.; Holy, A. Antiviral activity of phosphylmethoxyalkyl derivatives of purines and pyrimidines. *Antiviral. Res.* **1987**, *8*, 261-272.
288. Hitchcock, M. J. M.; Jaffe, H. S.; Martin, J. C.; Stagg, R. J. Cidofovir - a new agent with potent anti-herpesvirus activity. *Antiviral Chem. Chemother.* **1996**, *7*, 115-127.
289. (a) Yokota, T.; Konno, K.; Chonan, E.; Mochizuki, S.; Kojima, K.; Shigeta, S.; De Clercq, E. Comparative activities of several nucleoside analogs against duck hepatitis B virus *in vitro*. *Antimicrob. Agents Chemother.* **1990**, *34*, 1326-1330. (b) Yokota, T.; Mochizuki, S.; Konno, K.; Mori, S.; Shigeta, S.; De Clercq, E. Inhibitory effects of selected antiviral compounds on human hepatitis B virus DNA synthesis. *Antimicrob. Agents Chemother.* **1991**, *35*, 394-397.
290. Mendel, D. B.; Cihlar, T.; Moon, K.; Chen, M. S. Conversion of 1-[((*S*)-2-hydroxy-2-oxo-1,4,2-dioxaphosphorinan-5-yl)methyl]cytosine to cidofovir by an intracellular cyclic CMP phosphodiesterase. *Antimicrob. Agents Chemother.* **1997**, *41*, 641-646.
291. Weller, S.; Blum, M. R.; Doucette, M. Pharmacokinetics of the acyclovir pro-drug valaciclovir after escalating single and mutiple-dose administration to normal volunteers. *Clin. Pharmacol. Ther.* **1993**, *54*, 595-605.
292. Hostetler, K. Y.; Beadle, J. R.; Hornbuckle, W. E.; Bellezza, C. A.; Tochkov, I. A.; Cote, P. J.; Gerin, J. L.; Korba, B. E.; Tennant, B. C. Antiviral activities of oral 1-*O*-hexadecylpropanediol-3-phosphoacyclovir and acyclovir in woodchucks with chronic woodchuck hepatitis virus infection. *Antimicrob. Agents Chemother.* **2000**, *44*, 1964-1969.
293. Cheng, L. Y.; Hostetler, K. Y.; Ozerdem, U.; Gardner, M. F.; Mach-Hofacre, B.; Freeman, W. R. Intravitreal toxicology and treatment efficacy of a long acting anti-viral lipid prodrug of ganciclovir (HDP-GCV) in liposome formulation. *Invest. Ophtalmol. Vis. Sci.* **1999**, *40*, S872-S872.
294. Cheng, L.; Hostetler, K. Y.; Chaidhawangul, S.; Gardner, M. F.; Beadle, J. R.; Keefe, K. S.; Bergeron-Lynn, G.; Severson, G. M.; Soules, K. A.; Mueller, A. J.; Freeman, W. Intravitreal toxicology and

duration of efficacy of a novel antiviral lipid prodrug of ganciclovir in liposome formulation. *Invest. Ophtalmol. Vis. Sci.* **2000**, *41*, 1523-1532.

295. (a) Beadle, J. R.; Rodriguez, N.; Aldern, K. A.; Hostetler, K. Y.; Hartline, C.; Kern, E. R. Antiviral activity of alkoxyalkyl esters of cidofovir against drug-resistant strains of cytomegalovirus. *Antiviral Res.* **2002**, *53*, A60-A60. (b) Williams, S. L.; Kushner, N. L.; Hartline, C. B.; Harden, E. A.; Beadle, J. R.; Hostetler, K. Y.; Kern, E. R. Enhanced activity of alkoxyalkyl esters of cidofovir and cyclic cidoforiv against replication of herpesviruses *in vitro*. *Antiviral Res.* **2002**, *53*, A61-A61. (c) Aldern, K. A.; Ciesla, S. L.; Winegarden, K. L.; Hostetler, K. Y. 1-*O*-Hexadecyloxypropyl-[^{14}C]cidofovir: cellular uptake and metabolism in MRC-5 human lung fibroblasts, *in vitro*. *Antiviral Res.* **2002**, *53*, A62-A62. (d) Huggins, J. W.; Baker, R. O.; Beadle, J. R.; Hostetler, K. Y. Orally active ether lipid prodrugs of cidofovir for the treatment of smallpox. *Antiviral Res.* **2002**, *53*, A66-A66. (e) Winegarden, K. L.; Ciesla, S. L.; Aldern, K. A.; Beadle, J. R.; Hostetler, K. Y. *Antiviral Res.* **2002**, *53*, A67-A67.
296. Kumar, R.; Nath, M.; Tyrrell, D. L. J. Design and synthesis of novel 5-substituted acyclic pyrimidine nucleosides as potent and selective inhibitors of hepatitis B virus. *J. Med. Chem.* **2002**, *45*, 2032-2040.
297. Pockros, P. J. Developments in the treatment of chronic hepatitis C. *Exp. Opin. Inv. Drugs* **2002**, *11*, 515-528.
298. Sood, R. K.; Bhadti, V. S.; Fattom, A. I.; Naso, R. B.; Korba, B. E.; Kern, E. R.; Chen, H. -M.; Hosmane, R. S. Novel ring-expanded nucleoside analogs exhibit potent and selective inhibition of hepatitis B virus replication in cultured human hepatoblastoma cells. *Antiviral Res.* **2002**, *53*, 159-164.
299. Gill, J. K.; Wang, L.; Bretner, M.; Newman, R.; Kyprianou, N.; Hosmane, R. S. Potent *in vitro* anticancer activities of ring-expanded ("fat") nucleosides containing the imidazo[4,5-*e*][1,3]diazepine ring system. *Nucleosides Nucleotides* **2001**, *20*, 1043-1047.
300. De Clerq, E. Antiviral and antimetabolic activities of neplanocins. *Antimicrob. Agents. Chemother.* **1985**, *28*, 84-89.
301. Song, G. -Y.; Paul, V.; Choo, H.; Morrey, J.; Sidwell, R. W.; Schinazi, R. F.; Chu, C. K. Enantiomeric synthesis of D- and L-cyclopentenyl nucleosides and their antiviral activity against HIV and West Nile virus. *J. Med. Chem.* **2001**, *44*, 3985-3993.
302. Kusaka, T.; Yamamoto, H.; Shibata, M.; Muroi, M.; Kishi, T.; Mizuno, K. Streptomyces citricolor nov. sp. and a new antibiotic, aristeromycin. *J. Antibiot.* **1968**, *21*, 255-263.
303. Rajappan, V.; Schneller, S. W.; Williams, S. L.; Kern, E. R. The enantiomers of carbocyclic 5'-norguanosine: activity towards Epstein-Barr virus. *Bioorg. Med. Chem. Lett.* **2002**, *10*, 883-886.
304. Siddiqi, S. M.; Chen, X.; Schneller, S. W. Antiviral enantiomeric preference for 5'-noraristeromycin. *J. Med. Chem.* **1994**, *37*, 551-554.
305. Seley, K. L.; Schneller, S. W.; Korba, B. A 5'-noraristeromycin enantiomer with activity towards hepatitis B virus. *Nucleosides Nucleotides* **1997**, *16*, 2095-2099.

CHAPTER 2

CHIRAL SYNTHESIS OF ANTIVIRAL NUCLEOSIDES FROM CARBOHYDRATE TEMPLATES

GIUSEPPE GUMINA, SUREYYA OLGEN and CHUNG K. CHU

2.1. Introduction

When organic chemistry meets medicinal chemistry or natural products synthesis, stereoselectivity becomes a major challenge, because in most cases even simple chiral organic molecules cannot be prepared without a reasonable control of stereochemistry. In any case, there is an increasing need for organic synthesis to be stereoselective. This is particularly important in the field of drug discovery and development, whereas a racemic drug in a biological system behaves as a mixture of two different compounds and quite often only one has the desired properties, while the other is at least inactive or potentially harmful. Stereoselective synthesis can be a simple or difficult task, whether accomplished through a chiral auxiliary or by using a chiral starting material. Both strategies have advantages and limitations, depending on the starting materials and final target molecules. Chiral auxiliaries can be used with a number of molecules bearing the same functionality, but most times their action is only partially stereoselective. Chiral starting materials, on the other hand, can be available in very high optical purity. The major limitation of the latter strategy, of course, is that although thousands of optically active natural compounds are known, it is not always easy to select a starting material that is naturally abundant and will not require many steps to be converted to the target molecule.

Carbohydrates are among the most common and useful chiral starting materials, and most of them are commercially available and inexpensive. One drawback in using carbohydrates is that they tend to be "overfunctionalized" for many purposes, but this problem is usually solved elegantly by carbohydrate chemists. Furthermore, this "overfunctionalization" is often exploited in highly asymmetric reactions, and for this reason it provides carbohydrates great potentiality and versatility, particularly in the synthesis of carbohydrate-modified nucleosides.

Undoubtedly, carbohydrates have their greatest usefulness as synthons for compounds containing carbon chains with contiguous or noncontiguous secondary alcohols. However, they have been used as starting materials for the syntheses of a wide array of chiral compounds. As mentioned, one feature that has been taken advantage of, particularly in natural products synthesis, is the stereo-control that can be achieved in the manipulation of functionalities in small rings. Because carbohydrates often have ring structures that can be easily opened at a later stage, a common strategy is to carry out series of reactions on such rings to obtain, with excellent control of stereochemis-

Antiviral Nucleosides: Chiral Synthesis and Chemotherapy, Ed. by C.K. Chu. 77 — 189

try, compounds that sometimes only remotely resemble the parent sugar. Once reached the desired chirality, the structure can be manipulated by means of ordinary organic reactions to afford the product.

Nucleosides are among the most important carbohydrate-containing natural products, and carbohydrate-modified analogs are currently the cornerstone of antiviral therapy.[1] In this chapter, we will summarize a few important classes of nucleosides the chiralities of which were directly or indirectly originated from carbohydrates as chiral templates.

2.2. 4'-Thiofuranose nucleosides

A wide variety of carbohydrate-modified nucleoside analogs have been prepared and tested for antiviral activities. The first modification was the isosteric substitution of the furanose oxygen with a sulfur atom. The first nucleoside analogs containing such sugars were the adenine derivatives of 4'-thio-D-xylose and 4'-thio-D-arabinose.[2] 2'-Deoxy-4'-thionucleosides have shown potent anti-herpes virus activities, and several analogs, such as 4'-thiothymidine and 2'-deoxy-4'-thiocytidine, have potent cytotoxicity. The rationale for the synthesis of these classes of nucleosides was that the presence of sulfur in the sugar ring stabilized the *N*-glycosidic bond with respect to phosphorolysis.[3] Thus 4'-thioinosine was known to be resistant to phosphorolytic cleavage, which is a normal pathway in nucleosides catabolism. This is a major advantage of 4'-thio-nucleosides, since several "4'-oxy" antivirals have a drawback with regard to their metabolic stability caused by nucleoside phosphorylases. In addition, the potent antiviral activity and cytotoxicity of 4'-thionucleosides suggest that they are well recognized as substrates by both viral and host cell kinases. In most approaches for the syntheses of 4'-thionucleosides, the stereochemistry of the final products derives directly from the carbohydrate analogs by displacement of a leaving group with a sulfur-containing nucleophile, followed by ring closure or ring contraction,[4] acetolysis of an γ,γ-diethoxy episulfide, or ring closure of dialkyl dithioacetal.[5,6,7] Triiodoimidazole-triphenylphosphine and chlorodiphenylphosphine-iodine-imidazole have also been reported to effect ring closure of dialkyl dithioacetals to give the corresponding 4-thio furanoside derivatives.[8,9,10]

Dyson *et al.* reported the synthesis of (*E*)-5-(2-bromovinyl)-D-4'-thio-2'-deoxyuridine from D-2-deoxyribose using a double inversion strategy (Scheme 1).[11,12] D-2-Deoxyribose was kinetically cyclized to the furanose form **1-1** in the presence of a catalytic amount of HCl gas, and subsequently protected as the dibenzyl ether. Treatment of **1-2** with benzyl mercaptane in the presence of HCl gave the dithio derivative **1-3**. In order to obtain the desired configuration, the chirality at the C-4 position was inverted by the Mitsunobu reaction. Debenzoylation of the product **1-4**, followed by mesylation of the resulting alcohol gave compound **1-5**, which was cyclized to **1-6** with a second inversion of configuration and condensed with various pyrimidine bases.

Huang *et al.*[9] reported an improved method for the synthesis of 4'-thio-2'-deoxypyrimidine nucleosides using the triphenylphosphine/iodine/imidazole reagent system, which has been generally used for converting alcohols to deoxyiodo compounds.

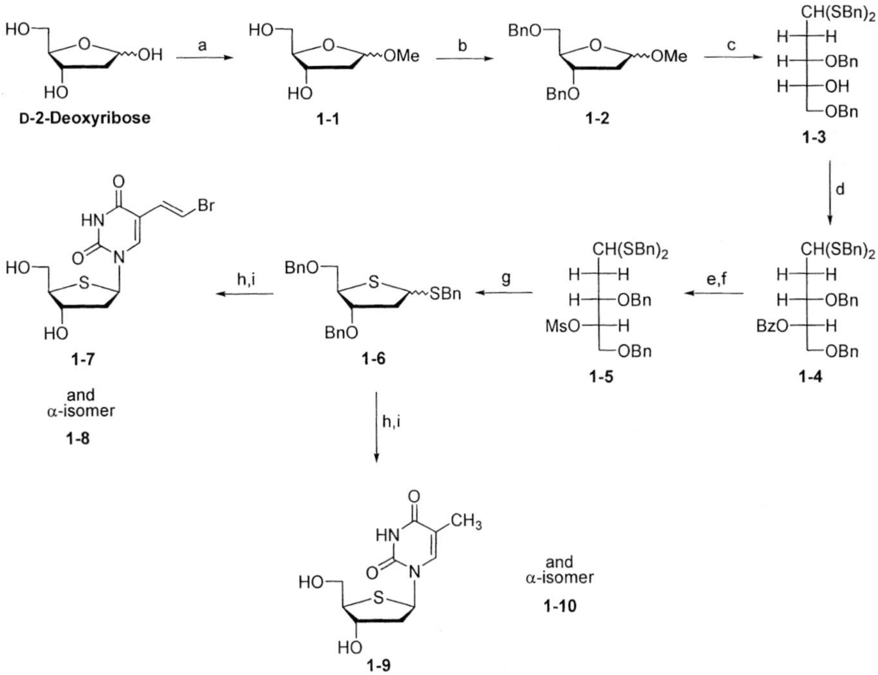

Reagents: a) MeOH, HCl, rt; b) NaH, TBAI, BnBr, THF; c) BnSH, conc. HCl; d) Ph_3P, Benzoic acid, DEAD, THF; e) NaOMe, MeOH; f) MsCl, Py; g) NaI, $BaCO_3$, acetone; h) silylated (E)-bromovinyluracil or thymine; i) NaOMe, MeOH, and then separation.

Scheme 1

Secrist III *et al.* also reported the synthesis of 2'-deoxy-4'-thio pyrimidine nucleosides from the methyl glycoside of 2-deoxy-4-thio-β-D-*erythro*-pentofuranose (Scheme 2).[13] The starting material **2-1** was prepared in several steps from L-arabinose by the literature procedure of Fu and Bobek[14] with several experimental modifications and was then converted to the 3,4-di-*O*-toluoyl derivative **2-2**. Direct coupling of **2-2** with pyrimidine bases failed to provide the desired nucleosides. Attempts to convert **2-2** to the more reactive glycosyl chloride using literature procedures were unsuccessful, apparently due to the instability of the chlorothiosugar. Experimental modification of conventional acetolysis conditions provided acetyl sugar **2-3** as a ~1:1 mixture of anomers, stable for some time at room temperature but reactive enough to provide the desired nucleosides.

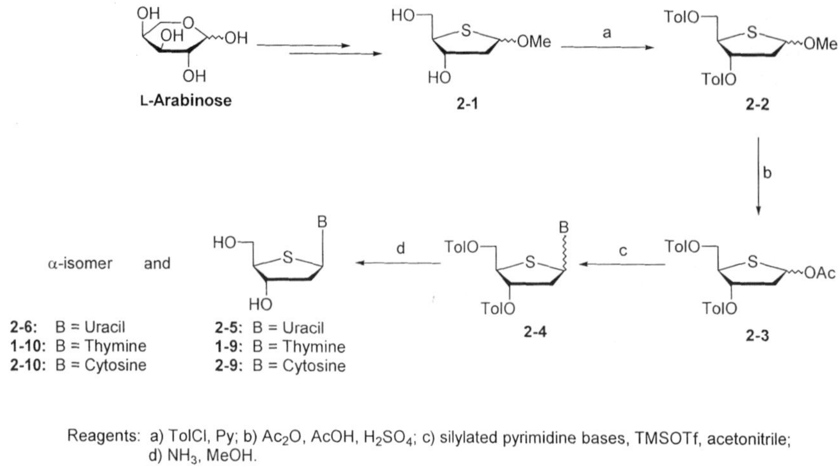

Scheme 2

Bellon *et al.* synthesized 4'-thio-D-ribofuranose nucleosides from L-lyxose (Scheme 3).[7] Their strategy was based on the introduction of a sulfur atom at the C-1 position followed by a nucleophilic displacement of an activated 4-hydroxylgroup with inversion of configuration. Thus, lyxose derivative **3-1** was treated with benzyl mercaptane, to give the dithiobenzyl acetal **3-2**. Mesylation of **3-2** gave the intermediate **3-3**, in which the nucleophilic substitution could be directly achieved either by heating the mesylation reaction in aqueous pyridine or by heating with sodium iodide in acetone or tetrabutylammonium iodide and barium carbonate in acetone. Treatment of **3-4** with mercuric acetate in acetic acid and condensation of the resulting intermediate **3-5** with silylated pyrimidines in the presence of TMSOTf gave the nucleoside derivatives. As expected when a 2'-non-participating group (-OBn) is present, anomeric mixtures were obtained. The anomers were separated by silica gel chromatography and deprotected independently using boron tribromide.

Leydier *et al.* synthesized 4'-thio-D-ribonucleosides starting from D-ribose using a double inversion strategy (Scheme 4).[15] The first S_N2 process in their synthesis was the Mitsunobu reaction on the D-ribose dithiobenzylacetal derivative **4-2** with inversion of the configuration at the C-4 atom, providing, after deprotection, the L-xylose dithiobenzyl acetal derivative **4-4**. This key intermediate has the stereochemistry needed to provide the 4-thio-D-ribofuranoside derivative **4-5** by means of another iodide-mediated S_N2 cyclization. Acetylation of **4-5** provided the 4-thio-D-ribofuranoside derivative **4-6**, from which 4'-thiocytosine and adenine derivatives were synthesized by appropriate glycosylation reactions.[16]

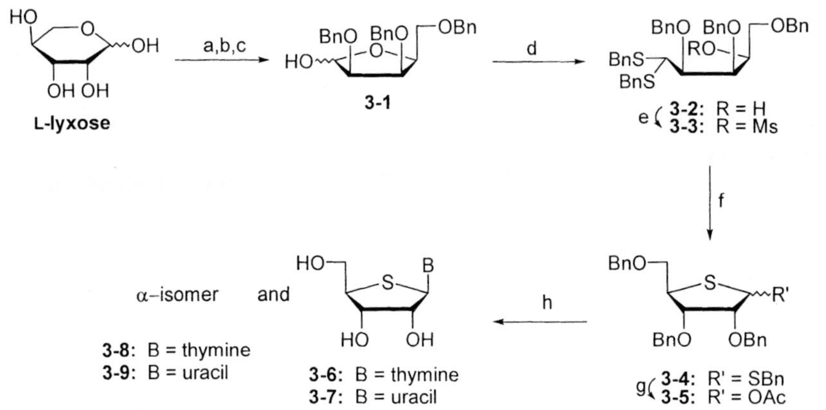

Reagents: a) HCl/MeOH; b) BnBr, KOH; c) HCl, H_2O/dioxane; d) BnSH, HCl; e) MsCl, Py; f) TBAI, $BaCO_3$, reflux; g) $Hg(OAc)_2$, AcOH; h) condensation with bases, separation, and deprotection.

Scheme 3

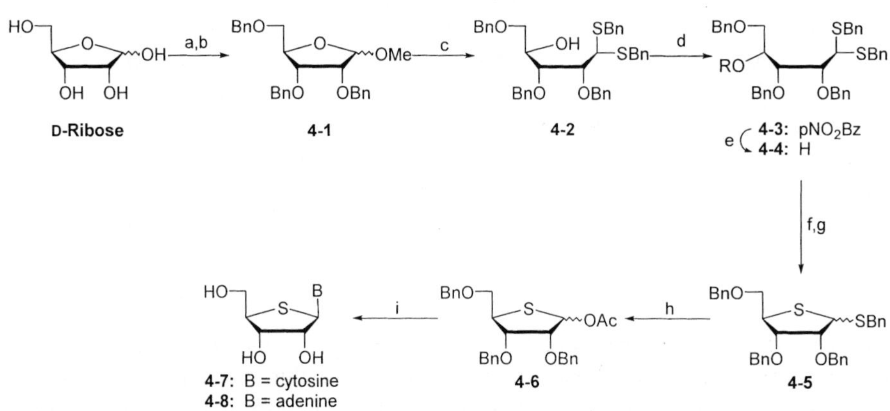

Reagents: a) MeOH/HCl; b) BnBr, KOH, THF; c) BnSH, $BF_3.Et_2O$; d) Ph_3P, DEAD, *p*-NO_2PhCO_2H, THF; e) K_2CO_3, MeOH; f) MsCl, Py; g) NBu_4I, $BaCO_3$, Py; h) $Hg(OAc)_2$, AcOH; i) condensation, and deprotection.

Scheme 4

Previously, Bobek *et al.* had synthesized 4'-thio analogs of the antibiotic toyocamycine[17] by condensation of 2,3,5-tri-*O*-acetyl-4-thio-D-ribofuranosyl chloride **5-2** with the chloromercury derivative of 4-acetamino-6-bromo-5-cyanopyrrolo-[2,3-d]pyrimidine **5-1** followed by removal of the protecting groups with MeOH-NH_3 and reductive debromination using H_2/Pd catalyst (Scheme 5). The key intermediate **5-2** was prepared from 1,2,3,5-tetra-*O*-acetyl-4-thio-α,β-ribofuranose, analogous to **4-5**. As a series, Bobek and co-workers synthesized 5-substituted 4'-thiouridine derivatives using the same method.[18]

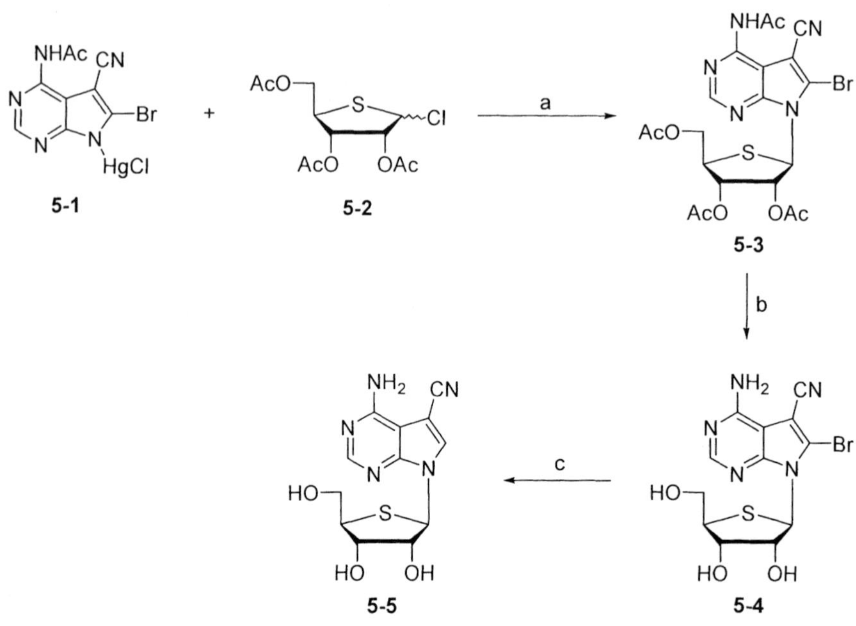

Reagents: a) H_2O, NaOH, EtOH; b) NH_3, MeOH; c)H_2/Pd

Scheme 5

Tber *et al.* reported the synthesis of D-3'-azido-2',3'-dideoxy-4'-thionucleosides from D-xylose (Scheme 6).[19] The starting material for this methodology is 2-deoxy-4,5-isopropylidene-D-*threo*-pentose diisobutyl dithioacetal **6-3**, prepared from D-xylose according to the procedure developed by Wong and Gray.[20] Using a Mitsunobu-type reaction, **6-3** was converted to the 3-azido-2,3-dideoxy-D-*erythro*-pentose derivative **6-4**, with inversion at the C-3 position. Hydrolysis of the 4,5-*O*-isopropylidene group and benzoylation at C-5 gave **6-5**, which has only the C-4 hydroxyl group unprotected. In a double inversion strategy, **6-5** was mesylated and, after deprotection/reprotection of the aldehyde functionality, converted to epoxide **6-8**, with inversion at C-4. The second inversion was obtained by treatment of **6-8** with thiourea to give the 4,5-epithio derivative **6-9**. Opening of the episulfide ring under conditions designed to give acyclic acetal **6-10** afforded the desired product in 48% yield, although some cyclization also occurred to give 33% of an anomeric mixture of methyl-*O*-acetyl-3-azido-2,3-dideoxy-4-thio-D-*erythro*-pentofuranosides **6-11**. The glycosyl donor **6-11**, obtained also from **6-10** by acidic hydrolysis with concomitant reacetylation, was condensed with various pyrimidine bases to give, after further elaboration, compounds **6-13~6-25**.

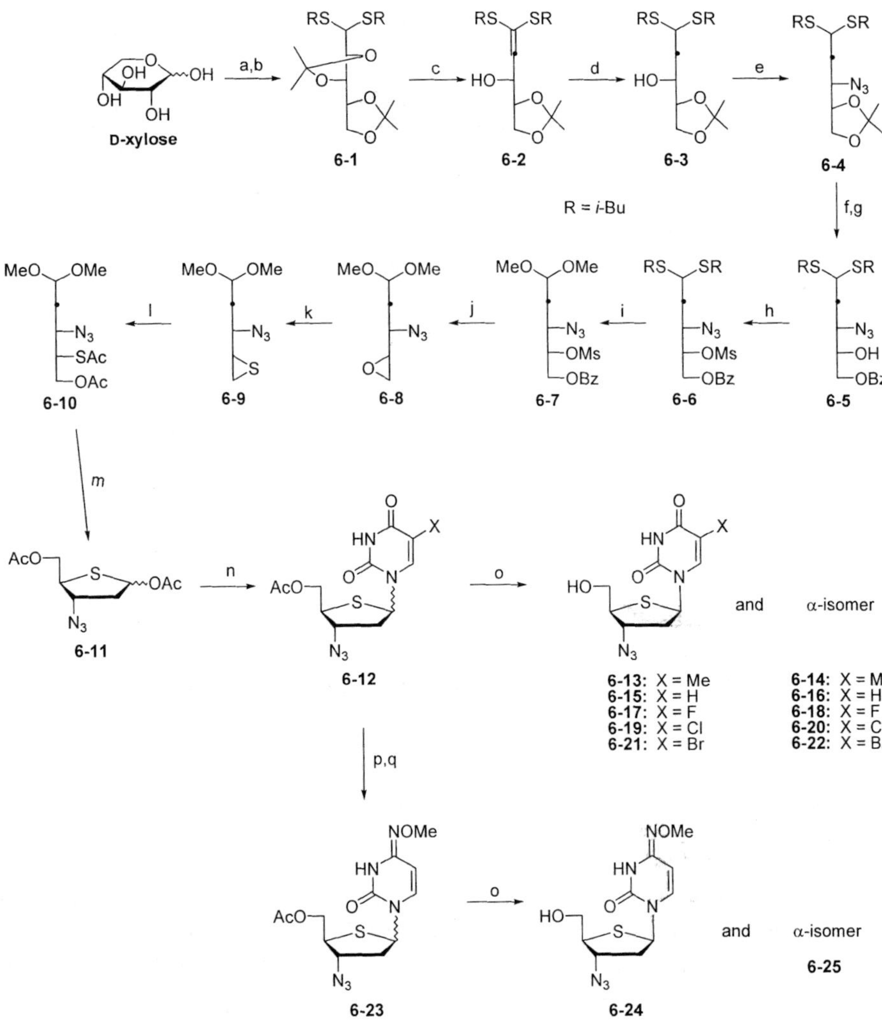

Reagents: a) *i*-BuSH, HCl; b) acetone, Dowex 50W-X8; c) KO*t*-Bu; d) LAH, THF; e) [Zn(N_3)$_2$.2Py], Ph_3P, DIAD, Tol; f) H^+; g) BzCl, Py; h) MsCl, TEA, CH_2Cl_2; i) HgO, $HgCl_2$, MeOH; j) MeONa, MeOH; k) $H_2NC(S)NH_2$; l) AcONa, Ac_2O, AcOH; m) Ac_2O, AcOH, H_2SO_4; n) persilylated pyrimidine, $SnCl_4$ or TMSOTf; o) NH_3/MeOH; p) K_2CO_3, TsCl; q) $MeONH_2 \cdot HCl$.

Scheme 6

Aitsir *et al.* subsequently devised another convergent synthesis of C3-substituted sugars *via* the 2-deoxy-4-thio-D-*erythro*-pentofuranose derivative **7-4** from L-arabinose in 10 steps (Scheme 7).[21] With an overall yield of 17%, this method may give some flexibility in the choices of protecting groups on the final product and has the advantage of using an inexpensive starting material.

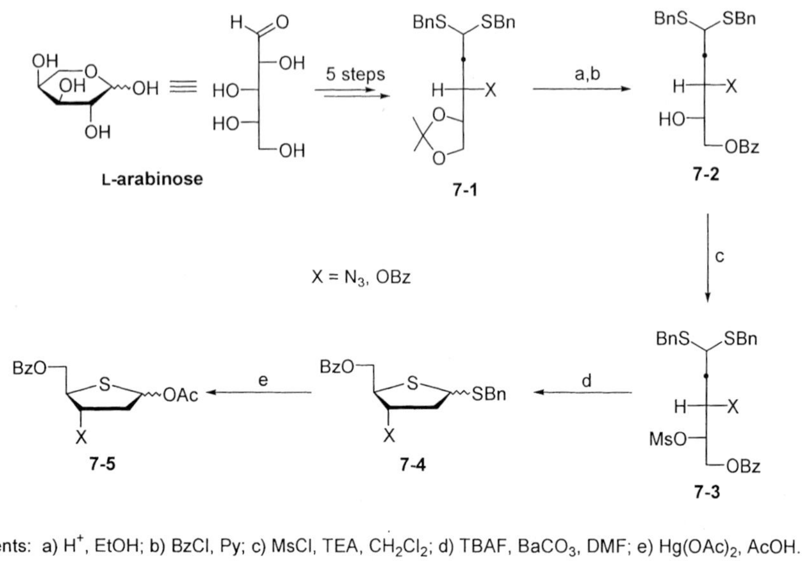

Scheme 7

Jandu and Selwood reported a specific preparation of the potent broad-spectrum antiviral agent 5-ethyl-4'-thio-2'-deoxyuridine that avoids the use of large quantities of thiol reagent and consecutive double inversion of the stereochemistry at C-4 of the sugar moiety (Scheme 8).[22] The key step uses the observation that 4-thiopyranoses **8-5** and **8-6** can be produced from thionocarbonates by radical-induced rearrangement, thus achieving the required double inversion in a single step.[23] The yield of 5-ethyl-4'-thio-2'-deoxyuridine was estimated in about 10% starting from 500 g of 2-deoxy-D-ribose.

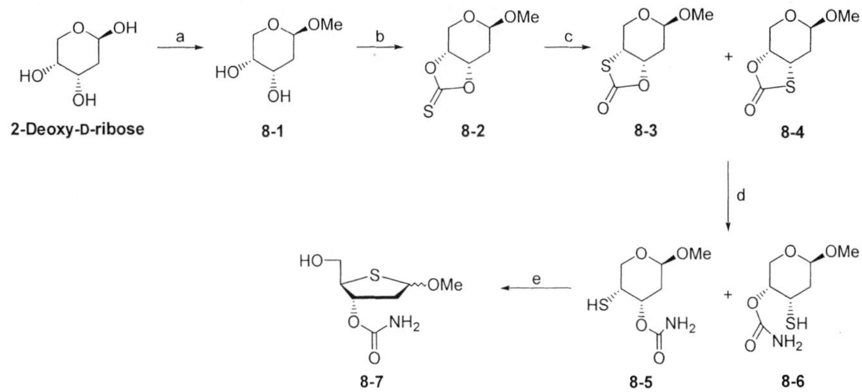

Reagents: a) HCl/MeOH; b) $CSCl_2$; c) t-$Bu_4N^+Br^-$/diglyme, 150 °C; d) NH_3/MeOH; e) Dowex H^+, MeOH.

Scheme 8

Rassu *et al.* reported the synthesis of 2',3'-dideoxy-4'-thiocytidines from *R* and *S*-2,3-isopropylidene glyceraldehydes (Scheme 9).[24] The synthesis of the cytidine derivative **9-11** (L-series) started from protected thiophene **9-1** and 2,3-*O*-isopropylidene D-glyceraldehyde **9-2**. Their diastereoselective addition in the presence of 1 equivalent of BF_3 etherate in CH_2Cl_2 resulted in preferential formation of the 4*R*-adduct **9-3** accompanied by less than 10% of 4*S*-diasteromer **9-4**. Hydrogenation of the major isomer gave thiolactone **9-5**. Selective deblocking of the acetonide protection of **9-5**, followed by exposure of the resulting crude triol to $NaIO_4/SiO_2$ in CH_2Cl_2, cleanly afforded aldehyde **9-6**. Reduction of the formyl functionality and subsequent silylation of the new primary hydroxyl functionality gave **9-7**, the immediate precursor of thiosugar **9-8**. Selective reduction of thiolactone carbonyl to thiolactol was achieved by careful treatment of **9-7** with $LiAlH_4$ in THF at -20 □C. The following acetylation of the anomeric hydroxyl group gave pure L-thiofuranose **9-8** as a 1:1 anomeric mixture in 37% overall yield based on **9-2**. The final coupling of **9-8** with cytosine was successfully conducted by a modification of the Vorbrüggen protocol. Separation of the individual anomers was carried out by preparative TLC allowing the synthesis of pure nucleosides **9-10** and **9-11**. For the synthesis of D-series, the same synthetic procedure was applied to protected L-glyceraldehyde **9-12** to obtain **9-13** and **9-14**.

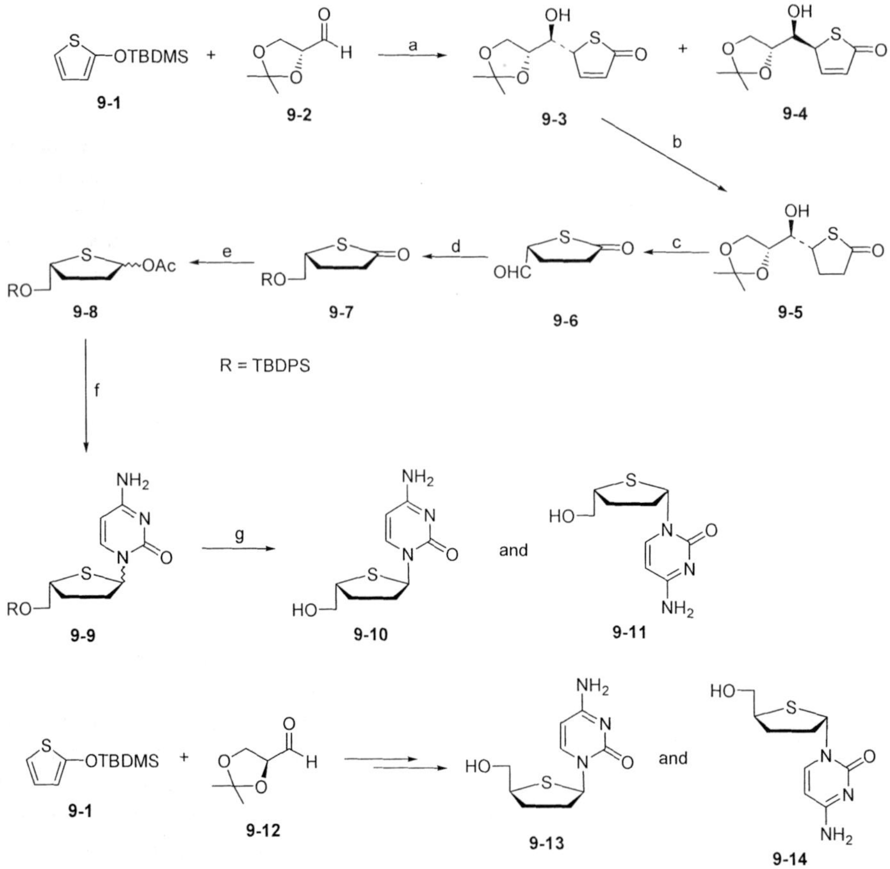

Reagents: a) TBSOTf, $BF_3 \cdot Et_2O$ (1.0 equiv), CH_2Cl_2, -90 °C, 5h; b) H_2, 10% Pd/C, NaOAc, THF, 20 °C, 12h; c) 80% aq. AcOH, 40 °C, 24h; then 0.65M aq. $NaIO_4$, SiO_2, CH_2Cl_2, 22 °C, 1h; d) $NaBH_4$, MeOH, 25 °C, 12h; then TBDPS-Cl, imidazole, DMF, 25 °C, 2h; e) $LiAlH_4$ (0.5 equiv), THF, -20 °C, 7h; then Ac_2O, py, DMAP, 25 °C, 2h; f) cytosine (1.5 equiv), nonafluorobutanesulfonate, CH_3CN, HMDS, TMS-Cl, 25 °C, 24h; g) preparative TLC, SiO_2, $CHCl_3$/MeOH (9:1) in NH_3 atmosphere; then TBAF, THF, AcOH, 25 °C, 20h.

Scheme 9

2,3-Isopropylidene D-glyceraldehyde **9-2** and its enantiomer **9-12** are important precursors in many syntheses leading to optically active nucleosides. The D-isomer **9-2** can be conveniently prepared from protected D-mannose **10-1** (Scheme 10) by oxidative cleavage using lead tetraacetate[25] or sodium periodate;[26] the L-isomer **9-12** can be prepared in the same way starting from L-gulonic acid-γ-lactone **10-2**.[27]

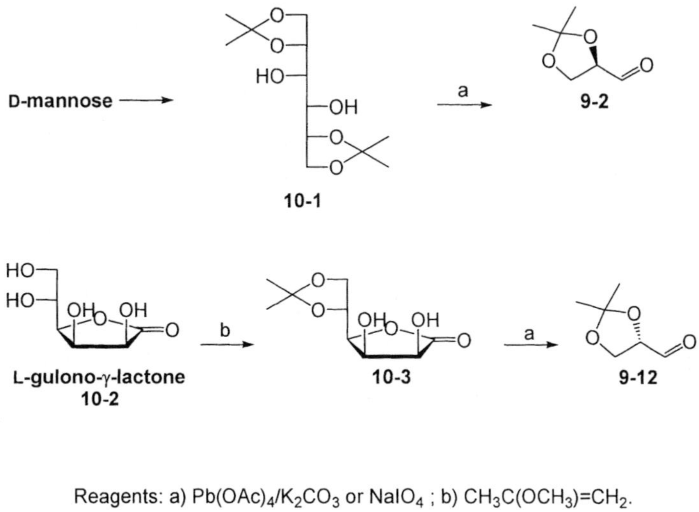

Scheme 10

Young *et al.* reported the enantioselective synthesis of 2',3'-dideoxy- and 2',3'-didehydro-2',3'-dideoxy-4'-thionucleosides from chiral thiolactones (Scheme 11).[28] The chiral key intermediate **11-4** was prepared starting from (*S*)-(–)-glycidol **11-1**. This was protected as its *tert*-butyldiphenylsilyl ether and reacted with thiourea in methanol[29] to give the (*R*)-thirane **11-2** with clean inversion of configuration. The conversion of **11-2** to the (4*S*)-2-methoxycarbonyl thiolactone **11-3** was achieved by reaction with dimethyl malonate and sodium hexamethyldisilazane in refluxing THF; careful control of concentration minimized side-products due to thiirane polymerization,[30] affording an enantiomeric excess between 82 to 91%. Similarly, (*R*)-(+)-glycidol **11-7** was converted to the (*R*)-lactone **11-8**. Each thiolactone was converted to the corresponding thiolactol acetate (i.e. **11-5** from **11-4**) by reduction with diisobutylaluminum hydride in toluene at -78 °C and subsequent acetylation in 90% yield. Glycosylation was achieved through standard methods, by reaction of the acetates with the silylated derivatives of a number of substituted uracils and cytosines in the presence of tin(IV) chloride in acetonitrile followed by deprotection to give the desired nucleosides. Better yields were obtained for cytidine glycosylations by using potassium nonaflate, trimethylsilyl chloride and hexamethyldisilazane.

Thiolactones **11-4** and **11-8** provided a useful entry to 2',3'-didehydro-2',3'-dideoxy-4'-thionucleosides, obtained by using the methodology developed for the corresponding oxa-series.[31] Thus, *O*-silyl-(*S*)-thiolactone **11-4** was phenylselenated at the 2 position *via* the TMS-enol ether, with better diastereofacial selectivity than the 4-oxa analog (Scheme 12). By optimizing the work-up conditions, this selectivity was further improved up to at least 15:1 α- to β-face selectivity.[31] Reduction and acetylation of **12-1** gave the desired glycosylating agent **12-2**.

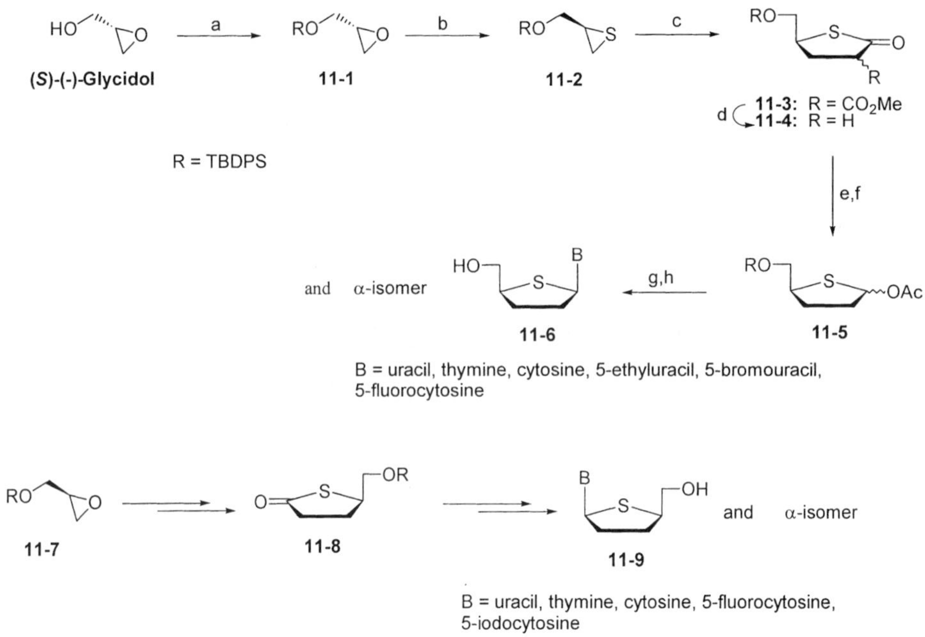

Reagents: a)TBDPSCl, DMAP, imidazole, DCM; b) thiourea, MeOH; c) $CH_2(CO_2Me)_2$, NaHMDS, then reflux; d) DMSO, brine, 160 °C; e) Dibal-H, Tol; f) Ac_2O, DMAP, DCM; g) silylated uracils; h) TBAF, THF.

Scheme 11

Coupling of **12-2** with silylated uracil in the presence of tin (IV) chloride in acetonitrile provided the glycosylated products **12-3** in 50 to 90% yield with 15:1 or better selectivity for the desired β-anomers. The 2'-selenyl functionality was removed either reductively with tributyltin hydride or by selective selenium oxidation with mCPBA. This latter method introduced regiospecifically the olefinic functionality of the novel 2',3'-didehydro-2',3'-dideoxy analogs in > 70% isolated yield. The enantiomeric compounds were similarly synthesized from the (*R*)-thiolactone **11-8**.[31]

Chu and co-workers recently reported the synthesis of L-2'-fluoro unsaturated 4'-thiocytidine starting from protected L-glyceraldehyde **9-12** (Scheme 13).[32,33] This was reacted under the Horner-Wadsworth-Emmons conditions with triethylfluorophosphonoacetate to give (*E*)-α,β-unsaturated fluoro ethylester **13-1**, together with the *Z* isomer as a minor product. Both isomers were hydrolyzed and protected as *tert*-butyldimethylsilyl ethers. Following the hydrolysis, the *E* isomer lactonized to give, after protection, lactone **13-2** which could be easily purified. Reduction of the α,β-unsaturated lactone gave only the 2-"up"-fluoro derivative **13-3**, which was converted to iodide **13-4** by saponification, methylation of the resulting carboxylate, and iodination in Mitsunobu fashion, the latter proceeding with inversion of configuration. A second inversion was then achieved by treatment with potassium acetate in dimethyl formamide to give thioacetate **13-5**.

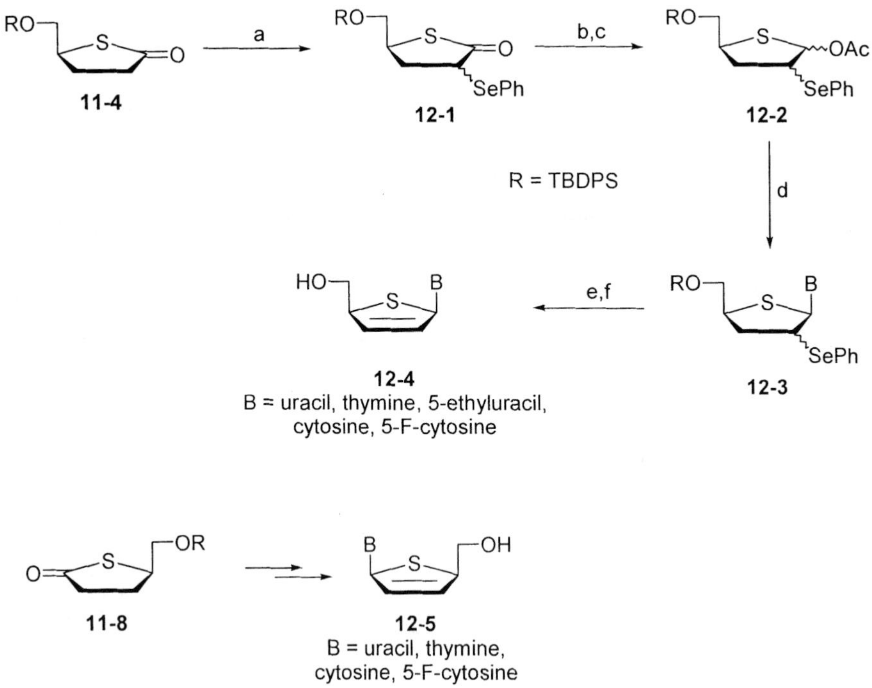

Reagents: a) LiHMDS, TMSCl, then, PhSeBr, all *in situ* -78 °C; b) Dibal-H, toluene; c) Ac_2O, DMAP, DCM; d) silylated uracil bases; e) m-CPBA, -30 °C; f) TBAF, THF.

Scheme 12

Treatment with DIBAL-H gave the thiolaldehyde intermediate, which cyclized to a thiolactol. *In situ* oxidation of the latter gave a stable thiolactone **13-6**. Phenylselenation proceeded stereoselectively to give almost exclusively α-selenide **13-7**, which was reduced and acetylated to the key intermediate **13-8**. Condensation with N^4-benzoylated cytosine gave protected nucleoside **13-9**, which was easily oxidized to the unsaturated analog **13-10** by mCPBA. Two deprotection steps afforded the target compound. D-Analogs were synthesized in the same way, starting from protected D-glyceraldehyde **9-2**.[34,35] Both isomers were endowed with potent anti-HIV activity.

Van Draanen *et al.*[36] synthesized various 2'-deoxy-4'-thiopurine nucleosides to study structure activity relationships by coupling a number of persilylated bases with key intermediate **14-1**, prepared from D-2-deoxyribose according to Dyson *et al.*[11,12] (Scheme 14). This synthetic procedure is an improvement over methods previously used to prepare purine 4'-thionucleosides. The compounds were tested against hepatitis B virus (HBV), human cytomegalovirus (HCMV), herpes simplex virus type 1 and 2 (HSV-1 and HSV-2), varicella zoster virus (VZV), and human immunodeficiency virus (HIV-1).

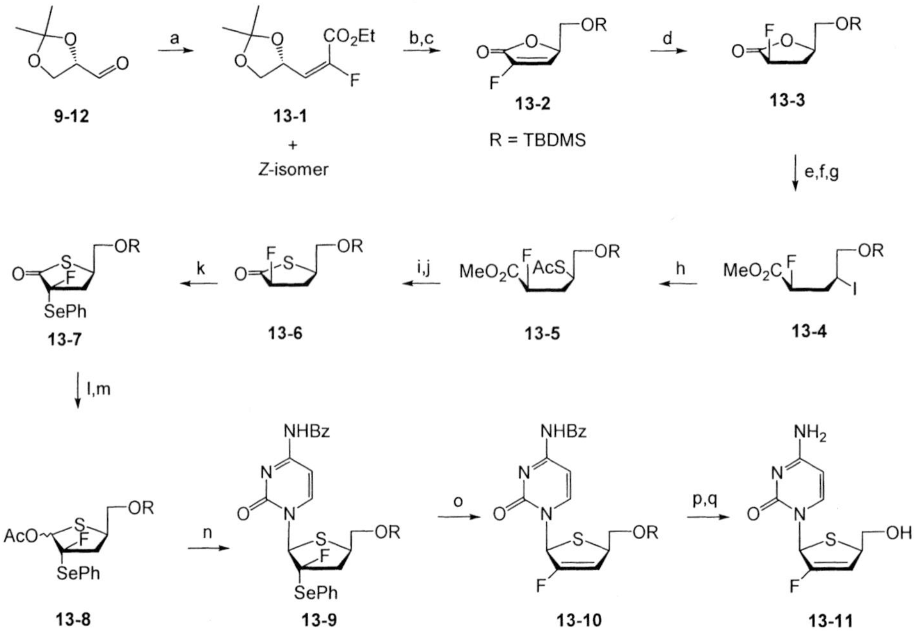

Reagents: a) $(EtO)_2P(O)CHFCO_2Et$/NaHMDS, THF; b) HCl, EtOH; c) TBDMSCl/imidazole, CH_2Cl_2; d) H_2, Pd/C, EtOAc; e) NaOH, EtOH; f) Me_2SO_4, DMSO; g) I_2, Ph_3P, imidazole, Tol; h) KSAc, DMF; i) DIBAL-H, Tol; j) Ac_2O, DMSO; k) LiHMDS, TMSCl, PhSeBr, THF; l) DIBAL-H, Tol; m) Ac_2O, TEA, CH_2Cl_2; n) silylated N^4-benzoylcytosine, TMSOTf, CH_3CN; o) *m*-CPBA, Py; p) TBAF, THF; q) NH_3, MeOH.

Scheme 13

Cytotoxicity was determined in a number of cell lines. Several compounds were extremely potent against HBV and HCMV and had moderate to severe cytotoxicity *in vitro*. The lead compound from the series, 2-amino-6-(cyclopropylamino)purine 2'-deoxy-4'-thioribose **14-13**, was the most potent and selective agent against HCMV and HBV replication *in vitro*; however, this analog was nephrotoxic when tested *in vivo*.

Rahim *et al.* also synthesized various 2'-deoxy-4'-thiopyrimidine nucleosides **15-2** to study the structure activity relationships (Scheme 15).[37] The key intermediates **1-6** and **2-3** were obtained according to the method reported by Dyson *et al.*[11,12] and Secrist *et al.*,[13] respectively (Schemes 1 and 2). A series of 5-substituted 2'-deoxy-4'-thiopyrimidine nucleosides was synthesized and evaluated as potential antiviral agents. A number of analogs such as 2'-deoxy-5-propyl-4'-thiouridine, 2'-deoxy-5-isopropyl-4'-thiouridine, 5-cyclopropyl-2'-deoxy-4'-thiouridine, 2'-deoxy-4'-thio-5-vinyluridine, and 5-(2-chloroethyl)-2'-deoxy-4'-thiouridine were found to be highly active against HSV-1 and VZV

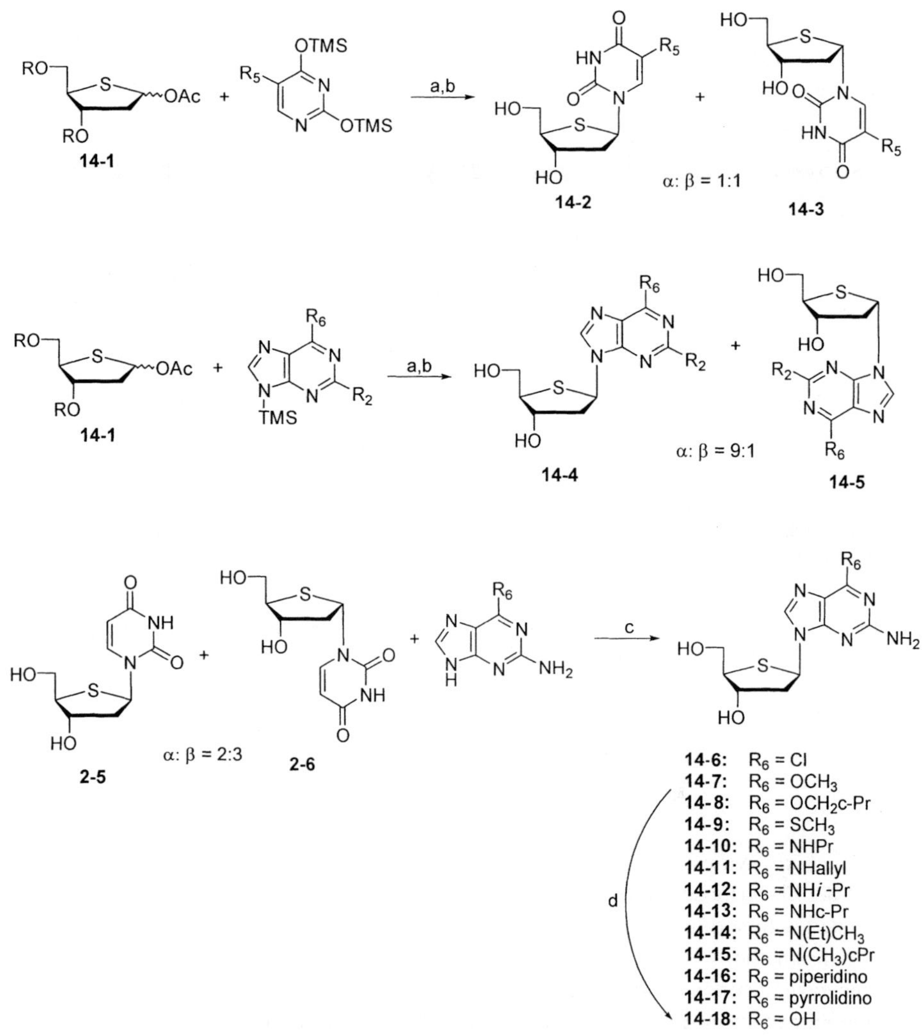

Reagents: a) Lewis acid; b) deprotection; c) *trans*-deoxyribosylase; d) adenosine deaminase.

Scheme 14

in vitro with no significant cytotoxicity. The compound with the broadest spectrum of activity was 2'-deoxy-5-ethyl-4'-thiouridine, which showed significant activity against HSV-1, HSV-2, and VZV.

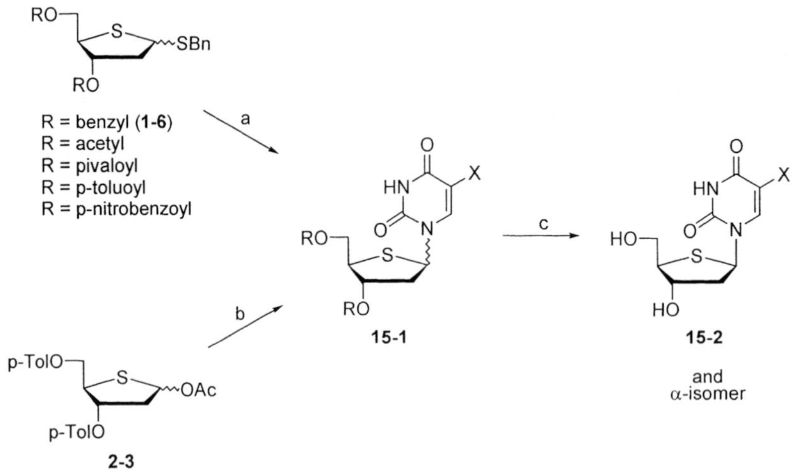

X = Et, *n*-Pr, *I*-Pr, *i*-Bu, (*E*)-vinyl, ethynyl, propynyl, methoxymethyl, 2-methoxyethyl, CF_3, 2-chloroethyl, 2-fluoroethyl, CH_2CF_3, $CH{=}CF_2$, (*E*)-CH=CHCl, $CF{=}CCl_2$, Cl, Br, I, CN, NO_2, OMe, acetyl, CH_2SMe, CH(OH)Me, 5,6-$[CH_2]_3$

Reagents: a) silyated bases, NBS or NIS, molecular sieve 4Å, acetonitrile; b) silyated bases,TBSOTf, acetonitrile; c) R= Bn: BCl_3, CH_2Cl_2, -78 °C; R = acyl: NaOMe, MeOH.

Scheme 15

Birk *et al.* sythesized (2-deoxy-4-thio-L-*threo*-pentofuranosyl)uracil and -thymine from 2-deoxy-D-ribose (Scheme 16).[38] The unprotected hydroxyl groups of **1-1** were deprotonated with sodium hydride in dimethylformamide and alkylated with iodomethane to give the methoxy derivative **16-1**. Acid-catalyzed reaction of **16-1** with benzyl mercaptane gave the open chain thioacetal **16-2**. The tosylate **16-3** did not need to be isolated but could be converted to the desired benzyl 2-deoxy-1,4-dithio-L-*threo*-pentofuranoside **16-4** without purification by refluxing with barium carbonate and sodium iodide in acetone. Coupling of **16-4** with 2,4-bis-*O*-(trimethylsilyl)uracil in the presence of mercuric bromide and cadmium carbonate according to Dyson *et al.* was not successful. 2'-Deoxy-4'-thio-L-*threo*-uridine could be successfully obtained using *N*-iodosuccinimide (NIS) in dry acetonitrile.

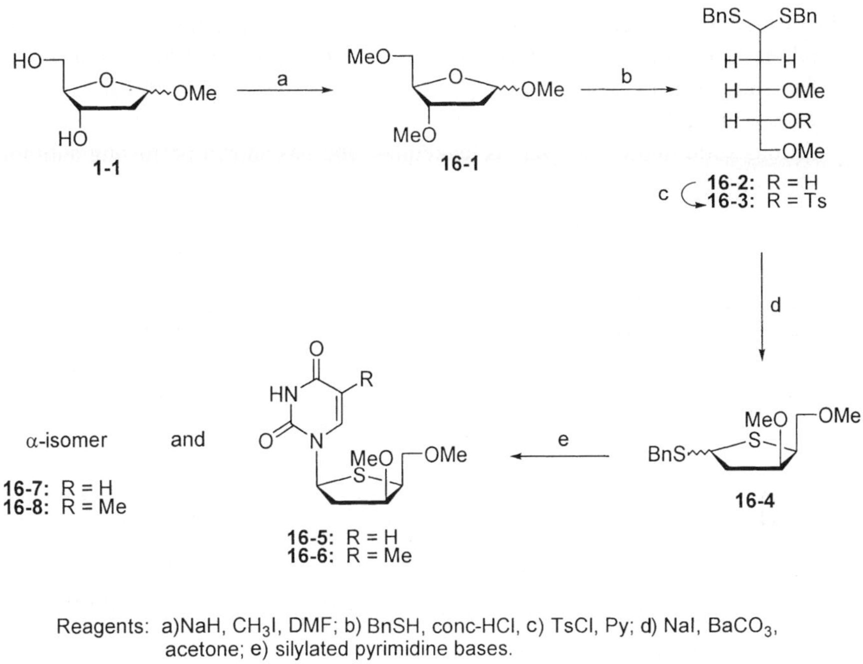

Scheme 16

Yoshimura *et al.* reported the synthesis of D-2'-modified-4'-thionucleosides from protected D-glucose **17-1** *via* anhydrothiosugar intermediate **17-8** using the Pummerer reaction (Scheme 17).[39,40] Hydrolysis of **17-8** and **17-9** followed by borohydride reduction gave 1,4-anhydro-4-thioarabitol **17-10**. The primary alcohol of **17-10** was selectively protected as *tert*-butyldiphenysilyl ether to give the common intermediate **17-11**. At first, to synthesize the 4'-thio analog of the antineoplastic agent 2'-deoxy-2'-methylenecytidine (DMDC), the secondary hydroxyl group of **17-11** was oxidized with DMSO/Ac_2O to give ketone **17-12,** which was treated without purification with methylenetriphenylphophorane to give **17-13**. Compound **17-13** was oxidized with mCPBA to the corresponding sulfoxide, which was treated with Ac_2O at 110 °C to give exclusively sigmatropically rearranged product **17-14**. Treatment of **17-14** with trimethylsilylated N^4-acetylcytosine and TMSOTf at 0 °C gave the 4'-thio-DMDC analog **17-15** along with its regioisomer **17-16**. Interestingly, the formation of the regioisomer was increased with a prolonged reaction time and an increase in the reaction temperature. Although several attempts were made to deprotect the benzyl group of **17-15**, the debenzylated product could not be obtained, due to the instability of **17-15** under the reaction conditions. Thus, it was necessary to remove the benzyl group prior to glycosylation. The reaction of **17-13** with boron trichloride proceeded smoothly to give the debenzylated product and the consequent mCPBA oxidation gave compound **17-17**. Compound **17-17** was treated with silylated *N*[4]-acetylcytosine and TMSOTf according to the method of O'Neil and Hamilton. The following deprotection and separation produced the 4'-thio-DMDC **17-18**. The same methodology was applied to the synthesis of 4'-thiogemcitabine[39] (*vide infra*), which, like DMDC, is endowed of prominent antineoplastic activities against various solid tumors, as well as leukemias, and has shown promising antitumor activities in clinical trials.[41]

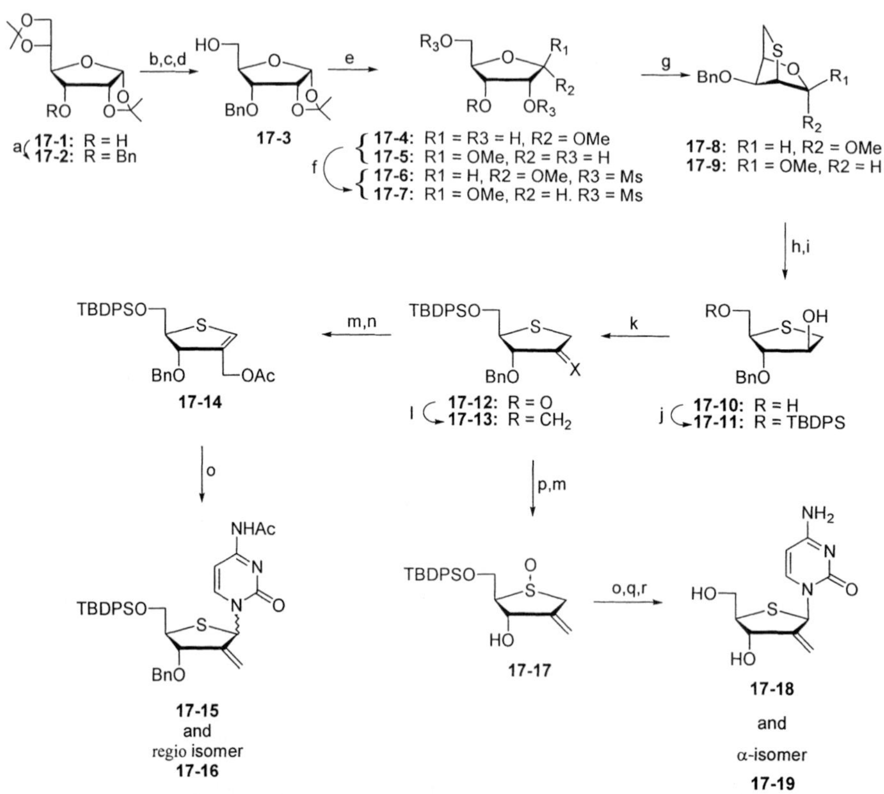

Reagents: a) BnBr, NaH, DMF, THF; b) 2 M HCl, THF. c) $NaIO_4$, H_2O, MeOH; d) $NaBH_4$, MeOH; e) 5% HCl/MeOH; f) MsCl, Py; g) Na_2S, DMF, 100 °C; h) 4 M HCl, THF; i) $NaBH_4$, MeOH; j) TBDPSCl,imidazole, DMF; k) Ac_2O, DMSO; l) $Ph_3P^+CH_3Br^-$, NaH; m) *m*-CPBA, CH_2Cl_2; n) Ac_2O, 110 °C; o) silylated *N*[4]-acetylcytosine, TMSOTf, DCE; p) BCl_3, CH_2Cl_2, -78 °C; q) TBAF, THF; r) NH_3/MeOH, then HPLC separation.

Scheme 17

In order to prepare 4'-thio-fluoroarabinocytosine **18-5**, a fluorine atom was introduced at the 2-position of **17-11** with retention of stereochemistry *via* double inversion strategy (Scheme 18). Marquez and co-workers reported that treatment of 1-(5-*O*-trityl-3-deoxy-4-thio-*threo*-pentofuranosyl)uracil with diethylaminosulfur trifluoride (DAST) gave the 2'-fluoro- 4'-thionucleoside with "*threo*" stereochemistry.[42] The endocyclic sulfur atom is known to play an important role in reactions of 4-thiosugars, such as the formation of ring-contracted products.[43] In this case, the episulfonium intermediate **18-1** was generated *in situ*. Attack of fluoride ion from the β-side allowed the formation of product **18-2**, having the same configuration of the starting alcohol. A Pummerer-type reaction was then successfully applied to **18-2** to obtain thioglycosyl donor **18-4**. This was treated with silylated N^4-acetylcytosine in the presence of $SnCl_4$ to give an anomeric mixture, which was deprotected and separated to 2'-deoxy-2'-fluoro-4'-thiocytidines **18-5** and **18-6**. Reaction of the intermediate **18-11** using diphenylphosphoryl azide (DPPA) efficiently allowed the introduction of an azido group at the 2-position. Surprisingly, the results of an NOE experiment of **18-12** revealed that the products had a *ribo* rather than *arabino* configuration, which indicates absence of sulfur participation. Compound **18-12** was further converted to *O*-acetoxy derivative **18-13** by a Pummerer-type reaction. The acetate **18-13** was subject *in situ* to Lewis acid catalyzed glycosylation, followed by deprotection and separation to give compound **18-14** and **18-15**, respectively. Newly synthesized 4'-thio- and 2'-fluoro-4'-thioarabinofuranosyl purine and pyrimidine nucleosides were compared with the corresponding 4'-oxo type arabinosyl nucleosides for anti-herpesvirus and anti-cell proliferative potencies. 4'-Thio- and 2'-fluoro-4'-thioarabinofuranosyl adenine exhibited biological activities similar to that of arabinofuranosyl adenine. Both 4'-thio-F-araG and 4'-thio-F-ara-2,6-diaminopurine had a 6-fold lower ED_{50} than ganciclovir against clinical isolates of HCMV. A ganciclovir-resistant isolate, obtained from a patient who had received long-term ganciclovir-treatment, was susceptible to 4'-thio-F-araG and 4'-thio-F-ara-2,6-diaminopurine.[44]

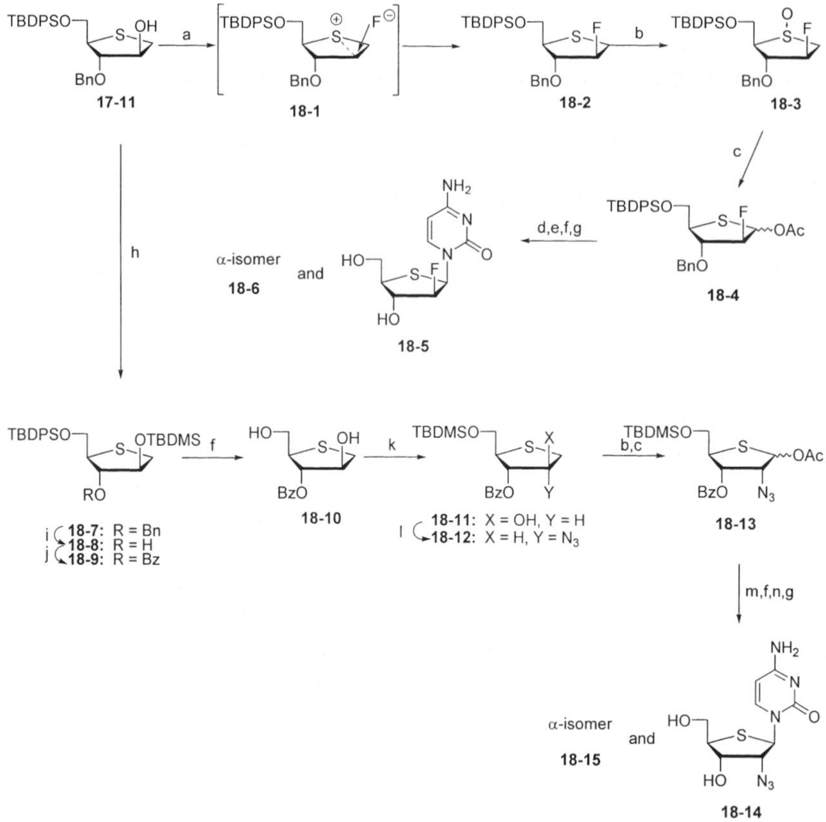

Reagents: a) DAST, CH_2Cl_2, -78 °C; b) m-CPBA, CH_2Cl_2, -78 °C; c) Ac_2O, 100 °C; d) silylated N^4-acetylcytosine, $SnCl_4$, CH_3CN; e) BBr_3, then MeOH, saturated $NaHCO_3$; f) $NH_4F.HF$, MeOH, CH_2Cl_2, 60 °C; g) NH_4OH, MeOH, then separation; h) TBSOTf, Py, CH_2Cl_2; i) BCl_3, CH_2Cl_2, -78 °C, then MeOH, Py; j) Bz_2O, TEA, DMAP, CH_3CN; k) TBSCl, imidazole, DMF; l) DPPA, DEAD, Ph_3P, THF; m) silylated N^4-acetylcytosine, TMSOTf, DCE; n) separation.

Scheme 18

DAST treatment of ketone **17-12** gave the 2,2-difluoro derivative **19-1** in 48% yield (Scheme 19).[45] A Pummerer-type glycosylation of **19-2** was used with 4'-thio-DMDC, since the 1-*O*-acetyl derivative resisted Lewis acid-mediated glycosylation due to the difluoro substituent at the 2-position. Reaction of **19-2** with silylated N^4-acetylcytosine in the presence of TMSOTf followed by deprotection and separation resulted in the desired 4'-thiogemcitabine **19-3**. 4'-Thio-DMDC showed potent antitumor activity *in vitro* (CCRF-HSB-2, IC_{50} = 0.0091 μg/ml; KB cells, IC_{50} 0.12 μg/ml), whereas 4'-thiogemcitabine showed only weak activity (CCRF-HSB-2, IC_{50} 1.5 μg/ml).[46]

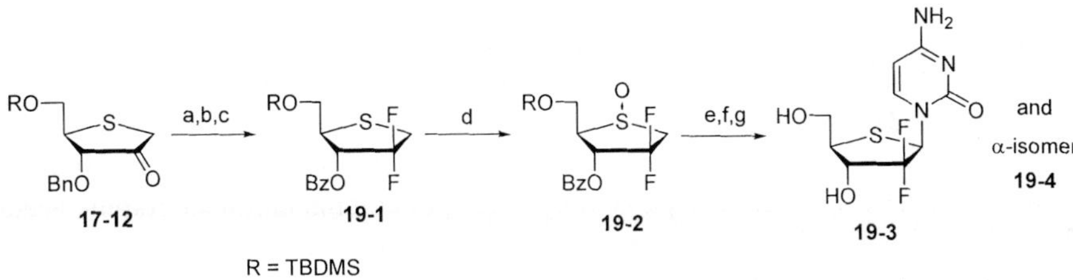

Reagents: a) DAST, benzene; b) BCl_3, CH_2Cl_2, -78 °C, then MeOH, Py; c) Bz_2O, TEA, DMAP, CH_3CN; d) mCPBA, CH_2Cl_2, -78 °C; e) silylated N^4-acetylcytosine, TMSOTf, DCE; f) TBAF, THF; g) NH_3.MeOH, MeOH, HPLC separation.

Scheme 19

In order to obtain other 2'-modified- 4'-thionucleosides, the intermediate **17-10** was fully benzylated to compound **20-1** (Scheme 20). This was subject to the Pummerer rearrangement to afford the thio-glycosyl donor **20-2**, which was condensed with various silylated pyrimidine and purine bases, followed by deprotection and HPLC purification to give the corresponding 4'-thioarabinonucleosides.[46,47] Among these, 5-methyl, -ethyl, -iodo, -chloro and -bromouridine analogs displayed anti-HSV-1 activity, with ED_{50} of 0.43 to 3.50 μg/ml.[46] The guanine and 2,6-diaminopurine derivatives also showed activity against a number of herpes viruses, particularly against HCMV (ED_{50} 0.010 and 0.022 mg/ml, respectively).[46] However, the 2,6-diaminopurine derivative was also found cytotoxic.[46]

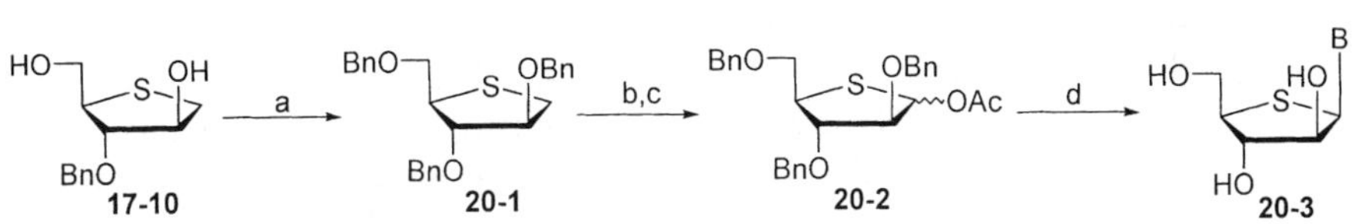

Reagents: a) BnBr, NaH, DMF; b) DMSO, Ac_2O; c) Ac_2O; d) silylayed pyrimidine bases, TMSOTf, DCE, then deprotection.

Scheme 20

In the examples so far reported, the syntheses of 4'-thionucleosides have been carried out by a Vorbrüggen-type condensation between an appropriate 4-thiosugar and a silylated nucleobase. However, a major drawback of this method is the lack of β-selectivity. Thus, in the synthesis of 2'-deoxy-4'-thionucleoside, the undesired α-isomer was obtained as a major product in many cases. Even in the case of 4'-thionucleosides, where neighboring group participation by an acyloxy group at the 2-position can be expected, the β-anomer is formed only in slight excess. Haraguchi *et al.* reported a highly stereoselective synthesis of 2'-deoxy-4'-thio pyrimidine nucleosides from 4-thio furanoid glycal **1-6**.[48] The thiosugar was converted to the corresponding *O*-(*t*-butyldimethylsilyl) derivative **21-1** by debenzylation with BBr_3 followed by silylation (Scheme 21). The dithioacetal moiety of **21-1** was subject to acetolysis with $Hg(OAc)_2$. Because a reagent-derived byproduct could not be separated by silica gel column chromatography, acetate **21-2** was hydrolyzed to hemiacetal **21-3**. This was mesylated in the presence of DMAP. Under these conditions, the desired elimination reaction occurred *in situ* to give 3,5-bis-*O*-(*tert*-butyldimethyl)-4-thio furanoid glycal **21-4**. By following the same reaction sequence, 2,5-*O*-(1,1,3,3-tetraisopropyldisiloxane-1,3-diyl)-4-thio glycal **21-5** was prepared from **21-3**. On the other hand, 2,5-di-*O*-t-butyl-4-thio furanoid glycal **21-7** could not be obtained by the same method. Therefore, this derivative was prepared from **21-6** through desilylation/re-silylation with di-*tert*-butylsilyl-bis(trifluoromethanesulfonate). In order to examine the electrophilic addition to the thioglycal, PhSeCl was added to a solution of TBDMS-protected gycal **21-4** in CH_3CN in the presence of bis-*O*-trimethylsilyluracil to give a mixture of stereoisomers β/α in 4:1 ratio as shown in entry 1 (see table in Scheme 21). In the case of **21-7**, the ratio of α-face attack by the electrophilic reagent increased and the desired β-2'-deoxy-4'-thionucleoside was obtained in 18:1 ratio (entry 2). As shown in entry 3, DTBS-protected glycal **21-5** gave β-isomer as the sole product in 88% yield. Using **21-5** as a substrate, the β-thymidine analog was also obtained in 62% yield stereoselectively. Instead of PhSeCl as an electophile, NIS also worked well to give 2'-deoxy-2'-iodo derivative (β-isomer).

De Valette *et al.* synthesized 2'-deoxy-4'-thio-L-nucleosides and their phosphotriester derivatives starting from D-xylose, in nine steps and 17% overall yield.[49] This synthesis involves the deoxygenation of the 2-hydroxyl group of protected D-xylofuranose, followed by a nucleophilic displacement[50] of the previously activated 4-hydroxyl group with only one inversion of configuration (Scheme 22). Deoxygenation[51,52] of **22-2** was performed according to the method of Barton,[53] using 1,1'-thiocarbonyldiimidazole in refluxing 1,2-dichloroethane *via* the intermediate thiourethane **22-3**. A radical reaction initiated with α,α'-azo-isobutyronitrile (AIBN) on **22-3** with tris(trimethylsilyl)silane in refluxing toluene led to methyl 2-deoxy-3,5-di-*O*-benzyl-D-xylofuranoside **22-4**. Dithioacetalization of **22-4** was performed with benzyl mercaptane and boron trifluoride etherate to give 2-deoxy-3,5-di-*O*-benzyl-1,1-dithiobenzyl acetal-D-xylose **22-5**. Treatment of the latter with mesyl chloride in pyridine, followed by the addition of barium carbonate and tetrabutylammonium iodide gave *S*-benzyl-2-deoxy-3,5-di-*O*-benzyl-4-thio-L-ribofuranoside **22-7**, which was subsequently treated with $Hg(OAc)_2$ in acetic acid to give acetate **22-8**. Coupling reactions between the thio-sugar **22-8** and various bases were performed by a modification of the Vorbrüggen method.[54] The two anomers of protected nucleosides were separated by flash silica gel column chromatography. The debenzylation of each anomer was achieved by treatment with boron trichloride in methylene chloride at -78 °C. No biological activity was detected for either the cytosine and adenine derivatives **22-9**, **22-11**, **22-12** and **22-14** or their bis(sate)phosphotriester prodrugs against a variety of DNA and RNA viruses.[49]

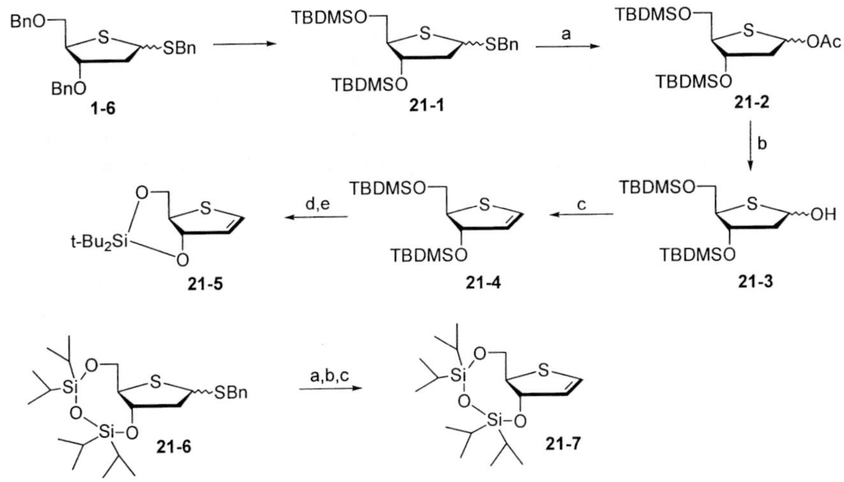

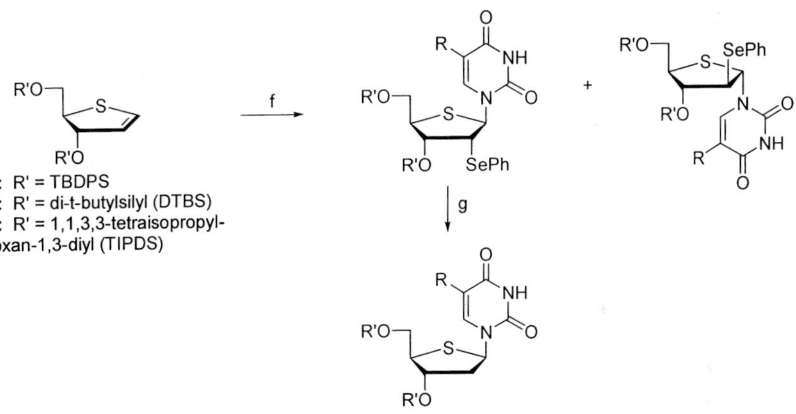

21-4: R' = TBDPS
21-5: R' = di-t-butylsilyl (DTBS)
21-7: R' = 1,1,3,3-tetraisopropyl-disiloxan-1,3-diyl (TIPDS)

Table. Electrophilc addition to 4-thio furanoid glycals.

Entry	glycal	electrophile (equiv.)	bases	Product ratio
1	**21-4**	PhSeCl (1.5)	Uracil	4 : 1
2	**21-7**	PhSeCl (1.5)	Uracil	18 : 1
3	**21-5**	PhSeCl (1.5)	Uracil	-
4	**21-5**	PhSeCl (2.3)	Thymine	-
5	**21-5**	NIS(1.5)	Uracil	-

Reagents: a) $Hg(OAc)_2$, AcOH; b) NH_3/MeOH; c) MsCl, DMAP, CH_2Cl_2; d) TBAF, THF; e) *t*-$Bu_2Si(OSO_2CF_3)_2$ (1.1 equiv.), DMAP (2.2 equiv.), DMF; f) PhSeCl or NIS, silylated uracil or thymine; g) Bu_3SnH, Et_3B, benzene, O_2, rt.

Scheme 21

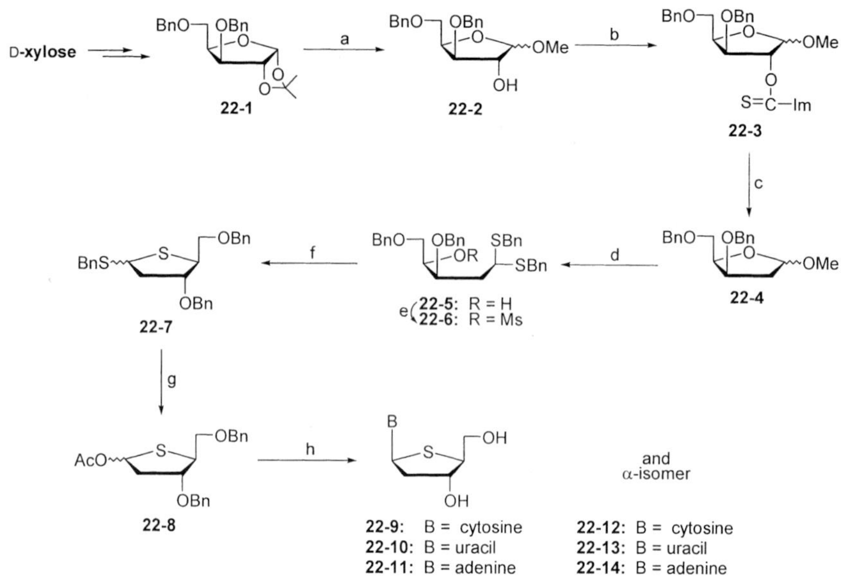

Reagents: a) CF_3COOH/H_2O, $MeOH/H_2SO_4$; b) $Im_2C{=}S$, DCE; c) $(Me_3Si)_3SiH$, AIBN, Tol; d) BnSH, BF_3/Et_2O; e) MsCl, Py; f) $BaCO_3$, NBu_4I; g) $Hg(OAc)_2$, AcOH; h) base condensation and deprotection.

Scheme 22

Secrist *et al.* reported the synthesis of 4'-thio-D-arabinofuranosylpurine nucleosides from L-xylose *via* conversion of L-*xylo* to D-*arabino* configuration (Scheme 23).[55] Conversion of L-xylose to methyl 2,3,5-tribenzyl-*O*-benzyl-L-xylofuranoside was first accomplished. Conversion to dibenzyl dithioacetal **23-3** employing benzyl mercaptane and stannic chloride proceeded in 57% yield. Cyclization at C-4 involving a single inversion, thus converting the L-xylo to the D-*arabino* configuration, was achieved using triphenylphosphine, iodine and imidazole. The final step, replacement of the benzylthio group at C-1 by an acetoxy group, was accomplished by treatment of **23-4** with mercuric acetate in acetic acid at room temperature. A series of purine nucleoside analogs were prepared through the coupling of **23-5** and 2,6-dichloropurine. A Lewis acid-catalyzed reaction utilizing $SnCl_4$ in acetonitrile was found to be an efficient method to achieve this coupling. The 2,6-diaminopurine and guanine derivatives **23-14** and **23-15** showed significant cytotoxicity. However, in the murine colon 36 tumor model, **23-14** did not show any selectivity.

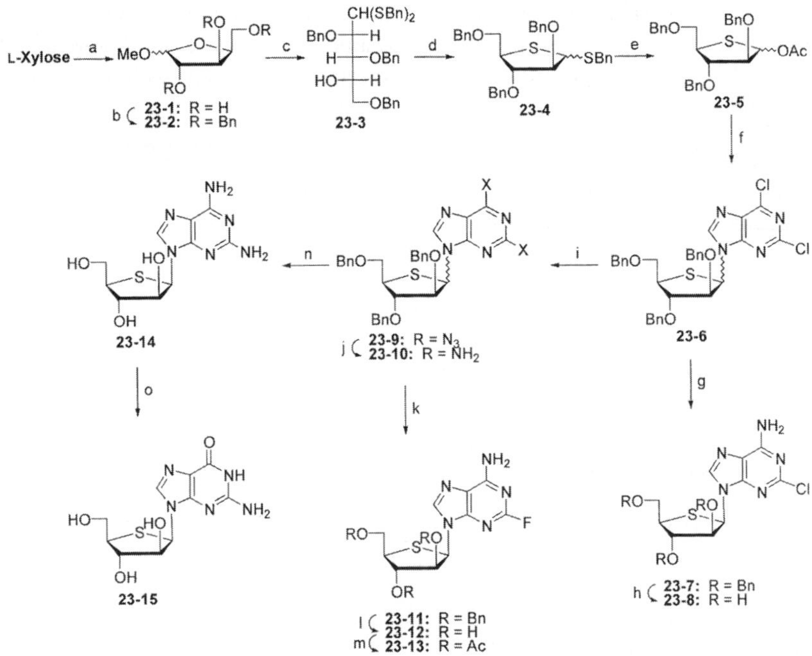

Reagents: a) MeOH, HCl, rt; b) BnBr, NaH, TBAI, THF, rt; c) BnSH, $SnCl_4$, CH_2Cl_2, rt; d) Ph_3P, imidazole, Tol, I_2, CH_3CN, 90 °C; e) $Hg(OAc)_2$, AcOH, rt; f) 2,6-dichloropurine, $SnCl_4$, CH_3CN, rt, 2 h; g) EtOH, NH_3, reflux, 2 h; h) BBr_3, CH_2Cl_2, -78 °C, rt, 0.5 h; i) NaN_3, 95%, EtOH, reflux; j) $SnCl_4$, CH_2Cl_2, MeOH, rt, 0.5 h; k) HF, Py, *t*-BuONO, -15 °C; l) BCl_3, CH_2Cl_2, -20 °C, 23 h; m) Py, Ac_2O, 20 h, rt; n) BCl_3, CH_2Cl_2, -20 °C, 16 h; o) ADA, H_2O, 17 h.

Scheme 23

Jeong *et al.* reported the synthesis of L-2'-disubstitued-4'-thionucleoside from D-xylose as a chiral template (Scheme 24).[56] D-Carbohydrate chirality was successfully used to synthesize L-configuration by conversion of a secondary hydroxyl group. Commercially available 1,2-isopropylidene-D-xylose **24-1** was dibenzylated. Deprotection of the isopropylidene group followed by benzoylation of the 2-hydroxyl gave compound **24-3**, which was treated with benzyl mercaptane and $BF_3{\cdot}Et_2O$ in CH_2Cl_2 at 40 °C to provide acyclic **24-4** in 84% yield. The latter was recyclized to the corresponding thiosugar *via* the methanesulfonate ester in the presence of n-Bu_4NI and $BaCO_3$.[15] Compound **24-5** was converted to the acetate after treatment with $Hg(OAc)_2$/AcOH following the method of Blumberg and co-workers,[57] and the successful removal of the acetate was subsequently achieved with Et_3SiH and TMSOTf to give the L-arabitol derivative **24-6**. Oxidation of **24-6** with DMSO/Ac_2O afforded the corresponding ketone, which was converted either to the difluoro (**24-7**) or to the methylene (**24-10**) derivative by Wittig reaction. The difluoro derivative **24-7** was oxidized to the sulfoxide with mCPBA, which was condensed with silylated N^4-benzoylcytosine and then deprotected to give the desired nucleoside **24-8**. Compound **24-11** was obtained by the same procedure. The synthesized compounds were tested against a number of tumor cell lines but none was found to be active.

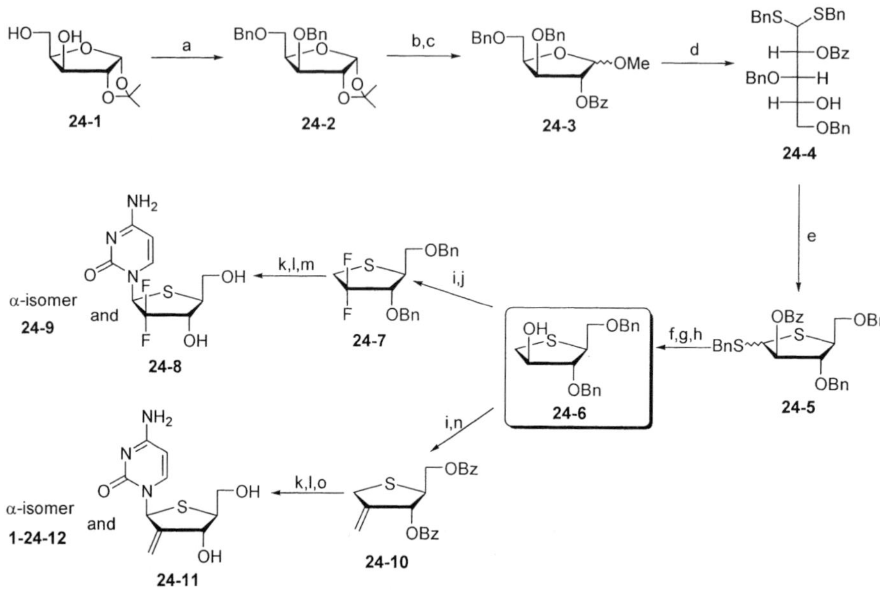

Reagents: a) NaH, TBAI, BnBr, THF; b) p-TsOH, MeOH, rt; c) BzCl, Py, rt; d) BnSH, $BF_3 \cdot Et_2O$, 40 °C; e) MsCl, Py, rt, then TBAI, $BaCO_3$, reflux; f) $Hg(OAc)_2$, AcOH, rt; g) Et_3SiH, TMSOTf, rt; h) NaOMe, MeOH; i) DMSO, Ac_2O; j) DAST, CH_2Cl_2, rt; k) mCPBA, CH_2Cl_2, 40 °C; l) silylated N^4-benzoylcytosine, TMSOTf, DCE; m) BBr_3, CH_2Cl_2, -40 °C, then NaOMe/MeOH; n) Ph_3PCH_3Br, NaH, *tert*-amyl alcohol, rt; o) NaOMe, MeOH.

Scheme 24

The key intermediate **24-6** was also used to synthesize the sulfur isoster of L-FMAU (Scheme 25). Reaction of **24-6** with DAST gave the 2-fluoro-4-thiosugar **25-2** with retention of configuration *via* intermediate **25-1** without any rearrangement or ring-contraction side products.[58] Compound **25-2** was reacted with mCPBA to give the corresponding sulfoxide, which was easily converted to the acetate by refluxing with acetic anhydride. Condensation of thioglycosyl donor **25-3** with silylated thymine in the presence of TMSOTf as the Lewis acid afforded the anomeric mixture **25-4**. To facilitate the separation of this mixture, the benzyl group was changed to benzoyl to obtain, after deprotection, L-SFMAU **25-6**. This was evaluated against HIV-1 and HSV types 1 and 2 and HBV, but no significant antiviral activity was found.

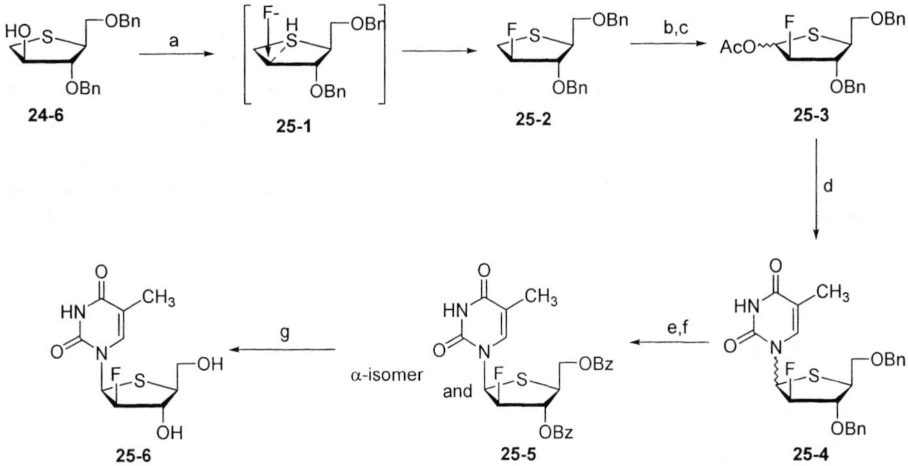

Reagents: a) DAST; b) *m*-CPBA; c) Ac_2O; d) silylated thymine, TMSOTf; e) BBr_3; f) BzCl, then separation; g) NaOMe, MeOH.

Scheme 25

Tiwari *et al.* synthesized 1-(2-deoxy-4-thio-α-L-*threo*-pentofuranosyl)thymine (Scheme 26).[59] The key intermediate, 3,5-di-*O*-benzyl-2-deoxy-D-*erythro*-pentose dithiobenzylacetal **26-1**, was prepared from 2-deoxy-D-ribose and converted to the 4-thiosugar **26-2** by treatment with triphenylphosphine, iodine and imidazole under Mitsunobu conditions. Further chemical manipulation afforded glycosyl donor **26-3**, which was condensed with silylated thymine and deprotected to give the target molecule **26-5**.

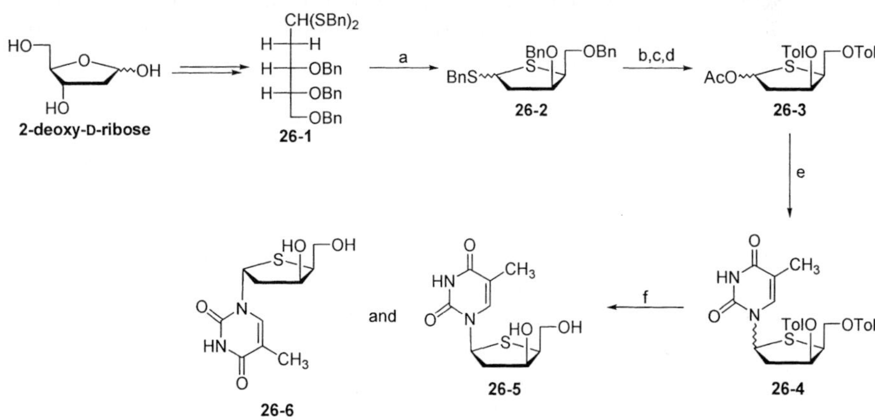

Reagents: a) Ph_3P, I_2, imidazole; b) BBr_3, CH_2Cl_2, -78 °C; c) p-Toluoyl chloride, Py; d) Ac_2O, AcOH; e) silylated thymine, TMSOTf; f) NaOMe, MeOH, then separation.

Scheme 26

2.3. Iso- and apionucleosides

The naturally occurring nucleosides and nucleotides are those in which the purine or pyrimidine bases are attached to the C-1 of ribose or its analogs. This linkage is quite susceptible to both hydrolytic and enzymatic cleavage. For many years, the design of congeners of these compounds has been based on the assumption that only nucleoside analogs with their bases attached in the β configuration to C-1 of D-furanoses were likely to fit the active sites of the anabolic enzymes. The same requirements were assumed to apply also to the incorporation of analogs into cofactors or macromolecules. Indeed, in the search for the effective and selective antiviral agents, a variety of strategies have been devised to design nucleoside analogs that interfere with viral replication without affecting cellular process. Several classes of nucleoside analogs containing significantly altered structures proved to be potent antiviral agents, and eventually it was established that, regardless their chemical nature or stereochemistry, the sugar templates of the modified nucleoside should allow a spatial orientation of the hydroxyl and nucleobase pharmacophores which mimics that found in natural nucleosides. An analog with an altered sugar template could be selectively recognized by the less discriminating viral enzymes, without affecting host cellular process. For this reason, iso- and apionucleosides (see Chapter 1) can display biological activity with improved metabolic stability.

Montgomery *et al.* synthesized purine and pyrimidine derivatives of 1,4-anhydro-2-deoxy-D-arabinitol from D-xylose (Scheme 27).[60] The method of Kjølberg for shortening the chain length of glycosides[61] followed by sodium borohydride reduction was used for the synthesis of 1,4-anhydro-D-xylitol **27-1**. The 1,4-anhydro-3,5-*O*-isopropylidene-D-xylitol **27-2** was prepared by the reaction of **27-1** with acetone containing 2,2-dimethoxypropane and 60% perchloric acid. Acetylation of **27-2** with pyridine-acetic anhydride gave 2-*O*-acetyl-1,4-anhydro-3,5-*O*-isopropylidene-D-xylitol **27-3**. Deprotection of **27-3** in 1 N ethanolic HCl gave 2-*O*-acetyl-1,4-anhydro-D-xylitol **27-4** which was subsequently acylated and activated as a mesylate to afford 1-*O*-acetyl-1,4-anhydro-3-*O*-mesyl-5-*O*-methoxycarbonyl-D-xylitol **27-5**. The latter was then cyclized to epoxide **27-6** in basic conditions.

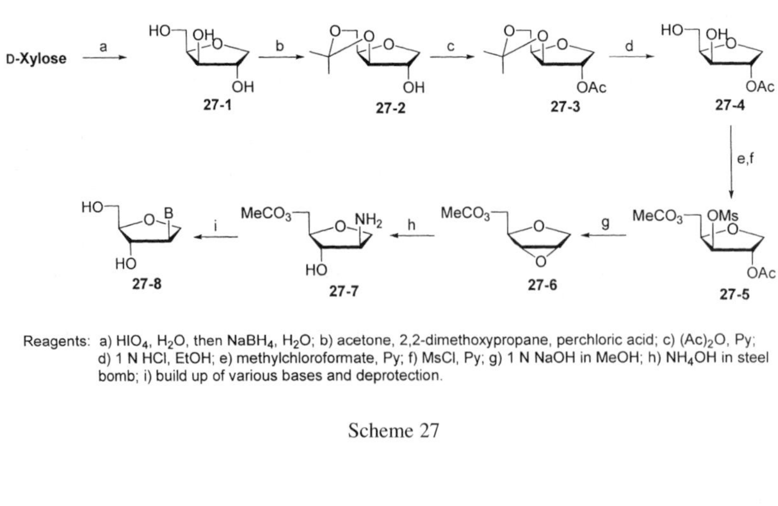

Reagents: a) HIO_4, H_2O, then $NaBH_4$, H_2O; b) acetone, 2,2-dimethoxypropane, perchloric acid; c) $(Ac)_2O$, Py; d) 1 N HCl, EtOH; e) methylchloroformate, Py; f) MsCl, Py; g) 1 N NaOH in MeOH; h) NH_4OH in steel bomb; i) build up of various bases and deprotection.

Scheme 27

The ring-opening reaction of **27-6** with concentrated ammonium hydroxide at 100 °C took place exclusively at C-2 to give 2-amino-1,4-anhydro-2-deoxy-D-arabinitol **27-7**, from which several nucleosides were prepared.

Bobek *et al.*[62] also synthesized another class of isonucleosides, 2'-deoxy-2'-fluoro-2'-pyrimidinyl-D-arabinopyranosides from methyl 3,4-*O*-isopropylidene-β-D-*erythro*-pento-pyranoside-2-ulose **28-2**, obtained by oxidation of protected D-arabinose **28-1** (Scheme 28). Fluorination of **28-2** with DAST proceeded with isomerization at the anomeric carbon to produce methyl 2-deoxy-2,2-difluoro-3,4-*O*-isopropylidene-α/β-D-*erythro*-pento-pyranoside **28-3**. The isopropylidene group of **28-3** was deprotected by treatment with 95% formic acid. Due to the sensitivity of the gem-difluoro group of **28-3** in acid environment, this hydrolytic step presented some difficulties, because long reaction times and/or higher temperature caused the elimination of fluorine and produced complex reaction mixtures. The labile free methyl glycosides **28-4a** (R' = H) were either acetylated to methyl 3,4-di-*O*-acetyl-2-deoxy-2,2-difluoro-α/β-D-*erythro*-pentopyranosides **28-4b** (R' = Ac), or treated with hexanoyl chloride in pyridine to give **28-4c** (R' = hexanoyl). Compounds **28-4** were readily converted to the corresponding 2-bromo-2-fluoro glycosides **28-5** by treatment with anhydrous HBr. While attempted condensation of **28-5** with silylated pyrimidines in the presence of $SnCl_4$ led to the decomposition of the starting material, condensation in the presence or HgO-$HgBr_2$ provided the protected nucleosides **28-6a/b~28-7a/b**. Alternatively, the 2,2-difluoro methyl glycoside **28-4b** was condensed with silylated pyrimidines in the presence of $BF_3.Et_2O$ to afford mixtures of nucleosides **28-6a/b~28-9a/b** in low yield along with recovered starting material. Acid hydrolysis of the compound **28-6a/b** in mild conditions resulted in the selective removal of the 3'-*O*-acetyl group to give **28-10a/b**. Similarly, porcine liver esterase (PLE) hydrolysis of the hexanoyl groups in **28-9a,b** gave the 3'-deprotected products **28-11a** and **28-11b**.

Huryn *et al.*[63] also reported the first synthesis of a series of isomeric 2',3'-dideoxy-nucleosides containing an isosugar unit (Scheme 29). The required isomeric sugar moiety **29-4** was prepared starting from tosylated D-2-deoxyxylose **29-1** by treatment with acetic acid in methanol. In these conditions, the methanolysis product **29-2** was in equilibrium with the open-chain acetal **29-3**, which cyclized to the desired structure **29-4**. Tosylation of the latter gave key intermediate **29-5**, from which iso-ddA was synthesized by condensation and deprotection.

The enantiomer of iso-ddA and its analogs were synthesized by Nair and Nuesca.[64] Protected D-xylose **30-1** was deoxygenated in two steps, and then subject to methanolysis to give methyl riboside **30-2** (Scheme 30). This was demethoxylated to alcohol **30-3**, which was tosylated to give the key intermediate **30-4**. Purine nucleosides could be obtained from **30-4** by direct coupling with silylated bases and further derivatization. In order to synthesize pyrimidines, intermediate **30-4** was converted to the β-amine, on which the pyrimidine ring was built. The nucleosides **30-5** were also used as intermediates for the synthesis of hydroxymethylated isodideoxynucleosides **30-7**, *via* oxidation to aldehydes **30-6** followed by hydroxymethylation.[65] The enantiomers of **30-7** were prepared combining this procedure with Huryn's synthesis of L-isodideoxynucleosides.[65]

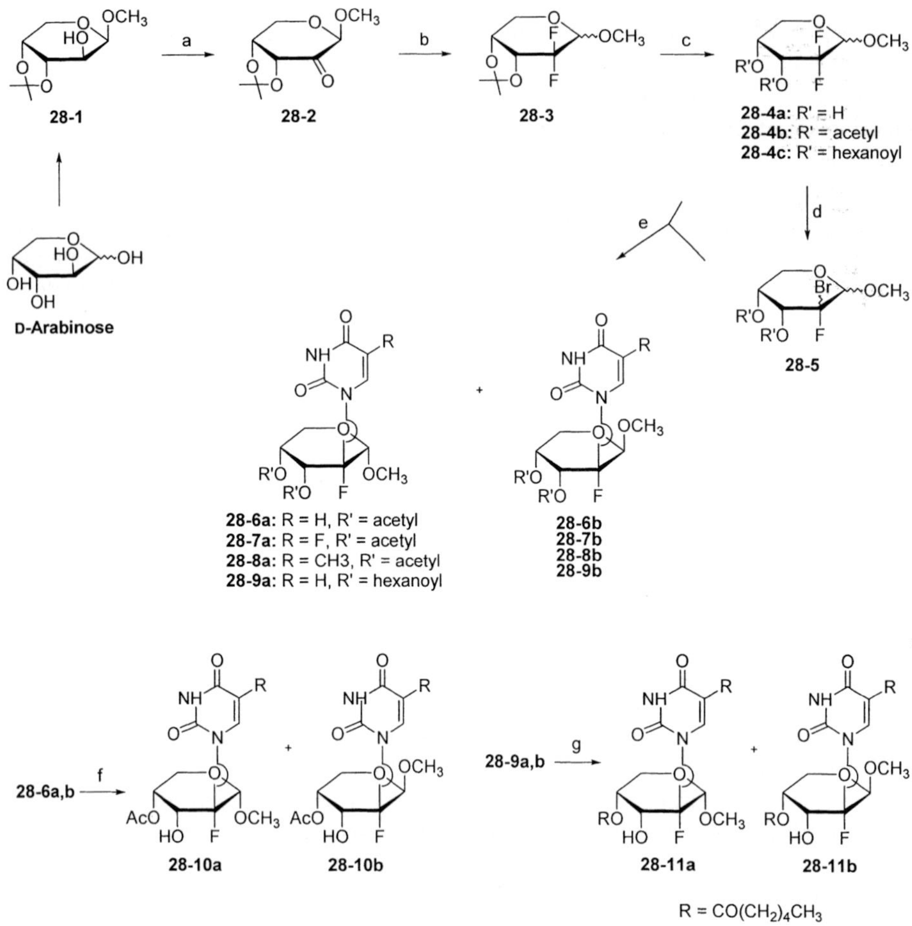

Reagents: a) CrO_3, Py, Ac_2O, CH_2Cl_2; b) DAST, benzene; c) 95% formic acid, and then $(Ac)_2O$, Py; d) HBr; e) silylated uracils, HgO-$HgBr_2$; f) HCl/MeOH; g) PLE.

Scheme 28

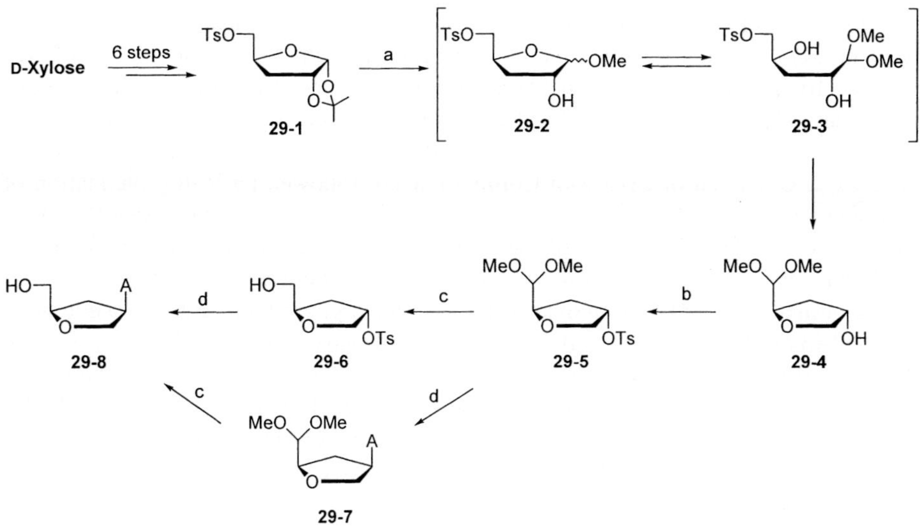

Reagents: a) 1%AcOH, MeOH; b) TsCl, Py; c) H_3O^+, then $NaBH_4$, H_2O; d) adenine, K_2CO_3, 18-crown-6.

Scheme 29

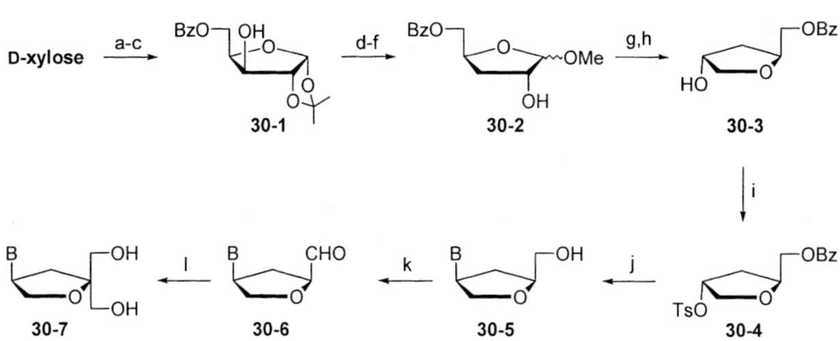

Reagents: a) acetone, H_2SO_4; b) 0.2 % HCl; c) BzCl, Py; d) $Im_2C(S)$, DCE; e) AIBN, Bu_3SnH, Tol; f) HCl, MeOH; g) HMDS, TMSCl; h) TMSOTf, Et_3SiH; i) TsCl, Py; j) ring construction (pyrimidines) or direct coupling (purines), then deprotection; k) DMSO, DCC, DCAA; l) 2 N NaOH, 37 % aq. $H_2C=O$, 1,4-dioxane.

Scheme 30

Tino *et al.* reported the synthesis of various isonucleoside analogs from D-xylose (Scheme 31).[66,67] Using modified literature procedures,[68] the known alcohol **31-7**[69] was prepared from commercially available 1,2-isopropylidene-D-xylofuranose. Reaction of the primary alcohol of **24-1** with tosyl chloride gave **31-1**, which was treated with the sodium salt of benzyl alcohol at 100 °C to afford **31-3** through the intermediate oxetane **31-2**. Oxidation of **31-3** with Collin's reagent followed by Wittig olefination of the crude ketone gave the exocyclic olefin **31-5**. Hydroboration of **31-5** with BH_3/THF afforded alcohol **31-6**. Protection of the primary hydroxyl group of **31-6** required harsh conditions. Treatment of **31-6** with sodium dimsylate followed by the treatment with benzyl bromide gave **31-7**. Removal of the acetonide group with aqueous acetic acid, followed by acetylation of the crude lactol-alcohol, gave diacetate **31-8** as a mixture of α- and β-anomers. Conversion of the anomeric acetates to the corresponding anomeric chlorides **31-9** with HCl in toluene and subsequent DIBAL reduction gave **31-10**. The key intermediate **31-11** was prepared by treatment of alcohol **31-10** with TsCl at low temperature. Tosylate **31-11** was coupled with various purine and pyrimidine bases to give isonucleosides **31-12**.

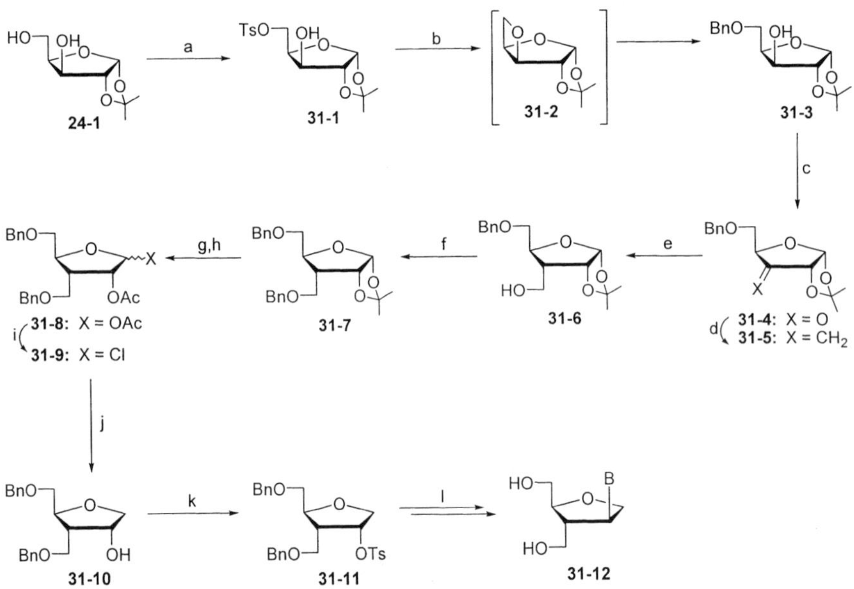

Reagents: a) TsCl, Py; b) Na, BnOH; c) CrO_3/Py; d) Ph_3PCH_2, THF, rt to 50 °C; e) (1) BH_3/THF, (2) NaOH, 30% H_2O_2; f) NaOH, DMSO, then BnBr; g) AcOH, H_2O; h) Ac_2O, Py; i) HCl, toluene; j) DIBAL, Tol; k) TsCl, Py; l) condensation with various bases.

Scheme 31

Franchetti *et al.* reported the synthesis of the aza and deaza analogs of isoddA as racemic mixtures (Scheme 32).[70] Nucleophilic substitution of (3*S*-*trans*)-tetrahydro-

5-(dimethoxymethyl)-3-furanol 4-methylbenzenesulfonate **29-5** with the appropriate heterocyclic bases in DMF in the presence of K_2CO_3 and 18-crown-6 afforded a mixture of regioisomers which were separated by flash silica chromatography. Hydrolysis of the dimethyl acetals, followed by reduction with $NaBH_4$, afforded the isonucleosides.

Varela *et al.* reported the synthesis of 3,4-dideoxyhexopyranosyl- and hex-3-eno-pyranoside-2-ulose-pyrimidine isonucleosides from 2-acyloxygylcals (Scheme 33).[71] Reaction of 2,3,4,6-tetra-*O*-acetyl-1,5-anhydro-D-*arabino*-hex-1-enitol **33-1** with TMS-uracil in the presence of tin(IV) chloride, followed by heating with methanol at reflux temperature for 1 h, led to 3-(6-*O*-acetyl-3,4-dideoxy-α-D-glycero-hex-3-enopyranoside-2-ulose)uracil **33-5** in 44% yield. The yield of **33-6** was considerably lower than that obtained with the glycal of *lyxo* configuration. The condensation of **33-1** or **33-2** with TMS-thymine led to the thymine ketoisonucleosides **33-7**, in 54 or 27% yield, respectively.

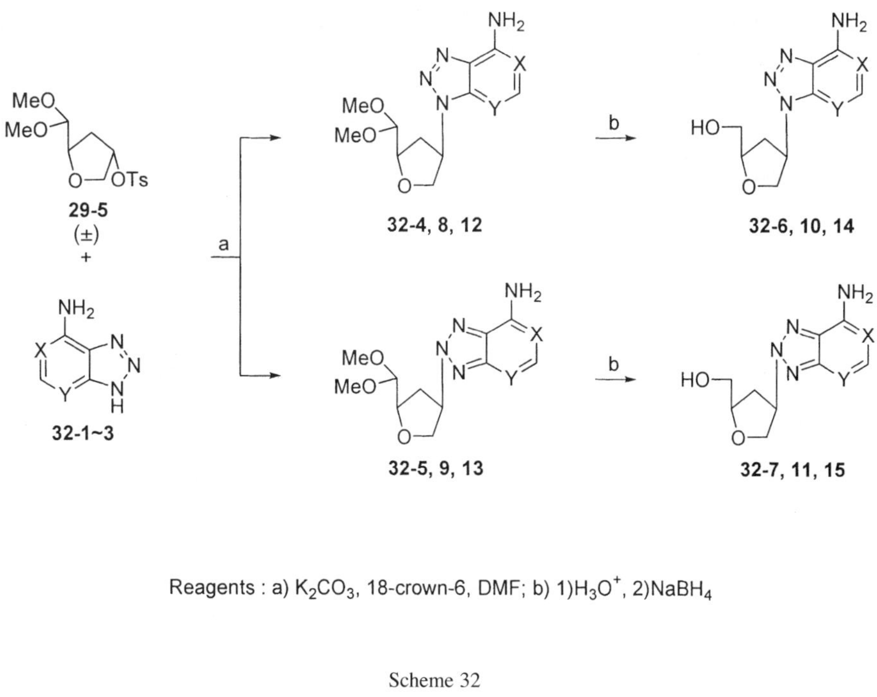

Reagents : a) K_2CO_3, 18-crown-6, DMF; b) 1)H_3O^+, 2)$NaBH_4$

Scheme 32

The high regio- and steroselectivities of the condensation and the reason why the N-3 position, being less reactive than N-1, is involved in the glycosidic linkage, may be explained by considering the sterochemical course of the allylic rearrangement of glycals. In the case of 2-acyloxyglycals, stabilization of the carbocation by anchimeric participation of the vicinal acyl substituent would lead to an acyloxonium intermediate I, having the β configuration, because of the stereoelectronic requirement of a positively charged anomeric substituent (reverse anomeric effect).[72] The more reactive N-1 of the pyrimidine will attack the acyloxonium carbon to give the species II, analogous to the orthoester proposed[73] as an intermediate in the $SnCl_4$-catalysed glycosylation of peracylated sugars with alcohols. A new allylic rearrangement would lead to the intermediate III, having the N-3 of the base suitably located for the attack of C-1' from the α face. Reduction of the α,β-unsaturated carbonyl system will readily produce 3',4'-dideoxyhexosyl pyrimidines, and it would also allow the confirmation of the anomeric configuration assigned for the ketoisonucleosides. Treatment of **33-6** with sodium borohydride, followed by deacetylation (sodium methoxide), afforded a single product **33-8**, whose ^{13}C NMR spectrum evidences the complete reduction of the enone. The configuration at 2'-C was determined by characterization of the sugar constituent of **33-8**, obtained by acetolysis of this nucleoside. As observed for the reduction of glycosuloses, the reduction of unsaturated ketoisonucleosides led to the corresponding nucleoside of a dideoxyhexose, with excellent diastereofacial selectivity. In both cases, the approach of the hydride ion took place from the β face of the molecule, opposite to the anomeric substituent. Similarly, Herscovici *et al.*[74] described that the reduction of 3'-substituted-3'-eno-pyranosid-2'-ulose drivative of theophilline gave a single product as result of the hydride addition from the less hindered face of the molecule. *anti* to the base.

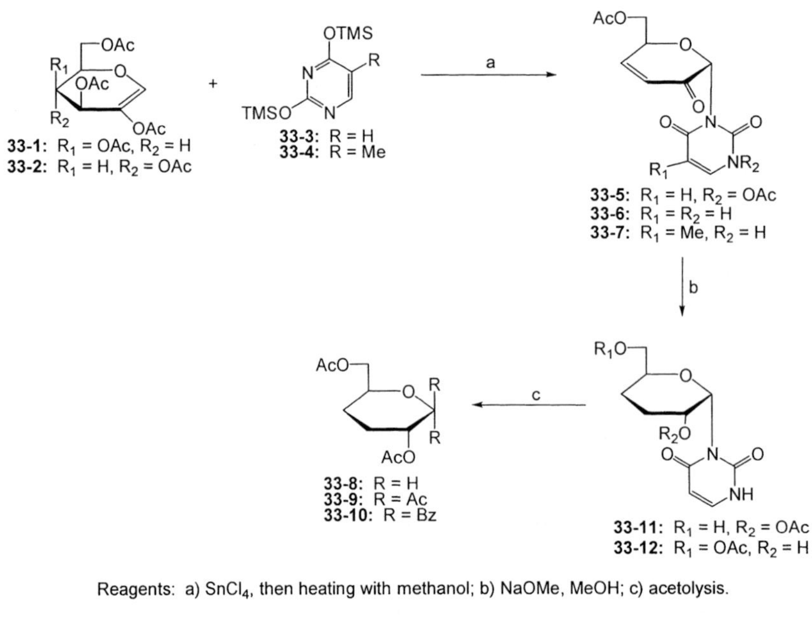

Reagents: a) $SnCl_4$, then heating with methanol; b) NaOMe, MeOH; c) acetolysis.

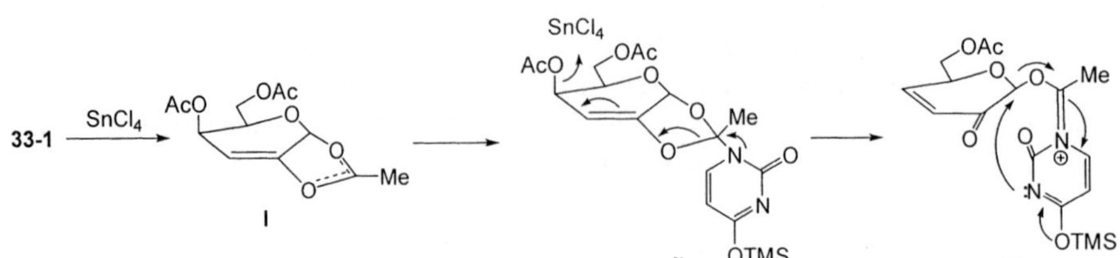

Scheme 33

Hammerschmidt *et al.* reported the synthesis of apio-β-D-furanosyl- and 2'-deoxy-apio-β-D-furanosyl nucleosides from D-mannose.[75] 2,3-*O*-Isopropylidene-D-apio-β-D-furanose **34-1**, available in 40% overall yield by modified literature procedures, could be benzoylated regioselectively to the 3'-*O*-benzoyl-derivative **34-2** (Scheme 34).

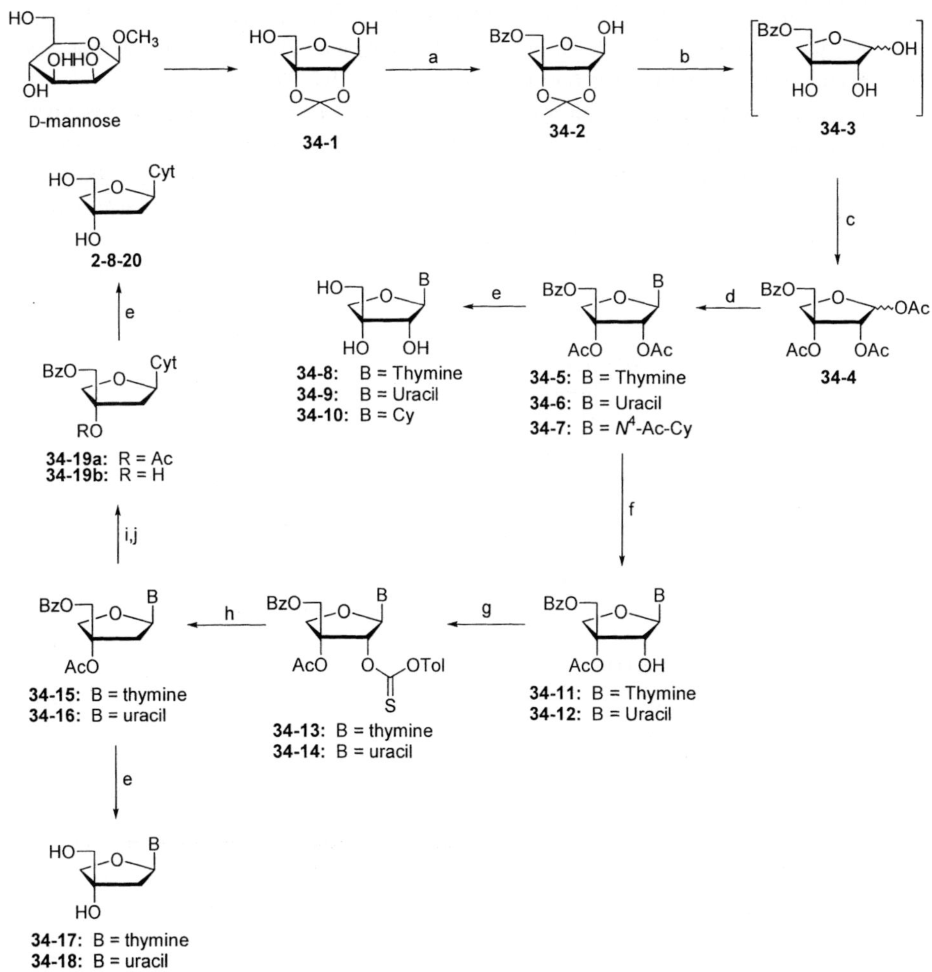

Reagents: a) BzCl, CH_2Cl_2, -78 °C; b) 80% water/CF_3COOH, CH_2Cl_2, 30%; c) Ac_2O, Py, DMAP; d) (1) Silylated bases, MeCN, Tol, TMSOTf, 80%; (2) $NaHCO_3$, H_2O; e) MeOH, NaOMe; f) $N_2H_4 \cdot H_2O$, AcOH, Py, 4 °C; g) Cl(C=S)OTol, DMAP, MeCN; h) $(Me_3Si)_3SiH$, AIBN, Tol, 60 °C; i) 1,2,4-triazole, $POCl_3$, MeCN, 0 °C; j) NH_4OH.

Scheme 34

This was converted to the anomeric triacetate **34-4** by cleavage of the isopropylidene protecting group followed by peracetylation. Reaction of compound **34-4** with silylated nucleobases and subsequent deprotection gave the D-apio-β-D-furanosyl nucleosides **34-8** ~ **34-10** in good yields. Regioselective hydrazinolysis of the protected nucleosides **34-5** and **34-6** afforded the 2'-hydroxy-derivative **34-11** and **34-12**, respectively, which were converted to the 2'-deoxyapio-β-D-furanosyl nucleosides **34-17** and **34-18** *via* deoxygenation of the corresponding thiocarbonates **34-13** and **34-14** with tris(trimethylsilyl)silane and subsequent deprotection. The 2'-deoxyapio-β-D-furanosyl cytosine **34-20** was prepared from the 2'-deoxyapio-β-D-furanosyl uracil derivative **34-16**.

Intermediate **34-2** was also used for the synthesis of 3'-deoxy-α-L-and 3'-deoxy-β-D-apionucleosides (Scheme 35).[76] Cyclic thiocarbonate **35-2**, easily available in three steps from **34-2**, was deoxygenated with n-Bu_3SnH to yield the 3'-deoxy epimers **35-3** and **35-7**. These were separated by chromatography and further elaborated to the 3'-deoxy-α-L-apionucleosides **35-6**, and the 3'-deoxy-β-D-apionucleosides **35-10**.

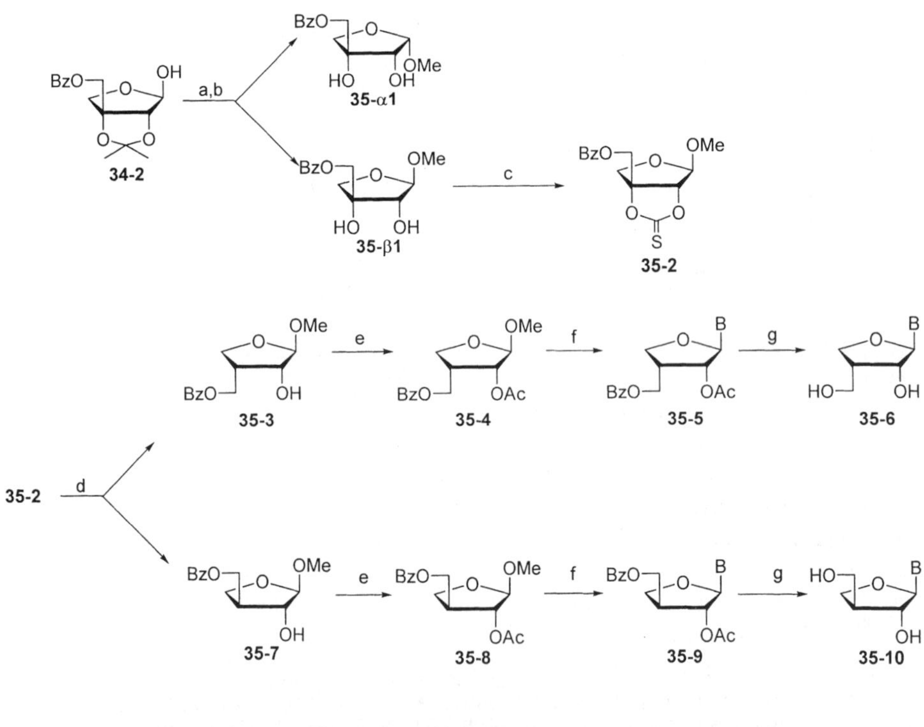

Reagents: a) 80% water, TFA, CH_2Cl_2, 30 °C; b) HCl/MeOH; c) $Im_2C{=}S$, CH_2Cl_2, 40 °C; d) n-Bu_3SnH, AIBN, Tol, 80 °C; e) Ac_2O, Py, DMAP; f) (1) silylated base, MeCN, TMSOTf, 80 °C; (2) $NaHCO_3$, H_2O; g) NaOMe, MeOH.

Scheme 35

Purdy *et al.* synthesized isonucleosides related to AZT and AZDU from D-ribose (Scheme 36).[77] The starting compound for the synthesis of the AZT analog was 1,4-anhydro-D-ribitol **36-3**.[78] Bennek and Gray had reported the synthesis of **36-3** *via* silylation of the methyl glycoside **36-1** followed by reductive cleavage in the presence of triethylsilane (TES) and TMSOTf.[79] Compound **36-6**, previously prepared from either the reduction of the 1-halogenated sugar or deoxygenation of **36-5** in the presence of pyridine-borane complex, appeared to be an attractive alternative.[80] High yields of **36-6** could be obtained by treatment of **36-5** with TES and BF_3 etherate.[81] Deprotection of **36-6** by catalytic hydrogenation afforded **36-3** in nearly quantitative yield. Selective protection of the hydroxyl functionalities at C-3 and C-5 was accomplished by treatment of **36-3** with 1,3-dichloro-1,1,3,3-tetraisopropyldisiloxane ($TIPDSCl_2$) in pyridine to afford **36-7** in 70-75% yields. Several attempts at the displacement of the mesylate of **36-8** by the appropriate pyrimidine bases under a variety of conditions led to poor overall yields of the N-1 alkylated product, accompanied by decomposition and formation of *O*-alkylated products. However, the displacement of the mesylate was readily accomplished with sodium azide in DMF to afford the protected 2-azido-2-deoxy-D-arabitol **36-9** in 80-85% yields. Construction of the pyrimidine ring was readily accomplished from the protected 2-amino-2-deoxy-1,4-anhydro-D-arabitol **36-10**, derived from **36-9** by catalytic hydrogenation.[82] Treatment of **36-10** with 3-methylacryloyl isocyanate, prepared in situ from 3-methoxy-2-methylacryloyl chloride and silver isocyanate,[83] afforded the intermediate acrylourea **36-11**. Cyclization of **36-11** with concomitant deprotection was readily accomplished under acid-catalyzed conditions to afford **36-12** in nearly quantitative yields.[84] After selective protection of the primary hydroxyl group and mesylation of the secondary, transformation of **36-14** into the new anhydro lyxitol analog **36-15** was accomplished by intramolecular cyclization in the presence of DBU in refluxing THF. Ring opening of protected anhydro nucleoside **36-15** with azide ions under a variety of conditions was difficult. Unlike the 6-membered ring anhydro nucleosides, the relatively strain-free tricyclic nucleoside **36-15** displays unexpectedly remarkable stability toward nucleophilic ring opening. This stability exceeds that of normal five-membered ring anhydro nucleosides and is comparable to anhydro *C*-nucleosides, which are also resistant to nucleophilic opening with azide ions.[85,86] However, opening of **36-15** was possible by alkaline hydrolysis, which afforded the deprotected 1,4-anhydro-D-lyxitol derivative **36-16**. Mesylation of this derivative gave intermediate **36-18**, which was smoothly converted to the protected azido derivative **36-19** upon treatment with lithium azide in DMF. Deprotection of **36-19** with methanolic ammonia followed by purification afforded the AZT isonucleoside **36-20**.

Kakefuda *et al.* reported the synthesis of purine and pyrimidine derivatives of 1,4-anhydro-2-deoxy-D-arabitol).[87] In order to synthesize the target nucleosides, a large quantity of **37-5** and **37-7** were required. The synthesis of 1,4-anhydro-3,5-di-*O*-benzyl-D-ribitol (**37-5**) from 1,2:5,6-di-*O*-isopropylidene-α-D-glucose **37-1** was effected by modifications of literature methods.[88] After hydrolysis of the isopropylidene group by aqueous 80% AcOH, **37-3** was obtained, which was treated with $NaBH_4$ in MeOH to afford the triol **37-4** in 74% yield from **37-2**. Cyclization was achieved by treatment of **37-4** with azodicarboxylate and triphenylphosphine in THF. Since this sequence to prepare **37-5** was rather lengthy and gave separation problems, the possibility of selectively reducing

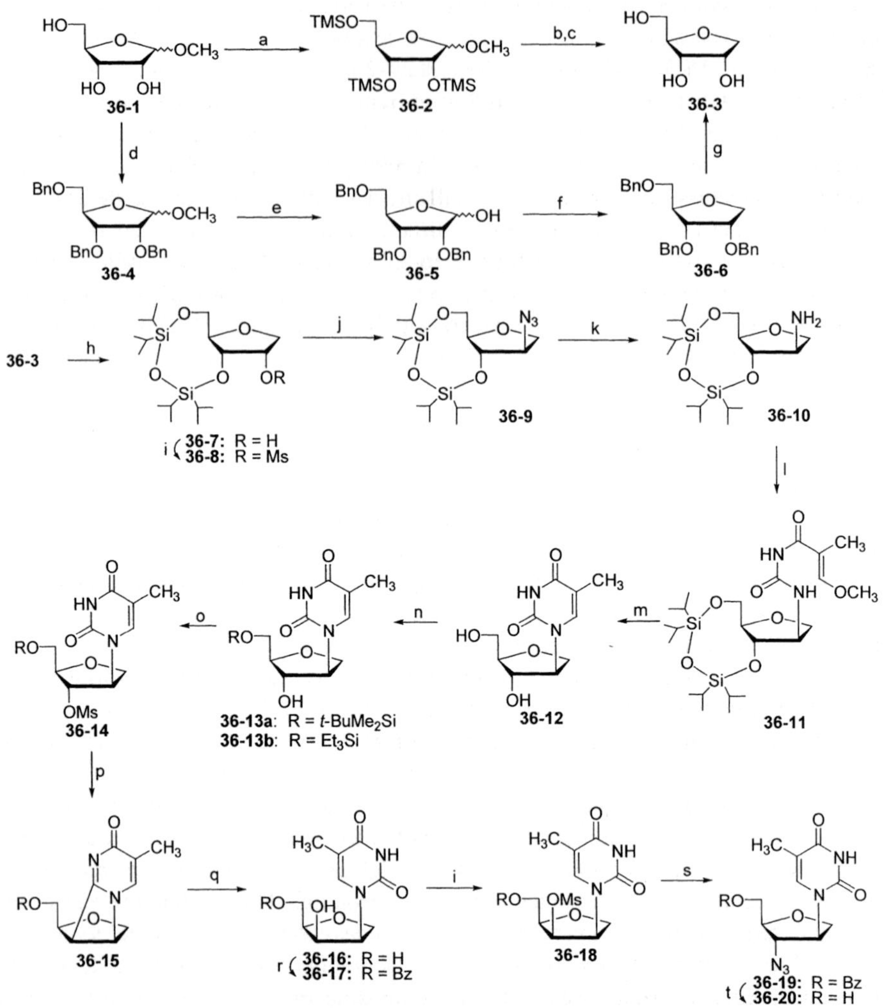

Reagents: a) HMDS, TMSCl, CH_3CN, reflux; b) TMSOTf, TES, CH_3CN; c) H_2O, Dowex OH; d) BnBr, NaH, DMF; e) H_2O,HCl, AcOH, reflux; f) TES, $BF_3 \cdot Et_2O$, CH_3CN; g) H_2, 10% Pd/C, EtOH; h) $TIPDSCl_2$, Py, -15 °C to rt; i) MsCl, Et_3N, CH_2Cl_2, 0 °C; j) NaN_3, DMF, reflux; k) 5% Pd/C, H_2, EtOH; l) $CH_3OCH{=}C(CH_3)C(O)N{=}C{=}O$, Tol, DMF, rt; m) dioxane, 2N H_2SO_4, reflux; n) RSiCl, DMAP, TEA, DMF; o) MsCl, Py, 0 °C to rt; p) DBU, THF, reflux; q) 1N NaOH, EtOH; r) BzCl, Py, 0 °C to rt; s) LiN_3, DMF, reflux; t) NH_3, MeOH, 0 °C.

Scheme 36

the anomeric position in **37-2** was examined. When **37-2** was treated with Et_3SiH[79] in the presence of TMSOTf in CH_2Cl_2 at 0 °C to room temperature, the desired tetrahydro-furan-3-ol **37-5** was obtained in 70% yield. Similarly, **37-7** was prepared from **37-6**, which was obtained from **37-1** according to literature methods[69,89] using Et_3SiH and TMSOTf. Mesylation of **37-5** and **37-7** gave the corresponding derivatives **37-8** and **37-9**, which were then treated with nucleobases in the presence of K_2CO_3 and 18-crown-6 in DMF. Reaction of **37-8** and **37-9** with adenine gave the desired nucleosides **37-10** in 50% yield and **37-11** in 37% yield, respectively. At lower temperature, the reactions did not proceed well and at higher temperature, the yields of the desired compounds were reduced. For the synthesis of the guanine derivatives **37-18** and **37-19**, attempted substitution reactions of **37-8** and **37-9** with 2-amino-6-chloropurine were unsuccessful. However, the reaction with 2,6-diaminopurine afforded **37-12** and **37-13** in 62% and 32% yields, respectively. The dibenzyl groups in both derivatives were removed similarly and the resulting **37-16** and **37-17** were treated with adenosine deaminase from calf intestine. Although the reaction catalyzed by the deaminase proceeded very slowly at 37 □C in phosphate buffer (pH 7.5), the desired guanine analogs **37-18** and **37-19** were obtained in 76% and 62%, respectively, after 10 days.

Ohrui *et al.* described the synthesis of an adenine substituted with a 1,4-anhydroribitol having a vicinal *cis*-oriented azide and an *O*-tosylate (Scheme 38).[90] The key intermediate for the synthesis of these analogs, compound **38-6**, was prepared from D-xylose-derived azide **38-3**, and it is noteworthy that the azide group is compatible under triethylsilane reduction conditions.

Yu *et al.* synthesized various 4'-hydroxy-5'-hydroxymethyl-tetrahydrofuranyl purines and pyrimidines from D-xylose (Scheme 39).[91] 3-(*R*)-Hydroxy-4-(*S*)-tosyl-5-(*S*)-dimethoxymethyl-tetrahydrofuran **39-1**, prepared from 1,2-*O*-isopropylidene-α-D-xylose **24-1**, was treated with potassium carbonate in methanol at room temperature to give 3,4-epoxy-5-(*S*, *trans*)-dimethoxymethyltetrahydrofuran **39-2**. The isonucleosides were obtained by the reaction of the epoxide **39-2** with different nucleobases in the presence of potassium *tert*-butoxide and crown ether. The yield of purine analogs was higher than that of pyrimidines, although guanine gave a particularly low yield due to its poor solubility. Two regioisomers were obtained and separated with **39-3~39-6** as the main products. The dimethoxy group in **39-3~39-6** was hydrolyzed in 3% TFA at 80 °C and reduced by $NaBH_4$ at room temperature to give the final nucleosides in good yields. Halogenated nucleosides **39-14** and **39-16** were prepared from **39-3** and **39-5** by general procedures. Uridine analog **39-5** was treated with $POCl_3$ and triazole followed by conc NH_4OH to give cytidine analog **39-15**.

Prévost *et al.* synthesized various pyranose isonucleosides by Michael additions on 2,5-disubstituted 2H-pyran-3(6H)-ones (Scheme 40).[92,93] Optically pure pyran-3-ones were obtained from commercially available tri-*O*-acetyl-D-glucal **40-1** *via* the Ferrier rearrangement as the key reaction.[94] Thus, **40-1** was converted to ethyl (or isopropyl) pyranoside **40-4~40-6** in four steps involving: a) iodine-catalyzed glycosylation with ethyl (or i-propyl) alcohol,[95] b) hydrolysis of the acetyl groups,[96] c) selective 6-OH protection with t-butyldimethylsilyl group or as a pivalate, and d) oxidation[97] of 4-OH. When pyranones were allowed to react in the presence of trimethylsilylated nucleobases (prepared in situ with an excess of *N,O*-bis(trimethylsilyl)acetamide in dry MeCN),[98] the Michael adducts were formed. Crude **40-10~40-16** were used as such in the subsequent step. Reduction of the carbonyl group of these adducts with $NaBH_4$ afforded a high diastereomeric excess in favor of the *cis*-C-4'/C-5' isomer.

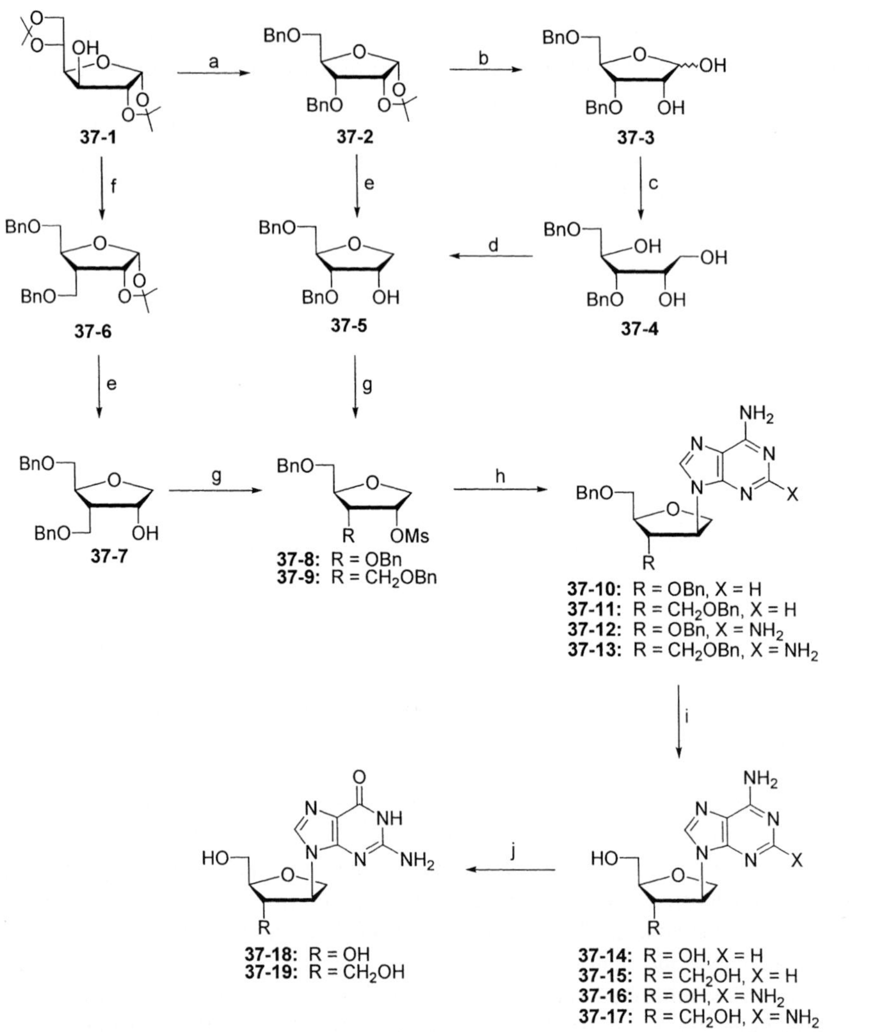

Reagents: a) known method, see text; b) 80% AcOH, 100 °C; c) $NaBH_4$, MeOH; d) Ph_3P, DEAD, THF, 60 °C; e) Et_3SiH, TMSOTf, CH_2Cl_2, -18 °C to rt; f) known method, see text; g) MsCl, TEA, DMAP, CH_2Cl_2, rt; h) nucleobase, K_2CO_3, 18-crown-6, DMF, 120 °C; i) BCl_3, CH_2Cl_2, -78 °C to -18 °C; j) adenosine deaminase, pH 7.5, 37 °C.

Scheme 37

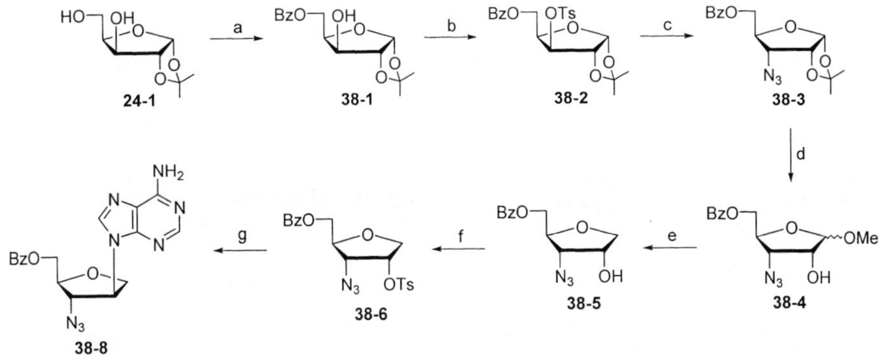

Reagents: a) BzCl, Py; b) TsCl, Py; c) NaN_3, DMF, H_2O; d) Dowex 50w x 8, MeOH, 90 °C; e) BSA, TMSOTf, Et_3SH, rt; f) TsCl, Py, rt; g) adenine, K_2CO_3, 18-crown-6, DMF, 115 °C.

Scheme 38

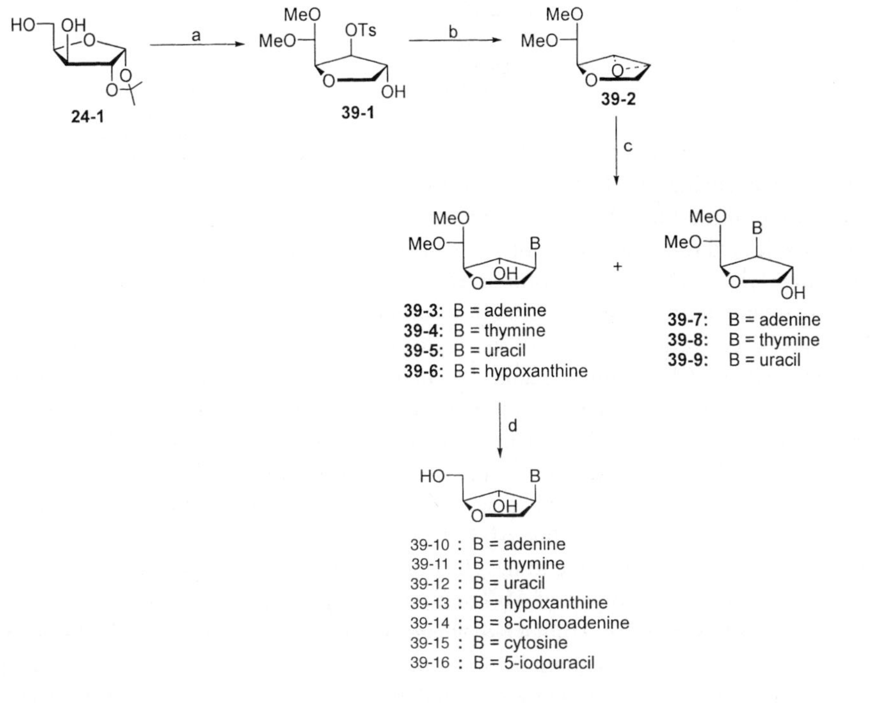

Reagents: a) (1) TsCl, 20 °C, 4 days; (2) 1% TFA, MeOH, 75 °C, 8h; b) K_2CO_3, MeOH; c) nucleobase, t-BuOK, 18-crown-6, DMF, 20 °C; d) (1) 3% TFA; (2) $NaBH_4$.

Scheme 39

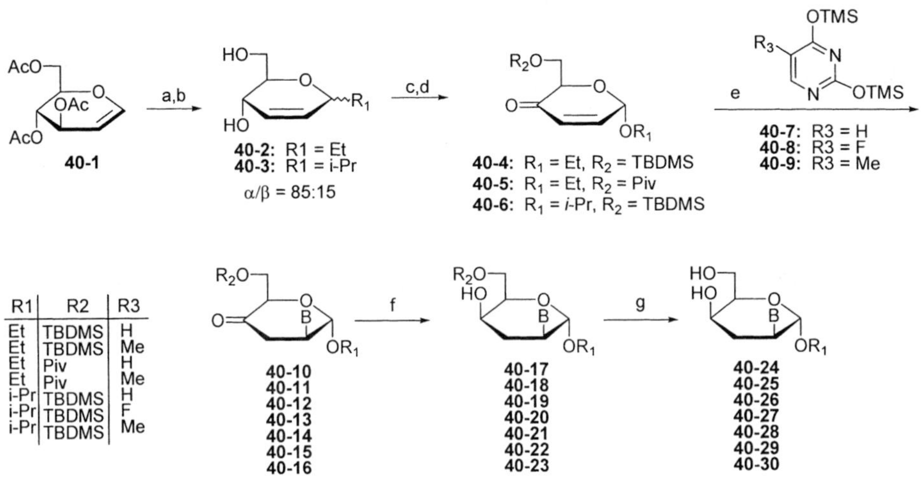

R1	R2	R3
Et	TBDMS	H
Et	TBDMS	Me
Et	Piv	H
Et	Piv	Me
i-Pr	TBDMS	H
i-Pr	TBDMS	F
i-Pr	TBDMS	Me

Reagents: a) Iodine-catalyzed glycosylation; b) Hydrolysis; c) TBDMSCl, TEA, DMAP or Pivaloylation; d) PDC; e) TMSOTf, MeCN; f) $NaBH_4$; g) TBAF or NaOH.

Scheme 40

Talekar *et al.* reported the syntheses of some pyrrolo[2,3-d]pyrimidine and 1,2,3-triazole isonucleosides (Scheme 41).[99] 1,2-*O*-Isopropylidene α-D-xylofuranose[100] was converted to its dimesylate **41-1**, which, upon refluxing in methanol containing 1% (v/v) of trifluoroacetic acid followed by treatment with anhydrous K_2CO_3, gave the acetal-epoxide **41-3** in high yield. The tetrahydrofuran **41-2** was assumed to be the intermediate after the acidic treatment, by analogy with other similar cyclization reactions.[101] The epoxide **41-3** could be converted cleanly to one regioisomeric azidoalcohol **41-4** by treatment with NaN_3 and NH_4Cl in refluxing aqueous ethanol. Catalytic hydrogenation of the latter gave the aminoalcohol **41-5**. The pyrrolopyrimidine ring could be built by reaction of amine **41-5** with the dichloropyrimidine (prepared from the known aldehyde[102] by treatment with NH_4Cl in ethanol at reflux) in the presence of Et_3N to give the substitution product **41-6**, which, on stirring in THF and aqueous HCl at room temperature, underwent smooth cyclization to the pyrrolopyrimidine **41-7**. The acetal functionality in **41-6** was stable in these conditions, presumably due to the electron-withdrawing effect of the oxygen of the tetrahydrofuran ring. Hydrolysis of **41-7** under more vigorous acidic conditions followed by treatment with $NaBH_4$ gave diol **41-8**, which could be converted to the adenosine analog **41-9** by treatment with methanolic ammonia at 100 °C. Chloro-derivative **41-8** could also be used as a precursor for the inosine analog **41-10** and for the thione **41-11**. Known methods were also used for the syntheses of isonucleosides **41-12**, **41-13**, and **41-14**.

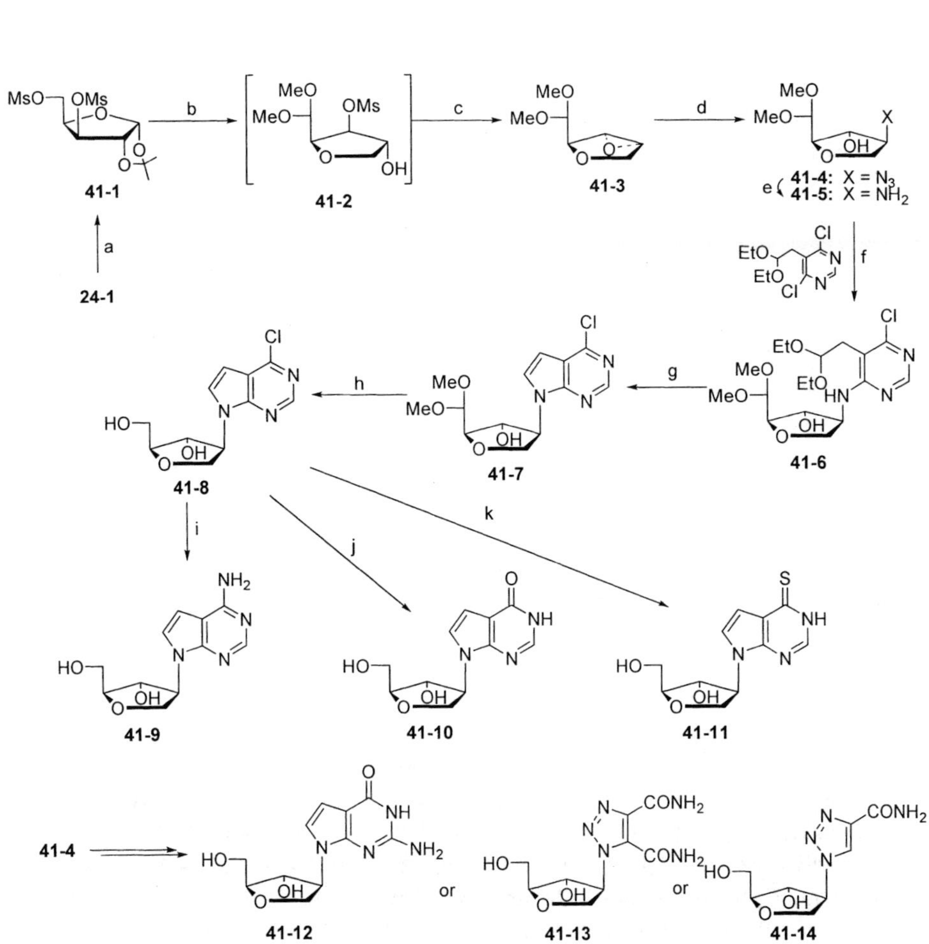

Reagents: a) MsCl, Py; c) MeOH, TFA, reflux; d) K_2CO_3, rt; c) NaN_3, NH_4Cl, H_2O/EtOH; e) H_2, PtO_2, EtOH; f) TEA, $EtOCH_2CH_2OH$, reflux; g) THF, HCl, aq., rt; h) TFA, THF/H_2O, 80 °C, then $NaBH_4$; i) NH_3, MeOH, 100 °C; j) NaOH, dioxane/ H_2O, reflux; k) thiourea, n-propanol, reflux.

Scheme 41

Yang *et al.* reported the synthesis of 4-deoxy-4-nucleobase-2,5-anhydro-L-mannitol derivatives from D-glucose (Scheme 42).[103] Fully protected **42-1** or **42-2** were hydrolyzed in acidic methanol to give **42-4** and **42-5**, respectively. Treatment of **42-4** or **42-5** with potassium carbonate in methanol generated epoxide **42-6** or **42-7**. Compound **42-7** reacted with nucleobases in the presence of potassium carbonate and crown ether to give the regioselective epoxide-opening products **42-11~42-13**. In the case of 5-fluorouracil, the N-3 alkylated side product was predominant.

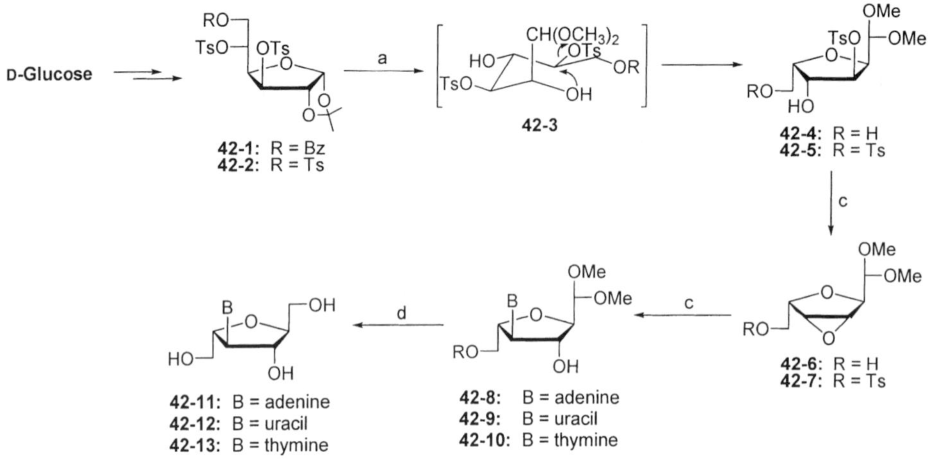

Reagents: a) 1% HCl in methanol; b) K_2CO_3 in methanol; c) nucleobases, K_2CO_3, 18-crown-6/DMF; d) (1) 0.5% TFA/H_2O, (2) $NaBH_4$.

Scheme 42

Yu *et al.* also reported syntheses of various types of isonucleosides from carbohydrate starting materials (Scheme 43).[104] Starting from L-xylose, they synthesized D-3'-hydroxy isonucleosides. Intermediate epoxide **43-1** was obtained from L-xylose in 91.3% yield by the same procedures reported for **42-6**. Due to the steric effect of the dimethylacetal group at C-5, the nucleobase preferably attacked the C-2 position of epoxide, and thus compounds **43-2** and **43-3** were obtained in 26:1 ratio. Deprotection of **43-3** followed by reduction afforded final isonucleosides **43-4**, **43-5**, and **43-6**.

The same authors reported the synthesis of *C*-isonucleosides in which the heterocyclic moiety is linked to the sugar by a C-C bond (Scheme 43). Thus, hydrolysis of epoxide **39-2**, followed by selective protection and Jones oxidation, gave ketone **43-8**, which was converted to cyanidrine **43-9**, in which the nitrile group is *trans* to the benzyl. Treatment with hydroxylamine gave intermediate **43-10**, which could be cyclized using different acyl anhydrides to give isonucleosides **43-11**.

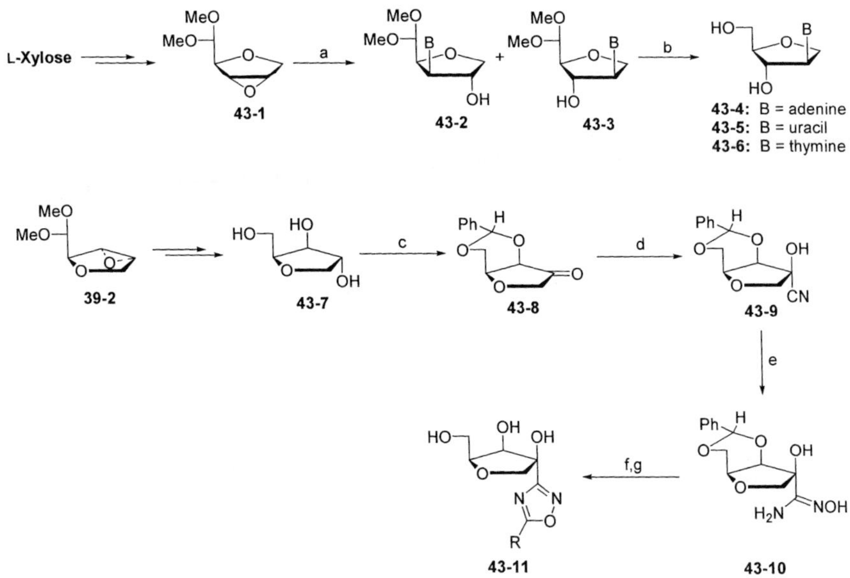

Reagents: a) Nucleobases/t-BuOK, 18-C-6, DMF, 75 °C, 30 min; b) 0.3% TFA, 80 °C, 3 h, $NaBH_4$, rt, 40 min; c) PhCHO, H^+, then CrO_3, Py, CH_2Cl_2, rt; d) KCN, EtOAc, H_2O, rt; e) NH_2OH, CH_3OH, 80 °C; f) $(RCO)_2O$, $CHCl_3$, 70 °C; g) 80% AcOH, 70 °C.

Scheme 43

Chu and co-workers reported the synthesis of enantiomeric 3'-fluoro-apionucleosides using Claisen rearrangement from 2,3-*O*-isopropylideneglyceraldhyde chirons (Scheme 44).[105] Intermediate **44-1**, prepared from protected D-glyceraldehyde **9-2**, was reduced and benzylated to **44-2**. The isopropylidene protecting group was then hydrolyzed to give diol derivative **44-3**, which was treated with di-*N*-butyltin oxide and benzyl bromide to give the dibenzyl allylic alcohol derivative **44-4** with high regioselectivity (10:1).[106] Alcohol **44-4**, suitable for a 1,3-chirality transfer in Claisen rearrangement conditions, was treated with triethyl orthoacetate and a catalytic amount of propionic acid to give the γ,δ-unsaturated tertiary fluoro ethyl ester **44-5** probably *via* a six-membered transition state in 86% yield with 90.4% ee. The double bond of **44-5** was ozonized to aldehyde **44-6**, which was subject to DIBAL-H reduction to give the lactol **44-7**. The apiose lactol **44-7** was then treated with acetic anhydride to give the intermediate **44-8**, which was condensed with silylated N^4-benzoyl cytosine and 6-chloropurine followed by corresponding treatment to give, after deprotection and derivatizations, the final apionucleosides **44-9** and **44-10**.

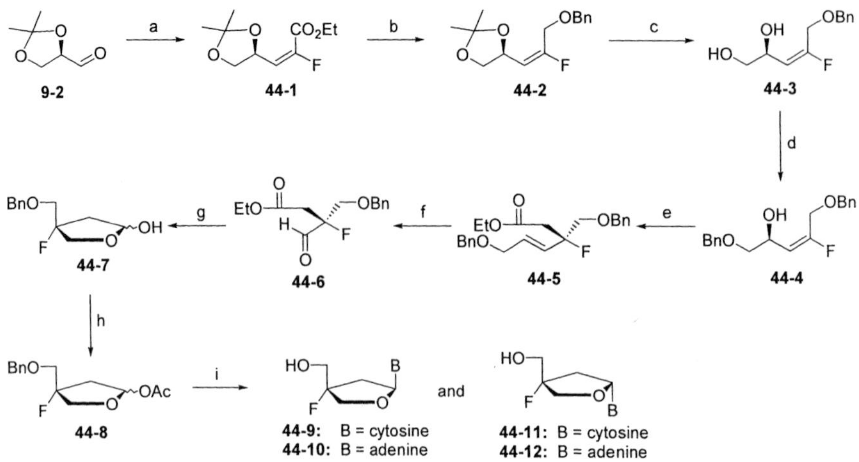

Reagents: a) $(EtO)_2P(O)CHFCO_2Et$/NaHMDS, THF; b) DIBAL-H, Tol, -78 °C, then BnBr, TBAI, Py; c) 2 N HCl, rt, 2 h; d) di-*n*-butyltin oxide, Tol, reflux, then BnBr, TBAI, 70 °C, overnight; e) $HC(OEt)_3$, $CH_3CH_2CO_2H$, 130 °C, 7 h; f) O_3/DMS; g) DIBAL-H, Tol, -78 °C; h) Ac_2O, Py, DMAP, rt, overnight; i) silylated nucleobases, TMSOTf, then deprotection and separation.

Scheme 44

The L-isomers were prepared using a similar approach.[107] Lactone **13-2**, synthesized from protected L-glyceraldehyde, was reduced in Luche conditions to diol **45-1** (Scheme 45). This was selectively protected with TBDMSCl to the allylic alcohol **45-2**, analog of **44-4**. The remaining steps are similar to the original procedure.

The same 3,3-Claisen rearrangement leading to a quaternary fluorinated carbon was exploited in the synthesis of 3'-*C*-methyl-4'-thio apionucleosides (Scheme 46).[108] Thus, 2,3-*O*-isopropylidene-D-glyceraldehyde was subject to a Wittig reaction with ethoxycarbonyl ethylidene triphenylphosphorane to give the trans-α,β-unsaturated ethyl ester **46-1**, which was reduced by DIBAL-H and benzylated to compound **46-3**. The isopropylidene group was then hydrolyzed to diol **46-4**, whose primary hydroxyl group was selectively protected with TBDMSCl. Subsequently, Claisen rearrangement of compound **46-5** gave chirality-transferred quaternary carbon chiron **46-6** with 98.5 % ee. The double bond of **46-6** was ozonized to aldehyde **46-7**, which was reduced to the apiose lactol **46-8**. This was treated with excess of benzyl mercaptane in the presence of $BF_3{\cdot}Et_2O$ as a Lewis acid. The resulting dithiane-protected alcohol was mesylated and cyclized to give the thio-glycosyl donor **46-10** as a diastereomeric mixture. Compound **46-10** was condensed with N^4-benzoylcytosine in the presence of NIS to give, after deprotection, the final thio-apionucleoside **46-11**.

Intermediate **46-8** was also used in the synthesis of various 3'-*C*-methyl apionucleosides (Scheme 47).[109]

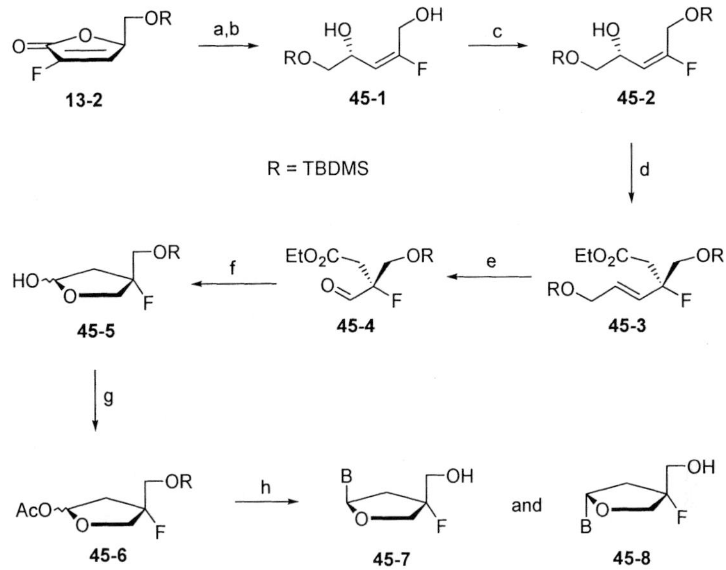

B = uracil, thymine, 5-F-uracil, guanine, inosine

Reagents: a) DIBAL-H, CH_2Cl_2; b) $NaBH_4$, $CeCl_3.7H_2O$, MeOH; c) TBSCl, imidazole, CH_2Cl_2; d) $HC(OCH_3)_3$, $CH_3CH_2CO_2H$; e) O_3, then DMS, MeOH; f) DIBAL-H, Tol, -78 °C; g) Ac_2O, TEA, DMAP h) silylated nucleobases, TMSOTf, deprotection, separation.

Scheme 45

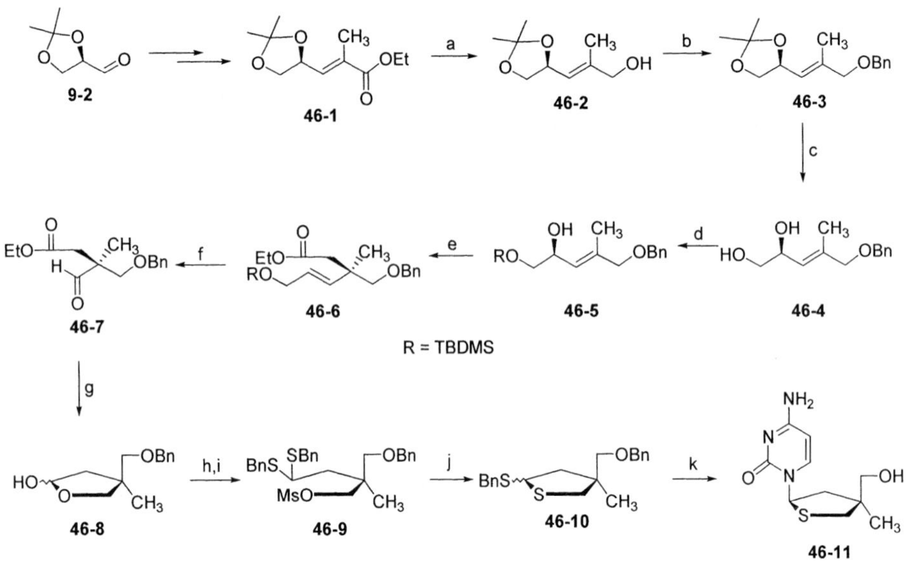

Reagents: a) Dibal-H, CH_2Cl_2, -78 °C; b) BnBr, Py, THF; c) 2 N HCl; d) TBDMSCl, imidazole, CH_2Cl_2; e) $HC(OCH_3)_3$, $CH_3CH_2CO_2H$, 135 °C; f) O_3, MeOH, then DMS, -78 °C; g) DIBAL-H, CH_2Cl_2, -78 °C; h) BnSH, $BF_3.Et_2O$, CH_2Cl_2; i) MsCl, Py; j) TBAI, $BaCO_3$, Py, reflux; k) nucleobases condensation, deprotection, and separation.

Scheme 46

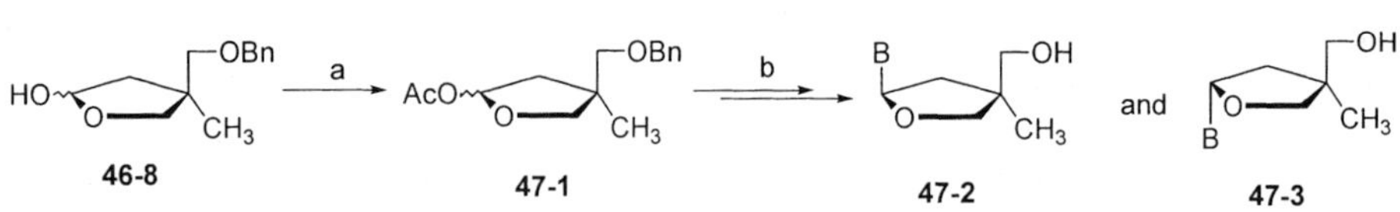

Reagents: a) Ac_2O; b) nucleobases condensation, deprotection, and separation.

Scheme 47

2.4. Oxathiolane and dioxolane nucleosides

Oxathiolane and dioxolane nucleosides are analogs in which the 3'-methylene is substituted by a sulfur or oxygen atom. This class has been extensively discussed in Chapter 3. Among them, important derivatives are 3TC, FTC, DAPD, L-OddC, L-OddFC, D-OddFC, L-I-OddU and L-BV-OddU (Figure 1).

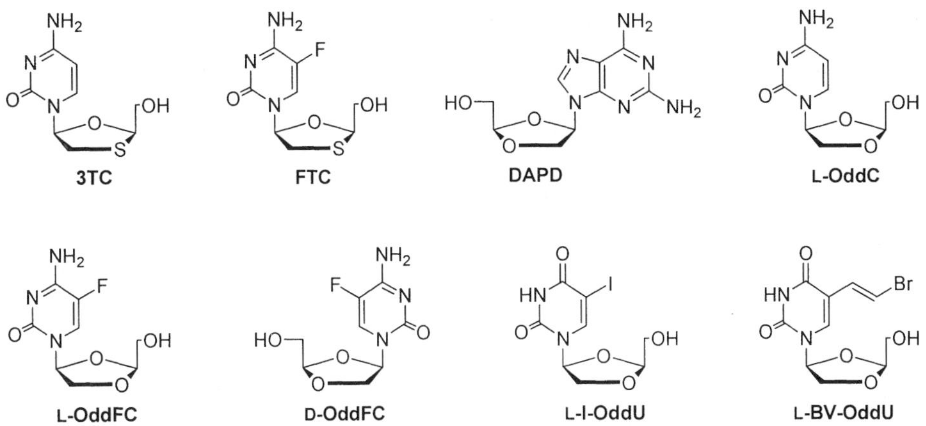

Figure 1

The early syntheses of compounds belonging to this category led to racemic mixtures. Most of the enantiomeric syntheses of oxathiolane and dioxolane nucleosides starting from carbohydrate templates have been reported by Chu and co-workers. These syntheses are briefly discussed here and described in greater details in Chapter 3.

The synthesis of enantiomerically pure (+)-(2'*S*,5'*R*)-1-[2-(hydroxymethyl)-1,3-oxathiolan-5-yl]-cytosine [(+)-BCH-189] was achieved using D-mannose as starting material *via* protected 1,6-thioanhydro-D-mannose **48-1** (Scheme 48).[110] This was subject to oxidative cleavage by lead tetraacetate followed by sodium borohydride reduction to give oxathiolane **48-2**, where the benzoyl group had migrated to the diol's primary hydroxyl. Further chemical manipulation afforded key intermediate **48-3**, which was condensed with silylated N^4-acetylcytosine to give, after deprotection, nucleosides **48-4** and **48-5**.

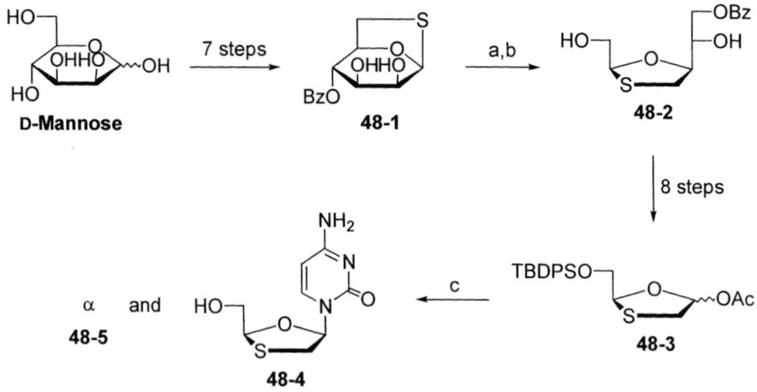

Reagents: a) $Pb(OAc)_4$; b) $NaBH_4$; c) silylated N^4-acetylcytosine, TMSOTf, then separation and deprotection.

Scheme 48

A more efficient methodology for the synthesis of (+)-BCH-189 starts from 1,6-thio-anhydro-D-galactose **49-1** (Scheme 49).[111,112] Oxidative cleavage of the *cis* diol in **49-1**, followed by reduction and protection afforded alcohol **49-2**, which was converted to the key intermediate **49-3** in 8 steps.

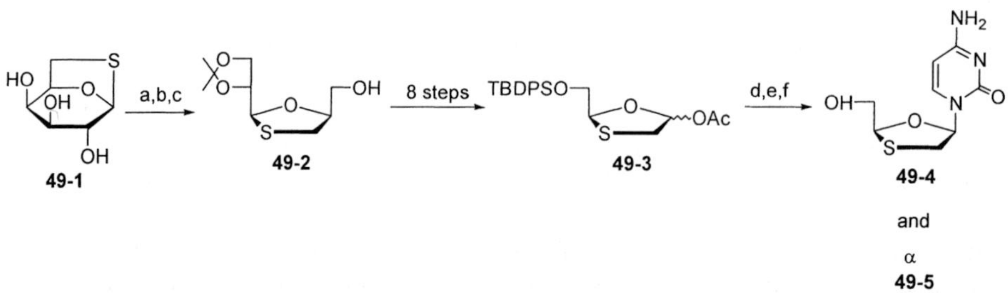

Reagents: a) $NaIO_4$, $MeOH/H_2O$; b) $NaBH_4$; c) $CH_3C(OMe)_2CH_3$, p-TsOH; d) silylated N^4-acetylcytosine, DCE, TMSOTf; m) NH_3, MeOH; n) TBAF, THF.

Scheme 49

3TC, the L- enantiomer of BCH-189, was synthesized by a similar strategy using 1,6-thioanhydro-L-gulose **50-1** as the starting material (Scheme 50).[113] Conversion of **50-1** to oxathiolane **50-2** was achieved in 3 steps, and 5 more steps allowed further manipulation to the key intermediate **50-2**, from which 3TC was synthesized by condensation and deprotection. Interestingly, it was found that the use of stannic chloride instead of TMSOTf as a Lewis acid during the condensation of **50-3** gave a racemic mixture, probably by the opening and reclosing of the oxathiolane ring under the reaction conditions.

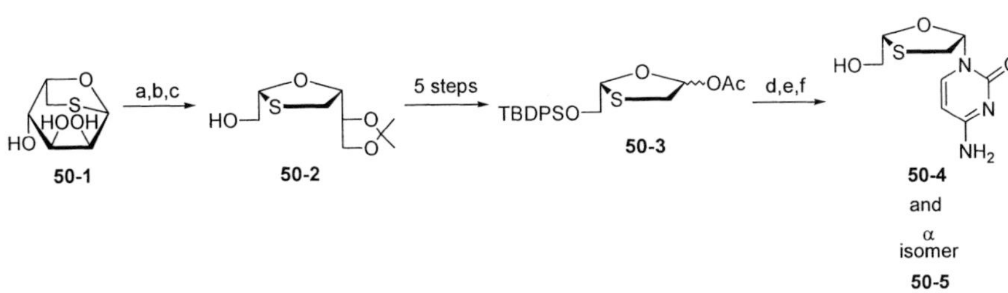

Reagents: a) $NaIO_4$, $MeOH/H_2O$; b) $NaBH_4$, MeOH; c) $CH_3C(OMe)_2CH_3$, *p*-TsOH; d) silylated N^4-acetylcytosine, DCE, TMSOTf; h) NH_3, MeOH; i) TBAF, THF.

Scheme 50

In order to study the structure activity relationships of various nucleobase derivatives, the same key intermediate **50-5** was also condensed with a number of pyrimidines and purines.[114] Upon evaluation of the anti-HIV activity of these nucleosides, the 5-fluorocytosine derivative was found to be the most potent.

Chu and co-workers[115,116] also reported the asymmetric synthesis of D-dioxolane nucleosides using D-mannose as a chiral starting material (Scheme 51). 1,6-Anhydro-D-mannose **51-1** was converted to protected derivative **51-2** in 3 steps. Oxidative cleavage followed by reduction in similar fashion as in the synthesis of oxathiolanes afforded dioxolane **51-3**, which was further elaborated to the key intermediate **51-4**. From this, a library of nucleosides was synthesized by coupling with natural and non-natural purine and pyrimidine bases, derivatization, and deprotection. Many of the synthesized compounds showed promising antiviral and antitumor activity.

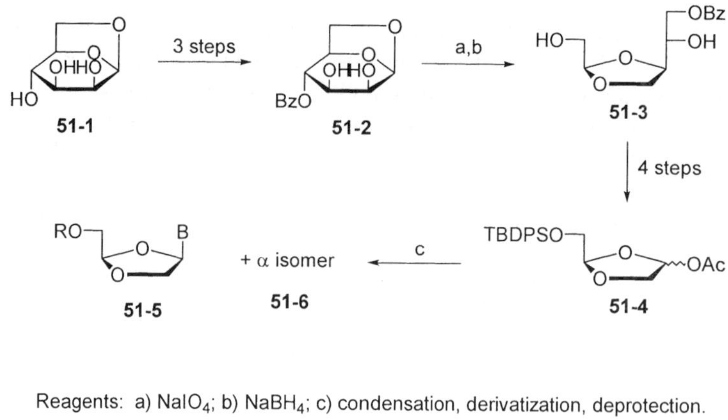

Scheme 51

L-Isomers of 1,3-dioxolane-pyrimidine nucleosides were synthesized starting from L-gulose (Scheme 52), which was dehydrated to 1,6-anhydro-L-gulose **52-1**. This was converted to dioxolane **52-2**, from which key intermediate **52-3** was derived in four steps. Also in this case, a comprehensive structure-activity relationships study was reported.[117,118]

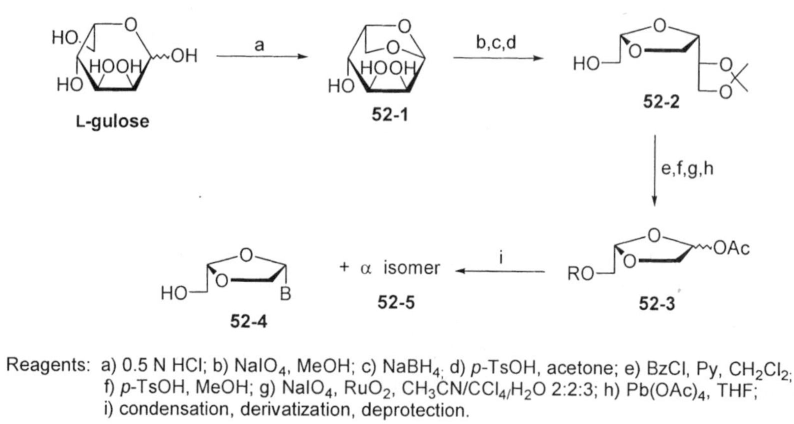

Scheme 52

Mansour and co-workers reported a divergent synthesis of dioxolanyl nucleosides using D-mannitol as the starting material (Scheme 53).[119] Condensation of D-mannitol and benzyloxymethylacetaldehyde gave the bis-acetal **53-1,** which was oxidized to give a 1: 1 mixture of acids **53-2** and **53-3**. These could be separated by silica gel chromatography and their oxidative decarboxylation afforded the key intermediates **53-4** and **53-5**, which were used in the synthesis of D- and L-dioxolanyl nucleosides, respectively.

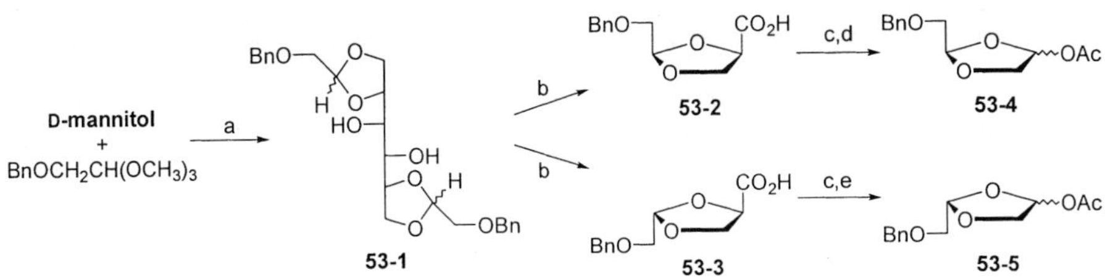

Reagents: a) 1.0 eq. $SnCl_2$; b) $RuCl_3$, $NaOCl_3$; c) flash chromatography 2% MeOH in CH_2Cl_2; d) $Pb(OAc)_4$, CH_3CN, Py; e) $Pb(OAc)_4$, CH_3CN/CH_2Cl_2, Py.

Scheme 53

Chu and co-workers[120] reported the synthesis of 1,3-dioxolanyl-*C*-nucleosides (Scheme 54). The chiralities of D-mannitol and L-γ-gulonolactone were exploited in the synthesis of thiazoles **54-1** and **54-4**, on which the sugar moiety was built by treatment with benzoyloxymethylacetaldehyde. Using similar reactions, the same authors synthesized selenazole and triazole *C*-nucleosides.[121] More details on the synthesis and pharmacology of oxathiolane and dioxolane nucleosides are discussed in Chapter 3.

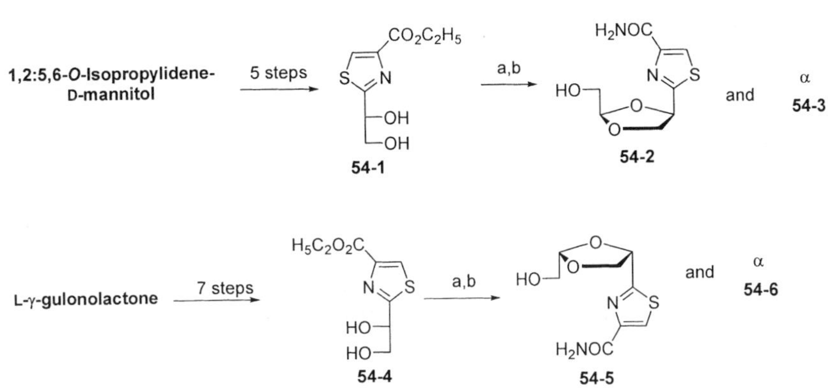

Reagents: a) $BzOCH_2CH(OMe)_2$, *p*-TsOH; b) NH_4OH.

Scheme 54

2.5. Cyclopentyl carbocyclic nucleosides

Although the first reported carbocyclic nucleoside (the racemic adenine analog) was a synthetic product,[122] the D-(-)-enantiomer of the same compound, aristeromycin[123] (Figure 2) was found to be a microbial product endowed with good antitumor activity. Other "antitumor antibiotics", the Neplanocins (of which neplanocin A is the prototype)[124] are cyclopentenyl derivatives which possess a wide range of biological activities.[125]

The biological activity of some carbocyclic nucleosides indicates that the isosteric substitution of the tetrahydrofuran system with a cyclopentyl ring does not prevent interaction with cellular and/or viral enzymes. On the other hand, the lack of a glycosylic bond makes these analogs resistant to enzymatic degradation by phosphorylases. Moreover, some derivatives, such as carbocyclic 2'-*ara*-fluoroguanosine,[126,127] are more active than their furanose counterpart. A number of carbocyclic adenosine analogs appear to exert their antiviral action through the inhibition of the enzyme *S*-adenosylhomocysteine hydrolase.[128] This mechanism might be exploited in combination therapy in association with nucleosides with different modes of action.

A number of carbocyclic analogs have been synthesized since the first reported synthesis; the most promising compounds are carbocyclic BVDU,[129] the potent anti-HBV agent BMS-200475[130] characterized by an *exo*-double bond, the anti-HIV agents (-)-BCA,[131] carbovir[132] and 1592U89 (abacavir[133,134]). Abacavir is one of the seven nucleosides/nucleotides approved by the FDA for the treatment of AIDS.

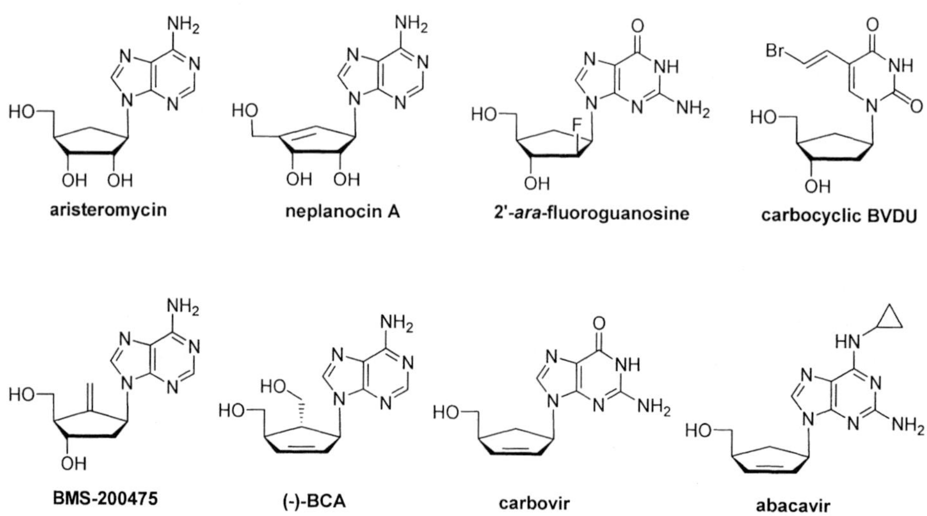

Figure 2

A number of different synthetic methods have proved effective in the syntheses of cyclopentane and cyclopentene carbocyclic nucleosides.[135,136,137] These methods utilize, as a precursor, a carbo- or heterocyclic bicyclic system or a functionalized cyclopentane/ene, and the chirality is introduced *via* chemical or enzymatic resolution. A smaller but not less important number of syntheses use carbohydrate templates as chiral starting materials. These syntheses, which produce enantiomerically pure nucleosides in stereoselective fashion, will be considered.

The first total synthesis of (-)-neplanocin A starting from a carbohydrate was reported by Marquez and co-workers[138,139,140] (Scheme 55), who obtained the final compound starting from protected D-ribonolactone **55-1** through 13 steps. Thus, **55-1** was coupled with lithium dimethyl methylphosphonate to afford the hemiketal **55-2**, which was transformed to the acyclic dibenzoate **55-3**. Debenzoylation and oxidation afforded diketophosphonate **55-5**, which was cyclized to **55-6** by treatment with potassium carbonate/18-Crown-6 in benzene under high dilution. Stereoselective reduction of **55-6** to **55-7** was achieved using $NaBH_4/CeCl_3 \cdot 7H_2O$. Mesylation and azidation with inversion of configuration followed by reduction afforded amine **55-10**, with the desired β-stereochemistry. From intermediate **55-10**, the 6-chloropurine derivative **55-12** was built by standard nucleoside chemistry. Ammonolysis and deprotection gave neplanocin A **55-14**.

Intermediate **55-7** was also used in a synthesis of conformationally constrained carbocyclic nucleosides (Scheme 56).[141] *Simmons-Smith* cyclopropanation afforded **56-1** as a single diastereomer, thanks to the directing effect of the allylic hydroxyl group. Cyclopentanol **56-1**, was subject to tosylation, azidation and reduction steps, to give intermediate **56-4**, on which the thymine moiety was built by conventional methods.

Kitagawa and co-workers reported the synthesis of (-)-aristeromycin from D-glucose in 21 steps (Scheme 57).[142,143] Configuration at C-3 of protected glucose **57-1** was inverted by Swern oxidation/$NaBH_4$ reduction, followed by benzylation to give fully protected derivative **57-2**. Further chemical manipulation gave ketone **57-4**, which was reacted with nitromethane to give isomeric nitrofuranose derivatives **57-5**. Acetylation-deacetoxyhydrogenation, followed by deprotection of the isopropylidene group gave diols **57-7**, which were oxidized to dialdehydes **57-8**. Treatment of **57-8** with potassium fluoride/18-Crown-6 in DMF resulted in an intramolecular condensation to give a mixture of cyclization products **57-9**. Dehydration of **57-9** resulted in only two epimers, a pseudo-L-lyxofuranose **57-10** and a pseudo-D-ribofuranose **57-11**, which were separated chromatographically. In the following step, an interesting Michael-type addition allowed highly stereoselective introduction of the purine moiety to give the β-nucleoside analog **57-12**. Removal of formyl, nitro and protecting groups afforded the target compound **57-14**.

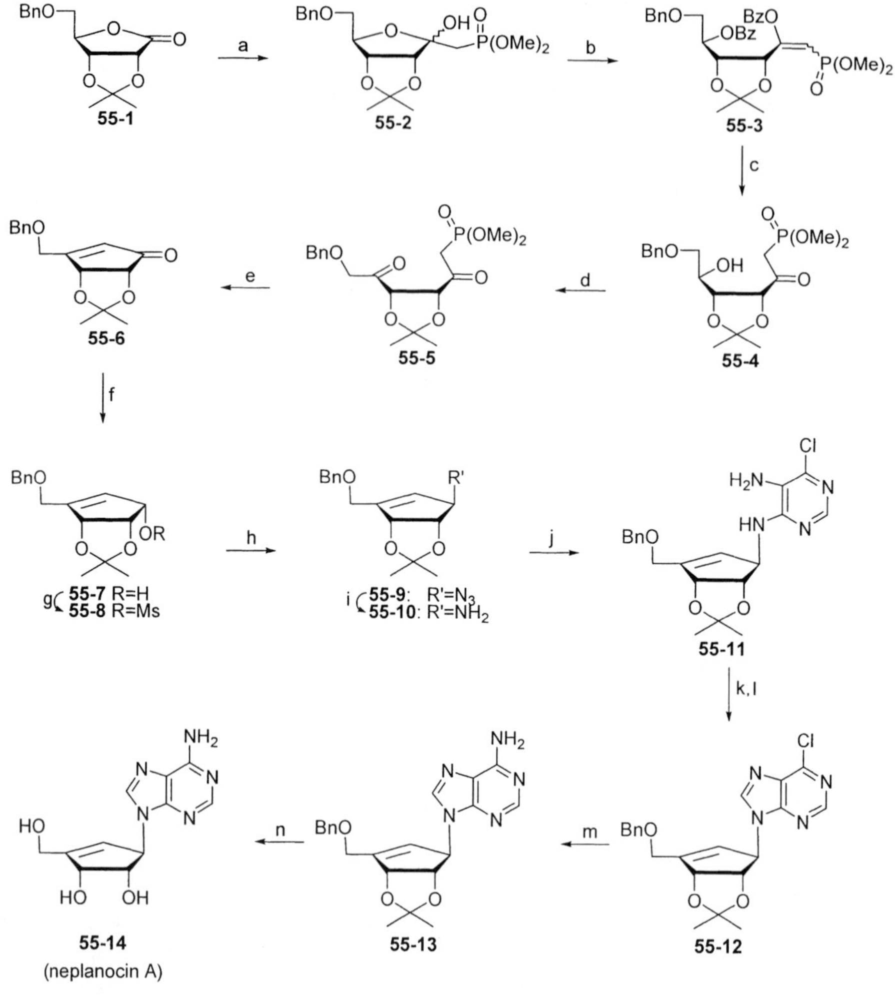

Reagents: a) $LiCH_2P(O)(OMe_2)$, THF; b) BzCl, Py; c) NaOMe, MeOH; d) $CrO_3{\cdot}2Py$, CH_2Cl_2; e) K_2CO_3/18-Crown-6, benzene; f) $NaBH_4/CeCl_3{\cdot}7H_2O$, MeOH; g) $MsCl/Et_3N$, CH_2Cl_2; h) LiN_3, DMSO/HMPA; i) $HS(CH_2)_3SH/Et_3N$, MeOH; j) 5-amino-4,6-dichloropyrimidine, *n*-BuOH; k) $HC(OEt)_3$, Ac_2O; l) HCl; m) NH_3, MeOH; n) BCl_3, CH_2Cl_2,.

Scheme 55

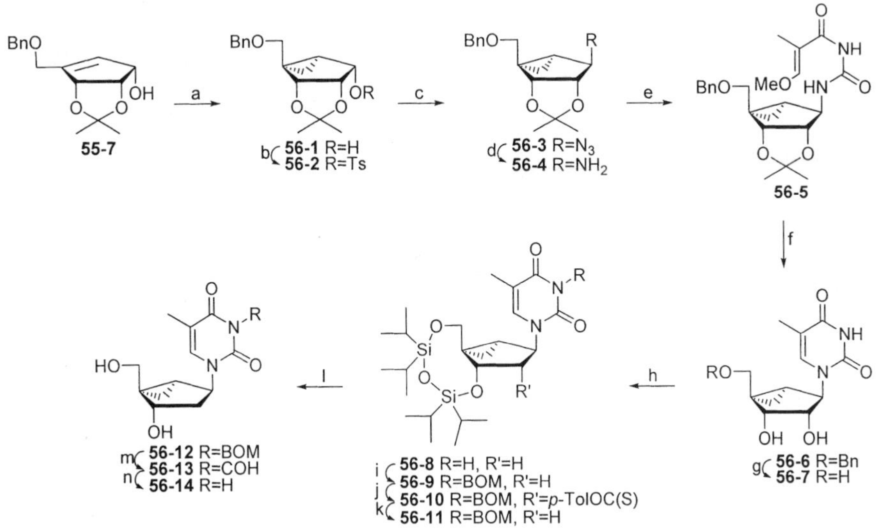

Reagents: a) Zn-Cu/CH_2I_2, Et_2O; b) TsCl/DMAP/Et_3N, CH_2Cl_2; c) NaN_3, DMF; d) H_2-Lindlar catalist; e) $CH_3OCH=C(CH_3)CONCO$, CH_2Cl_2; f) 0.2 N HCl, EtOH/H_2O 9:1; g) H_2/10%Pd-C, AcOEt/ MeOH 1:1; h) $TIPSiCl_2$/imidazole, DMF; i) BOMCl/DBU, CH_3CN; j) *p*-TolOC(S)Cl/DMAP/Et_3N, CH_2Cl_2; k) Bu_3SnH/AIBN, DME; l) TBAF, THF; m) H_2/10% Pd-C; n) NaOMe, MeOH.

Scheme 56

A shorter way to obtain aristeromycin has been described by Borchardt and co-workers,[144,145,146] who synthesized the target compound in 9 steps starting from D-ribono-lactone **58-1** (Scheme 58). Protection of **58-1**, followed by oxidation-deformylation and protection of the resulting L-erythruronolactone derivative **58-2**, afforded intermediate **58-3**. This was reacted with lithium dimethyl methylphosphonate to give cyclopente-none **58-4** *via* a ring opening-closure mechanism (Scheme 59). Conjugated addition of lithium di-(*tert*-butoxymethylene)cuprate to **58-4** gave cyclopentanone **58-5**, which was stereoselectively reduced to the corresponding α-alcohol. This was converted to triflate **58-7**, intermediate for the condensation with adenine. Acidic hydrolysis of the protecting groups afforded aristeromycin **57-15**.

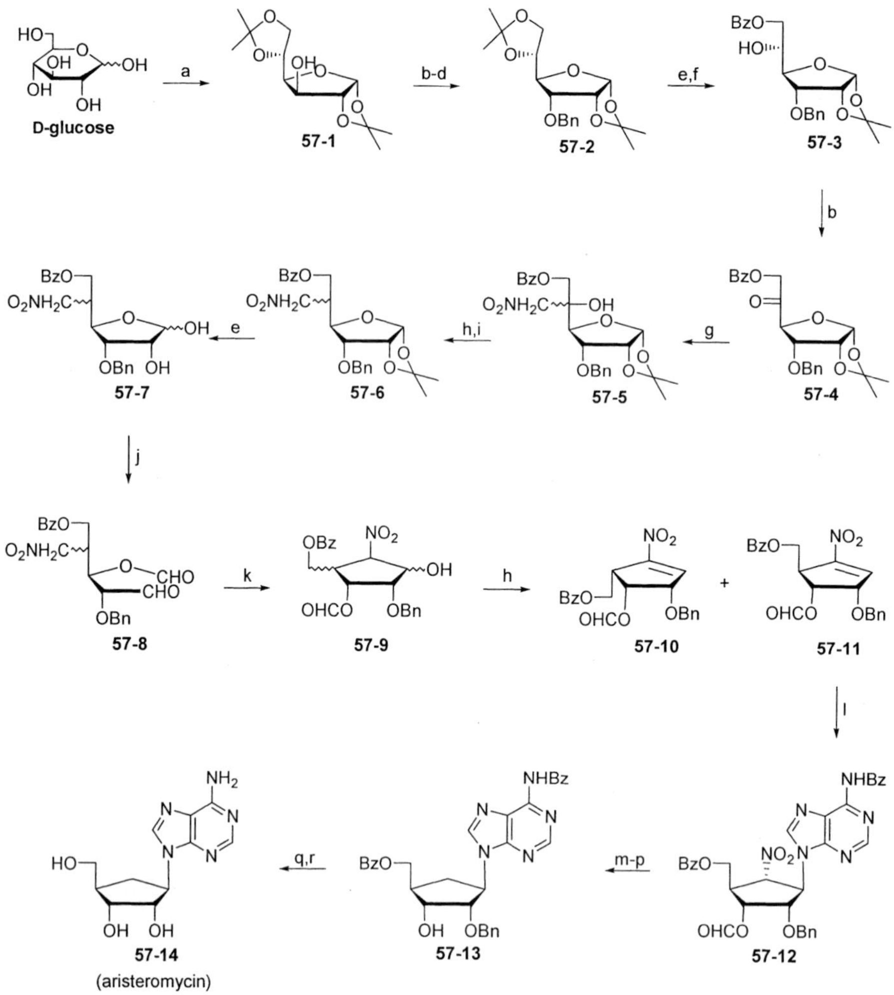

Reagents: a) H_2SO_4, acetone; b) $(COCl)_2$/DMSO/Et_3N, CH_2Cl_2; c) $NaBH_4$, 95% aq. EtOH; d) BnCl/NaH, DMF; e) 80% aq. AcOH; f) BzCl/Py, CH_2Cl_2; g) CH_3NO_2/KF/18-Crown-6, DMF; h) Ac_2O, *p*-TsOH·H_2O; i) $NaBH_4$, 95% aq. EtOH; j) $Pb(OAc)_4$, benzene; k) KF/18-Crown-6, DMF; l) N^6-benzoyladenine/KF/18-Crown-6, THF; m) 28% aq. NH_4OH, 95% aq. EtOH; n) ethyl vinyl ether/CSA, CH_2Cl_2; o) n-Bu_3SnH/AIBN, toluene; p) 10% aq. AcOH, acetone; q) 5% NaOMe, MeOH; r) Na/liq. NH_3, THF.

Scheme 57

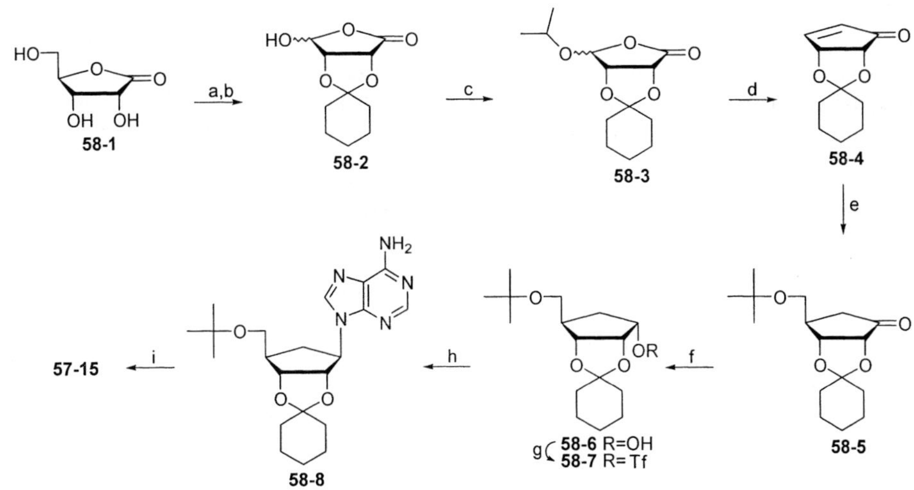

Reagents: a) Cyclohexanone, $FeCl_3$; b) $NaIO_4$, $NaOH/H_2O$; c) PPTS, 2-propanol; d) $CH_3PO(OCH_3)_2$/*n*-BuLi, THF; e) $[(CH_3)_3COCH_2]CuLi$; f) DIBAL-H, CH_2Cl_2; g) $(CF_3SO_2)_2O$/Py, CH_2Cl_2; h) adenine/NaH/18-Crown-6, DMF; i) CF_3COOH, H_2O.

Scheme 58

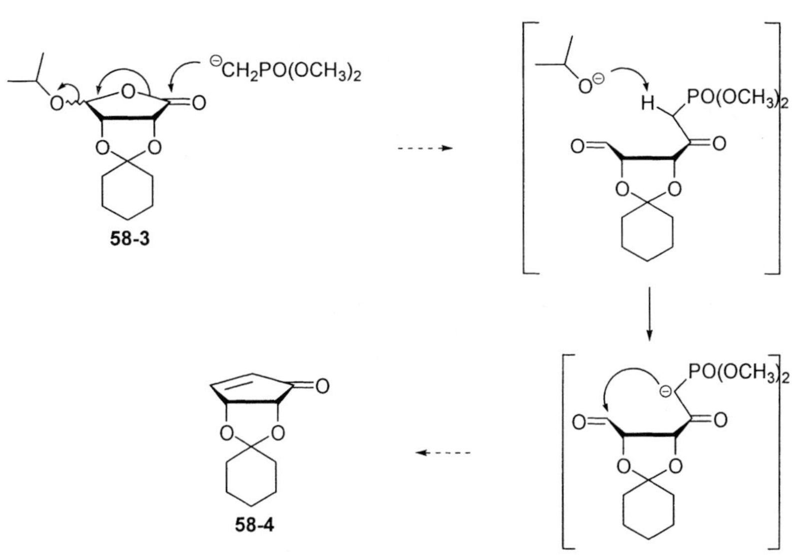

Scheme 59

Enone **58-4** is an important intermediate which had previously been used in a synthesis of nor-C-5 analogs of neplanocin A **60-5** and **60-6** (Scheme 60).[145] In that case, **60-4** had been stereoselectively reduced to alcohol **60-1**, which had been tosylated and coupled to adenine or 3-deazaadenine.

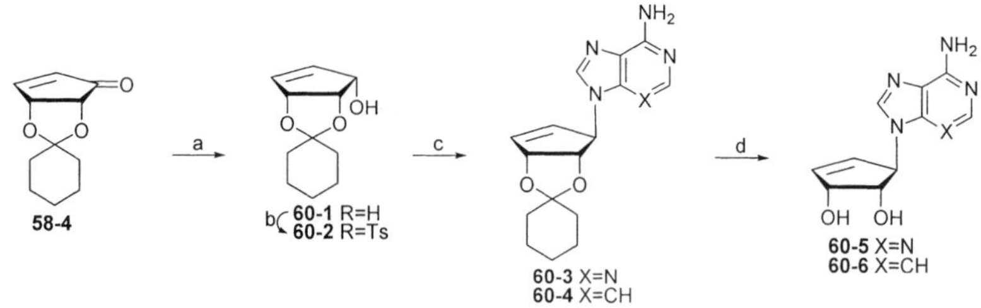

Reagents: a) $NaBH_4/CeCl_3 \cdot 7H_2O$, MeOH; b) $TsCl/Et_3N$, CH_2Cl_2; c) adenine or 3-deazaadenine/NaH, DMF; d) dilute HCl.

Scheme 60

The enantiomer of **58-4** has been prepared starting from D-mannose **61-1** (Scheme 61),[145] which was protected as the bis-cyclohexylidene and oxidized to **61-2**. Selective deprotection and oxidative cleavage of the resulting diol gave **61-3**, which, after protection and ring opening-closure, afforded the enantiomeric enone **61-5**.

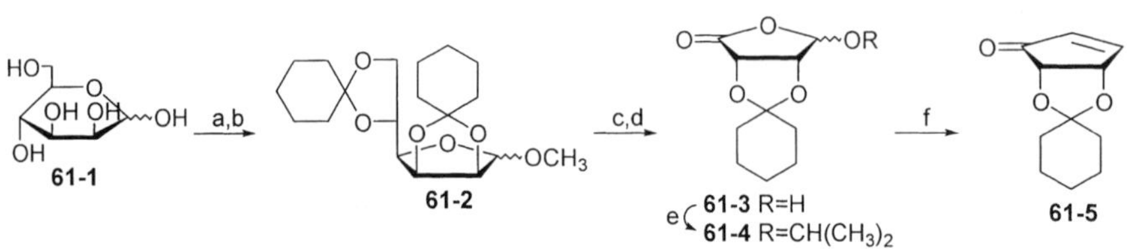

Reagents: a) Cyclohexanone, H_2SO_4; b) Collins reagent; c) Dowex 50W, H_2O/EtOH; d) $NaIO_4$, $NaOH/H_2O$; e) PPTS, 2-propanol; f) $CH_3PO(OCH_3)_2$/*n*-BuLi, THF.

Scheme 61

Enone **62-5**, analog of **61-5**, was prepared by Bélanger and Prasit[147] starting from D-ribonolactone **58-1** (Scheme 62). This was fully protected to tosylate **62-2**, which was converted to the *exo*-alkene **62-3** *via* an intermediate iodide. Treatment with lithium tri-*tert*-butoxyaluminum hydride afforded hydroxycyclopentanone **62-4**, from which the target molecule **62-5** was obtained by mesylation/elimination.

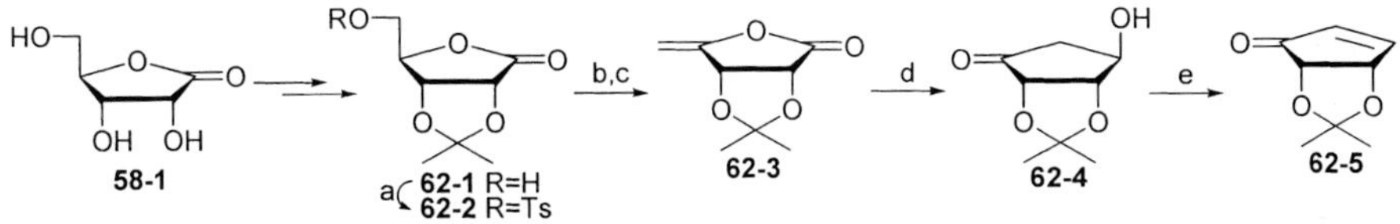

Reagents: a) TsCl/Py, CH_2Cl_2; b) NaI, acetone; c) DBU, benzene; d) $LiAlH(Ot\text{-}Bu)_3$, THF; e) MsCl/Py, CH_2Cl_2.

Scheme 62

Enone **62-5** was also synthesized by Borchardt and co-workers through an efficient 3-step synthesis starting from D-ribose. The same procedure was used for the synthesis of the enantiomer **63-5**, analog of **61-5**, starting from D-lyxose (Scheme 63).[148] The key reaction was the unusual pyridinium chlorochromate (PCC) oxidation-deformylation of **63-1** or **63-3** to lactones **63-2** or **63-4**. Target compounds were obtained by the ring opening-closure reaction discussed above (Scheme 59).

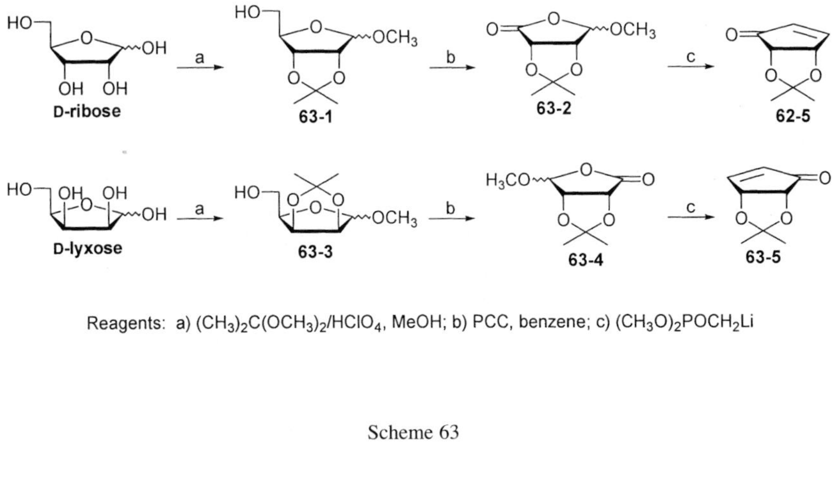

Reagents: a) $(CH_3)_2C(OCH_3)_2/HClO_4$, MeOH; b) PCC, benzene; c) $(CH_3O)_2POCH_2Li$

Scheme 63

Enone **62-5** was used by Chu and co-workers as starting material in the synthesis of (+)-L-aristeromycin and its analogs, enantiomers of the natural products (Scheme 64).[149,150,151,152] Conversion of **62-5** to alcohol **64-2** was accomplished by the conjugated addition of lithium di-(*tert*-butoxymethylene)cuprate[144,145] followed by stereoselective reduction of the carbonyl group. Compound **64-2** was, then, converted to the triflate, which was reacted with the sodium salt of adenine to give, after deprotection, L-aristeromycin **64-5**. The thymidine analog **64-10** was obtained by the same route. In this case, however, because of the poor yield of the condensation step, it was found more convenient to build the heterocyclic ring onto the carbocyclic moiety.

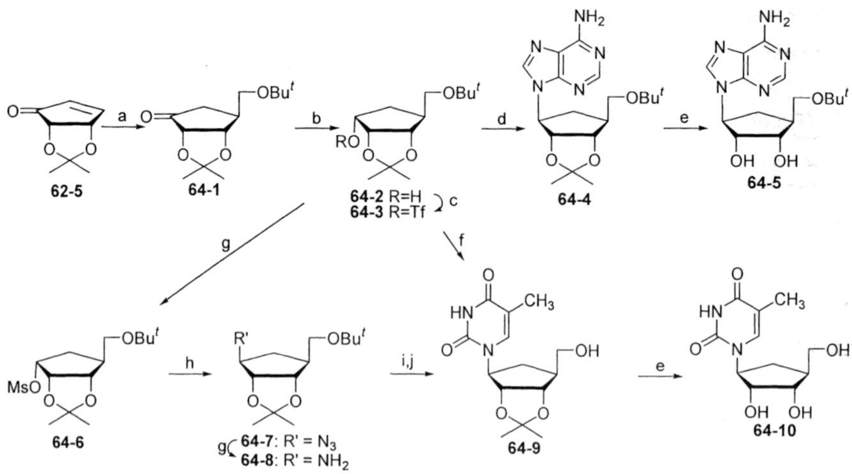

Reagents: a) (t-BuOCH$_2$)CuLi/t-BuOMe, THF; b) DIBAL-H, CH_2Cl_2; c) Tf_2O, Py; d) adenine/NaH/18-crown-6, DMF; e) CF_3CO_2H, H_2O; f) thymine/K_2CO_3/18-crown-6, DMF; g) MsCl/Et_3N, CH_2Cl_2; h) LiN_3, DMF; i) $CH_3OCH{=}C(CH_3)CONCO$, DMF; j) 30% NH_4OH, EtOH.

Scheme 64

Enones **62-5** and **63-5** were used by Chu and co-workers for the synthesis of a complete series of D- and L-neplanocin analogs (Scheme 65).[153] The starting enones **63-5** or **62-5** were reacted with *tert*-butyl methyl ether to give, *via* 1,2-addition, alcohols **65-1** or **65-7**, respectively. These were acetylated and subject to palladium-catalyzed rearrangement to give acetates **65-3** and **65-9**. Deprotection of the acetates gave key intermediates **65-4** and **65-10**. Intermediate **65-4** could also be obtained from enone **62-5**, *via* 1,4-addition of *tert*-butyl methyl ether, selenation of the resulting cyclopentanone **64-1**, oxidative deselenation and Luche reduction of enone **65-6**.

Alcohols **65-4** and **65-10** were reacted with protected bases in Mitsunobu fashion to afford, after derivatization and/or deprotection, L- and D-neplanocin analogs, respectively. Among the synthesized nucleosides, the cytidine and 5-fluorocytidine analogs displayed anti-HIV and anti-West Nile Virus activity, but also severe cellular toxicity.[153]

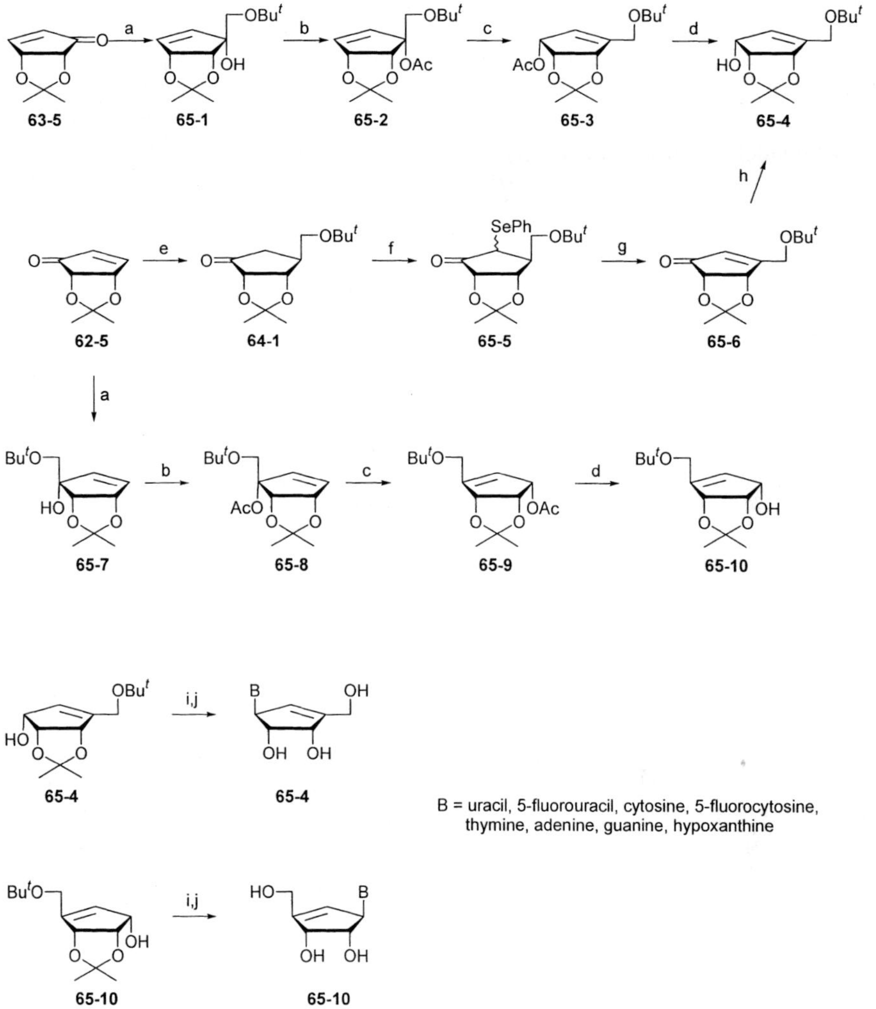

Reagents: a) *t*-BuOMe, *t*-BuOK, *sec*-BuLi, THF; b) Ac_2O, Et_3N, DMAP, CH_2Cl_2; c) $PdCl_2(CH_3CN)_2$, *p*-benzoquinone, THF; d) K_2CO_3, MeOH; e) $(t\text{-}BuOCH_2)_2CuLi$, *t*-BuOMe, THF; f) PhSeBr, LDA, THF; g) H_2O_2, H_2O; h) $CeCl_3{\cdot}7H_2O$, $NaBH_4$, MeOH; i) PPh_3, DEAD, (protected) base; j) derivatization and/or deprotection.

Scheme 65

Another important intermediate in the synthesis of aristeromycin is cyclopentylamine **67-15**. This compound has been synthesized, by Tadano *et al.*,[154,155,156,157] through a 24-steps scheme. The key intermediate **66-8** was prepared starting from D-glucose, which was converted to the D-erythrose derivative **66-2** in 2 steps, according to MacDonald and co-workers[158] (Scheme 66). Full protection of **66-2** afforded intermediate **66-4**, from which the aldehyde group was regenerated in **66-5**. Malonic condensation followed by reduction and deprotection afforded **66-8**.

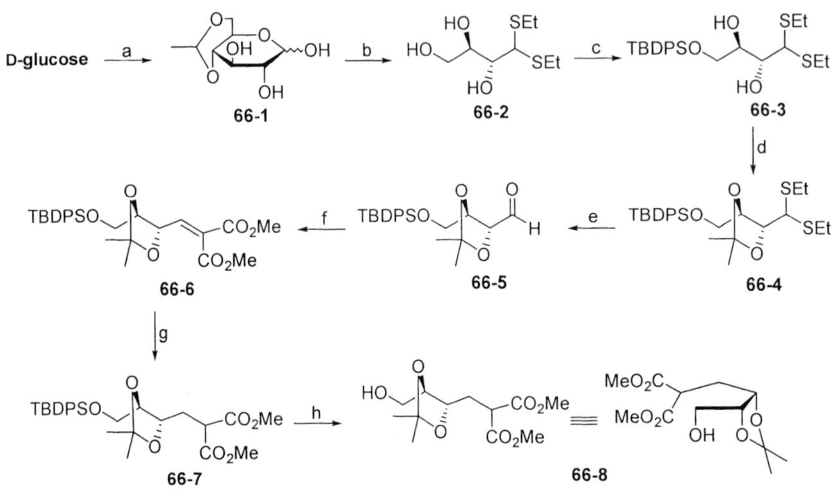

Reagents: a) paraldehyde, conc. H_2SO_4; b) EtSH, H^+; c) TBDPSCl/imidazole, DMF; d) $(CH_3)_2C(OCH_3)_2$/TsOH, acetone; e) $HgCl_2/CaCO_3$, CH_3CN/H_2O; f) $(CH_3O_2C)CH_2/Ac_2O$, Py; g) $NaBH_4$, MeOH; h) TBAF, THF.

Scheme 66

PCC oxidation of **66-8** produced an aldehyde that spontaneously underwent intramolecular aldol condensation to give isomeric alcohols **67-1** (Scheme 67). These were acetylated to **67-2**, which were thermally decarbomethoxylated to the α,β-unsaturated ester **67-3**. Reduction, oxidative hydroboration, and acetylation gave diacetate **67-5** along with a small amount of isomer **67-6** from which **67-5** was purified chromatographycally; subsequent deprotection, monoprotection as silyl ether, and oxidation gave ketone **67-8**, which was smoothly epimerized to **67-9** by silica gel catalysis. Stereoselective reduction of the ketone, followed by deprotection-protection and mesylation gave intermediate **67-12**, which was reacted with sodium azide in S_N2 fashion to afford, after deprotection steps and reduction of the azido group, target compound **67-15**.

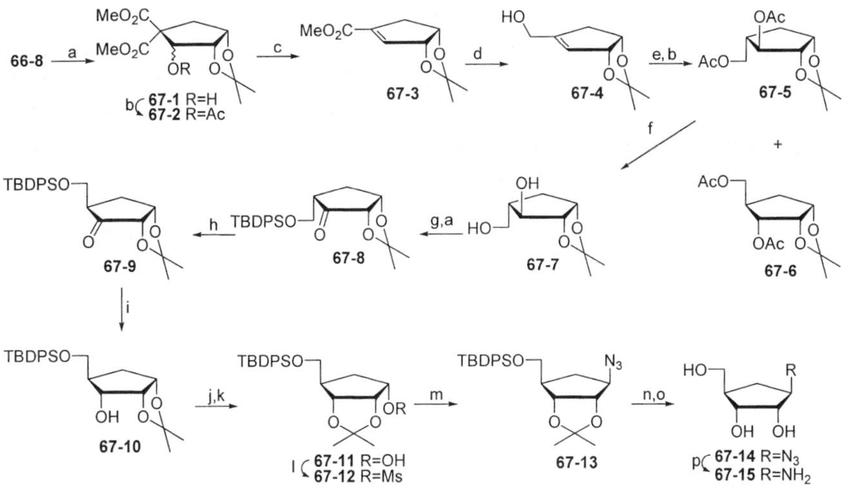

Reagents: a) PCC/4Å mol. sieves, CH_2Cl_2; b) Ac_2O, Py; c) NaCl, H_2O/DMSO; d) DIBAL-H, Tol; e) $BH_3 \cdot THF$; f) MeONa, MeOH; g) TBDPSCl/ imidazole, DMF; h) silica gel, CH_2Cl_2; i) $NaBH_4$, MeOH; j) AcOH, H_2O, MeOH 20:4:1; k) $(CH_3)_2C(OCH_3)_2$/CSA,DMF; l) MsCl, Py; m) NaN_3, DMF; n) TBAF, THF; o) 80% AcOH; p) H_2/Raney Ni, MeOH.

Scheme 67

Vandewalle and co-workers reported a total synthesis of (-)-neplanocin A from L-ribulose (Scheme 68).[159,160] An interesting trans-orthoesterification with concomitant methylation of the 5-OH produced only the β-epimer **68-2**. This was converted to the versatile intermediate **68-5** through a series of protection-deprotection steps. Olefination of the crypto-carbonyl function, followed by Swern oxidation of the originated primary alcohol, led to the key intermediate **68-7**. This was treated with benzylhydroxylamine to produce the intermediate nitrone which, upon heating, cyclized to bicyclic isoxazolidine **68-8** in 1,3-dipolar cycloaddition mode (Scheme 69). Reductive cleavage of the N-O bond in **68-8** followed by catalytic hydrogenation gave aminoalcohol **68-10**, on which the adenine moiety was built by standard reactions to afford (-)-neplanocin A **55-14**.

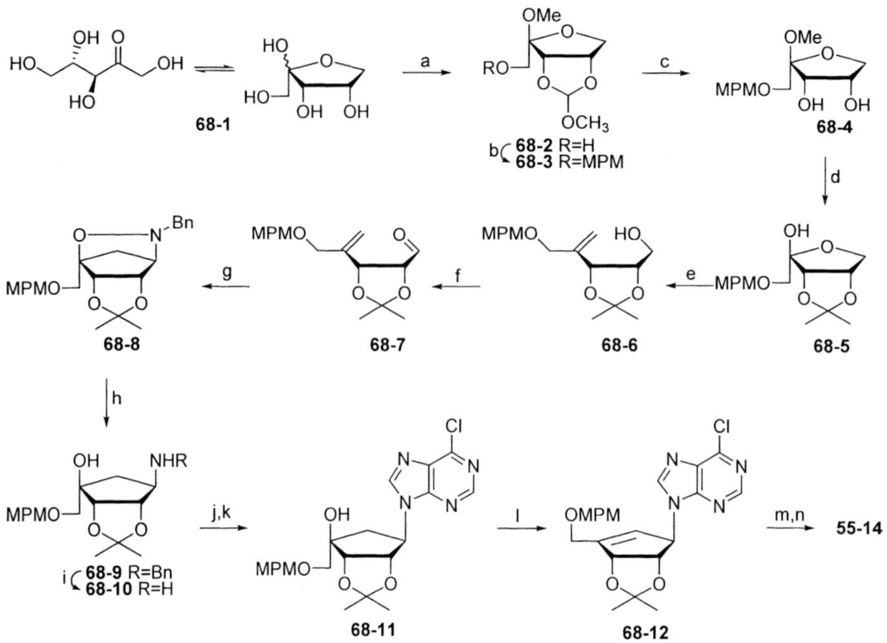

Reagents: a) $HC(OMe)_3$/TsOH, CH_3OH; b) [MPM]; c) 10% HCl, THF; d) $(CH_3)_2C(OMe)_2$, TsOH; e) $Ph_3P{=}CH_2$/12-Crown-4; f) DMSO/$(COCl)_2$, CH_2Cl_2, then Et_3N; g) BnNHOH, Tol; h) Zn/AcOH, Et_2O; i) H_2/Pd-C (10%)/AcOH, EtOAc; j) 5-amino-4,6-dichloropyrimidine/Et_3N, *n*-BuOH; k) $HC(EtO)_3$, TsOH; l) $POCl_3$/4-DMAP, CH_2Cl_2; m) liq. NH_3; n) BCl_3, CH_2Cl_2.

Scheme 68

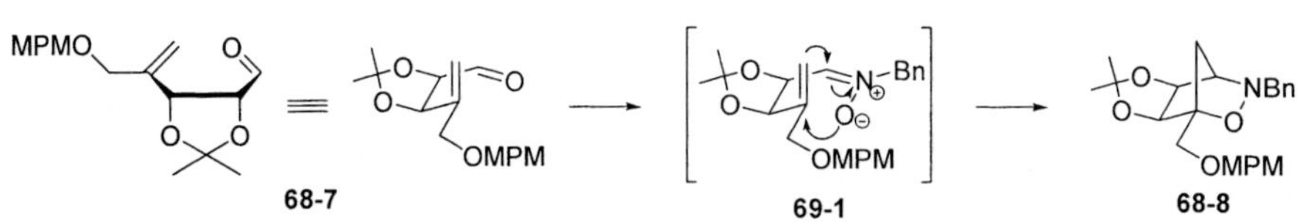

Scheme 69

The Dieckmann cyclization is a tool that cannot be neglected in the synthesis of cyclopentyl rings. In fact, Rapoport and co-workers[161,162] devised a synthesis of carbovir precursor **70-26** based on the well-known reaction (Scheme 70). Diester **70-14**, the intermediate for the cyclization step, was synthesized through a 13-step sequence starting from D-glucono-δ-lactone **70-1**. The most notable of this steps is the protection of amine **70-5**, synthesized in four steps from **70-1**, with a *N*-(9-phenylfluoren-9-yl) group (Pf in Scheme 70), the favorable stereoelectronic characteristics of which drive more than one step through the desired regio- and stereoselectivity. Thus, derivative **70-6** was deprotected, oxidized and methylated to diester **70-9**, which was deoxygenated at the 4- and 5-hydroxylfunctionalities in four steps to give the key intermediate **70-14**. Upon basic treatment, **70-14** cyclized to cyclopentanone **70-15** with very high regioselectivity (12:1 diastereomeric ratio). Separation of the desired diastereomer was achieved after stereoselective reduction of the carbonyl group, to give alcohol **70-16**. The latter was deoxygenated through a three-step sequence to ester **70-19**, which was converted to the unsaturated compound **70-24** in four more steps. Final deprotections afforded **70-26** as the trifluoroacetate.

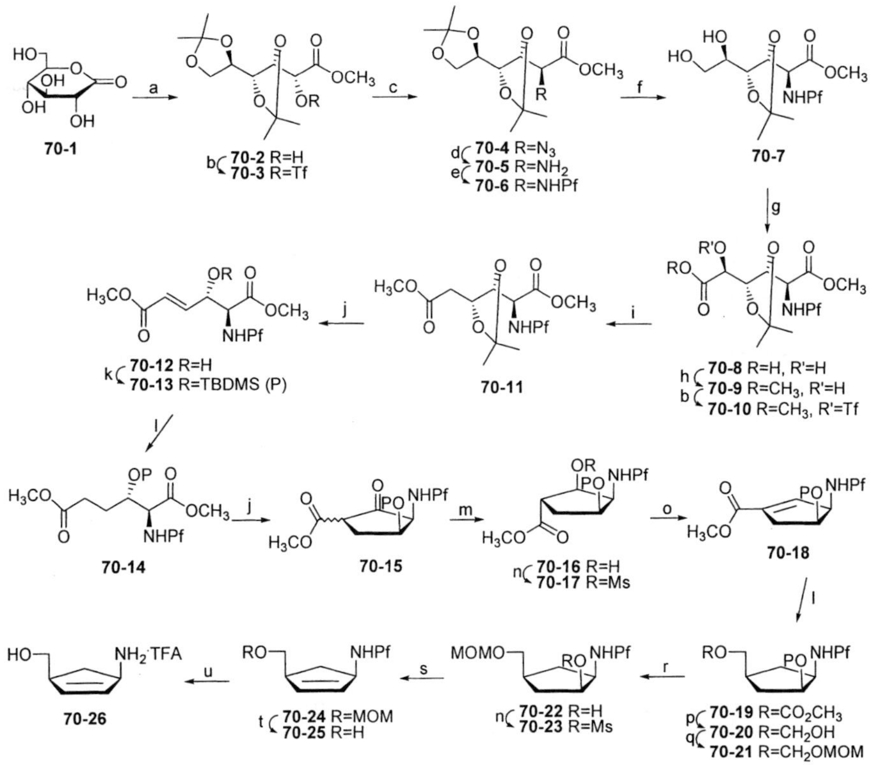

Reagents: a) $CH_3CH(OCH_3)_2$/acetone/TsOH·H_2O, MeOH; b) Tf_2O/Py, CH_2Cl_2; c) Bu_4NN_3, CH_3CN; d) H_2-Pd/C, EtOAc; e) PfBr/$Pb(NO_3)_2$/Et_3N, CH_2Cl_2 ; f) HCl/Et_2O, H_2O; g) O_2-Pt, *i*-PrOH/H_2O/EtOAc; h) CH_3I/K_2CO_3, DMF; i) HAc/LiI·$3H_2O$, THF; j) KHMDS, THF; k) TBDMSCl/imidazole, DMF; l) H_2-5%Pt/C, EtOAc; m) $NaBH_4$, *i*-PrOH, then separation from diastereomer; n) MsCl/Et_3N, THF; o) NaH, THF; p) $LiAlH_4$,THF; q) MOMCl/(*i*-Pr)$_2$NH, CH_2Cl_2; r) TBAF, THF; s) NaH/$CH_3O(CH_2)_2OH$, DMF; t) PPTS, *t*-BuOH; u) TFA, CH_2Cl_2.

Scheme 70

The *arabino* analog of aristeromycin, (+)-cyclaradine **71-15** was prepared by Murakami and co-workers starting from D-arabinose (Scheme 71).[163] Full protection/mono-deprotection of the starting material led to alcohol **71-4**, which underwent the Swern oxidation to ketone **71-5**. Nitromethane addition produced the alcohol **71-6**, which was acetylated and reduced to acetal **71-8**, the deprotection of which regenerated the aldehyde group. An aldol-type ring closure catalyzed by cesium fluoride gave an epimeric mixture of nitrocyclopentanes **71-10**. Tosylation and elimination afforded the α,β-unsaturated nitro-derivative **71-11**, with which N^6-benzoyladenine reacted in a Michael type addition to give adduct **71-12**. This was denitrohydrogenated with tributyltin hydride and 2,2'-azobisisobutyronitrile (AIBN), and deprotected in two steps to (+)-cyclaradine. Reaction of **71-11** with persilylated uracil, followed by analogous reductive and hydrolytic steps, afforded the synthesis of the uridine analog.

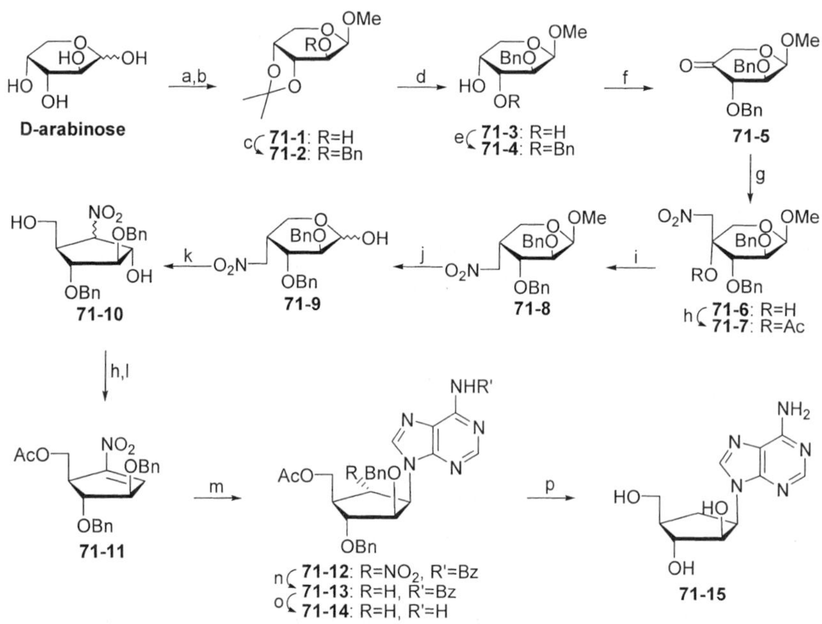

Reagents: a) HCl, MeOH; b) $(CH_3)_2C(OCH_3)_2$/TsOH, DMF; c) BnCl/NaH, DMF; d) 80% AcOH; e) Bu_2SnO, Tol, then BnBr, CsF, DMF; f) DMSO, $(COCl)_2$, CH_2Cl_2, then Et_3N; g) CH_3NO_2, KF, 18-crown-6, DMF; h) TsOH, Ac_2O; i) $NaBH_4$, EtOH; j) conc. HCl, AcOH; k) CsF, DMF; l) Py; m) N^6-Bz-adenine, CsF, DMF; n) Bu_3SnH, AIBN, Tol; o) MeONa, MeOH; p) H_2-Pd black, 10% AcOH, EtOH.

Scheme 71

Another interesting scheme leading to (-)-neplanocin A is based on the C-H insertion of an alkylidenecarbene (Scheme 72).[164] Ketone **72-4** was synthesized in three steps from protected ribose **72-1**. Treatment of **72-4** with lithium trimethylsilyldiazomethane generated the reactive alkylidenecarbene **72-5**, which evolved into cyclopentene derivative **72-6**. Further elaboration to **55-14** was accomplished by conventional methods.

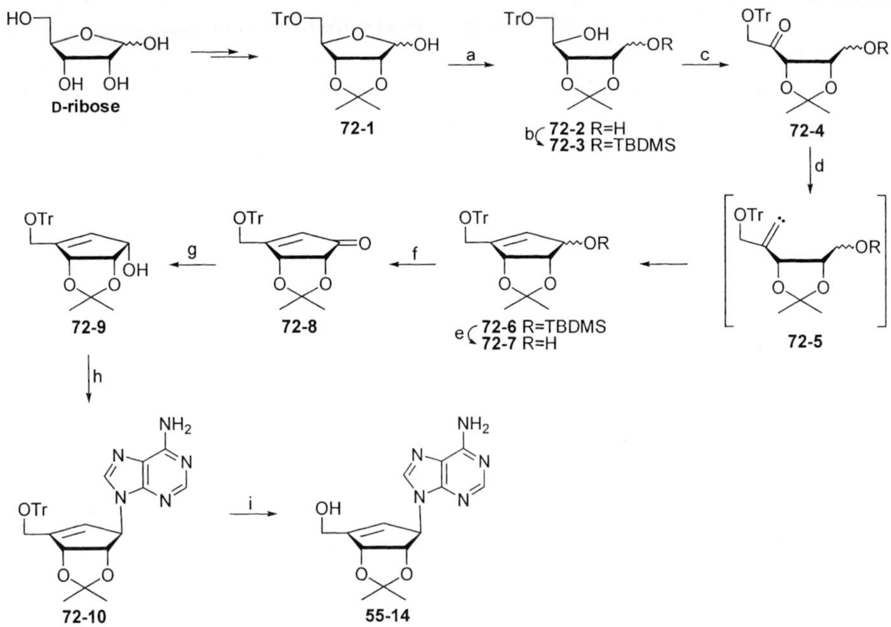

Reagents: a) $LiAlH_4$, Et_2O; b) TBDMSCl/imidazole, DMF; c) DMSO/$(COCl)_2$, then Et_3N; d) $TMSC(Li)N_2$, THF; e) TBAF, THF; f) PDC, CH_2Cl_2; g) $LiAlH_4$, THF; h) adenine/DEAD/Ph_3P, THF; i) HCl, MeOH.

Scheme 72

Chu and co-workers described a divergent synthesis of carbocyclic 4'-fluoro-2',3'-dideoxynucleosides.[165,166] Intermediate **73-1**, prepared in the same way as its enantiomer **45-3**, (Scheme 45), was reduced to **73-2**, common intermediate of the divergent scheme (Scheme 73). In the synthesis of D-nucleosides, **73-2** was oxidized to aldehyde **73-3**, which was reacted with vinylmagnesium bromide or vinyllithium to give alcohol **73-4**. This was cyclized *via* ruthenium-catalyzed ring-closing metathesis, using Grubb's catalyst. The unsaturated cyclization product **73-5**, obtained as an epimeric mixture, was hydrogenated to give key intermediate **73-6**, which could be separated chromatographycally from its epimer. Condensation of the key intermediate with several nucleobases in Mitsunobu conditions, followed by deprotection reactions, afforded D-nucleoside analogs **73-7**.

In order to obtain L-analogs, the common intermediate **73-2** was benzylated to the fully protected triol **73-8**, which was subject to ozonolysis to give aldehyde **73-9**. This was reacted in Horner-Emmons-Wadsworth fashion to give α,β-unsaturated ester **73-10**. Saturation of the double bond and debenzylation were accomplished in one step by catalytic hydrogenation. The resulting alcohol **73-11** was mesylated to **73-12**, whose treatment with sodium hydride in THF allowed an intramolecular nucleophilic substitution to give epimeric cyclopentyl carboxylic esters **73-13**. Saponification of the latter, followed by radical oxidative iododecarboxylation, gave iodides **73-15**. Surprisingly, hydrolysis of the epimeric iodides (ratio 1:1) gave only α-cyclopentanol **73-16** in 40% yield, probably because the α-iodide was not reactive under the hydrolytic conditions. Alcohol **73-16** was the key intermediate for the synthesis of L-nucleosides. Its reaction with 6-chloropurine in Mitsunobu conditions, followed by ammonolysis and deprotection, afforded L-4'-fluoro-2',3'-dideoxyadenosine **73-17**.

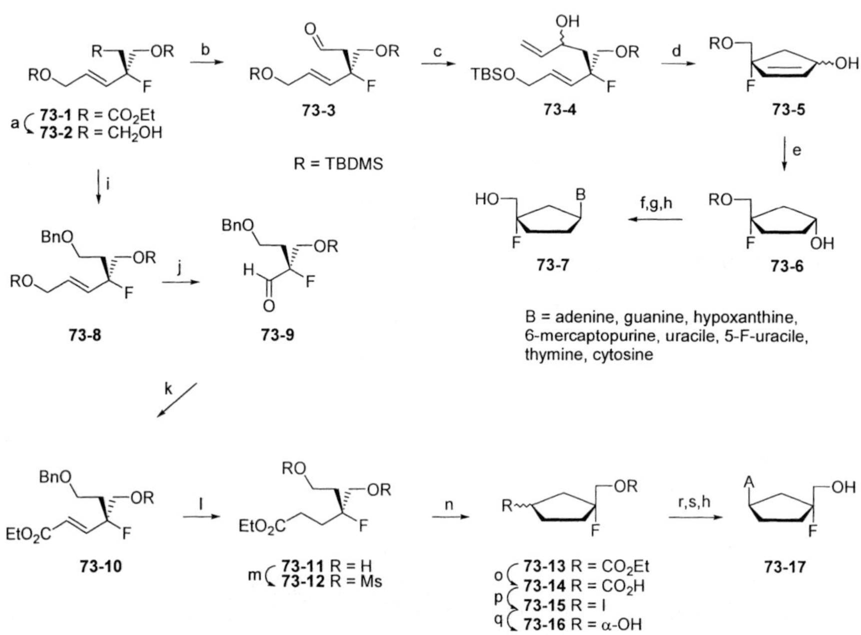

Reagents: a) LAH, THF; b) PCC/4Å mol. sieves, CH_2Cl_2; c) CH_2=CHMgBr or CH_2=CHLi, THF; d) Grubb's catalyst, CH_2Cl_2; e) H_2, 10% Pd/C, cyclohexane, rt, 6 h, then chromatographic separation; f) (protected) base, [PPh_3/DEAD], THF; g) base modification/deprotection; h) TBAF, THF; i) NaH, THF, then BnBr, TBAI; j) O_3, MeOH, then Me_2S; k) [$(EtO)_2P(O)CH_2CO_2Et$/NaHMDS], THF; l) H_2, 10% Pd/C, cyclohexane; m) MsCl, Py, CH_2Cl_2; n) NaH, THF; o) NaOH/H_2O, EtOH; p) $Pb(OAc)_4$, CCl_4, hν, then I_2, CCl_4, hν; q) $NaHCO_3$, 15% (v/v) water/HMPA; r) 6-Cl-purine, [PPh_3/DEAD], THF; s) NH_3/MeOH.

Scheme 73

2.6. Cyclopropyl carbocyclic nucleosides

Structure-activity relationships studies of acyclic nucleosides showed that, if the side chain could be frozen into the optimal conformation for enzyme interaction, a superior biological activity could result. The first asymmetric synthesis of optically pure (1'S, 2'R)-cyclopropyl carbocyclic nucleosides has been first accomplished by Chu and co-workers[167,168,169,170,171,172] (Scheme 74). Protected L-glyceraldehyde **9-12** was subject to Wittig olefination, followed by reduction to an alcohol, which was protected with *tert*-butyltrimethylsilyl bromide to give **74-2**. Protected derivative **74-2** was treated with $Zn(Et)_2/ICH_2Cl$ at 0□C to give desired cyclopropyl derivative **74-3**. Deprotection of this cyclopropyl sugar gave the alcohol **74-4**, which was treated with $RuO_2/NaIO_4$ to give the acid, whose treatment with chloroethylformate followed by sodium azide afforded compound **74-5**. This was subject to Curtius rearrangement conditions to give the isocyanate intermediate, which was treated with ammonia to afford the urea derivative **74-6**. The urea derivative was then reacted with β-methoxymethacryloyl or β-methoxyacryloyl chloride in pyridine to provide an intermediate that was cyclized to thymine **74-7** or uracil **74-8** derivatives by treatment with ammonium hydroxide in ethanol. The isopropylidene group was removed by acidic hydrolysis to give the diols **74-9** and **74-10**. The diol nucleosides were oxidized to aldehydes, which were directly reduced to the desired nucleosides **74-11** and **74-12**. The synthesis of cytosine derivative **74-13** was accomplished by treatment of **74-12** with 1,2,4-triazole and chlorophenyl phoshorodichloridate. Subsequent hydrolysis with ammonium hydroxide afforded the cytidine derivative. L-Cyclopropyl carbocyclic nucleosides **74-14~74-16** were synthesized by following the same procedure starting from protected D-glyceraldehyde **9-2**.

The purine nucleosides were synthesized from the key intermediate cyclopropylamine **75-1**, obtained by hydrolysis of **74-6**. Coupling reaction with 4,6-dichloroformamidopyridine in the presence of triethylamine provided **75-2** (Scheme 75). Formation of the imidazole ring by heating of **75-2** in diethoxymethylacetate gave 6-chloropurine derivative **75-3**. The adenine **75-8** and hypoxanthine **75-9** derivatives were obtained by known procedures from 6-chloropurine. The guanine derivative **75-13** was prepared by coupling cyclopropylamine **75-1** with 2-amino-4,6-dichloropyrimidine in the presence of triethylamine to give **75-10**. The amino derivative **75-11** was treated with triethylorthoformate in the presence of conc HCl to afford the 6-chloropurine derivative, which was hydrolyzed by 2N HCl to provide guanosine analog **75-12**. Cleavage of the diol in **75-12** gave cyclopropyl nucleoside **75-13**. Likewise, enantiomers **75-15** to **75-17** were synthesized starting from **75-14**, derived from protected D-glyceraldehyde **9-2**.

D- and L-2'-hydroxyethylcyclopropylmethyl nucleosides **76-4** to **76-6** and **76-8** to **76-10** were synthesized from intermediates **9-12** and **9-2**, respectively (Scheme 76). In the synthesis of the D-series, deprotection of **74-3** followed by oxidative cleavage of the resulting diol gave an aldehyde that was treated with methyltriphenyl phosphonium bromide to obtain the olefin **76-1**. This was readily converted to the hydroxyethyl group by hydroboration and the resulting primary hydroxylgroup was protected with the methoxymethyl (MOM) group. Deprotection of the silyl group provided the key intermediate **76-2**. Oxidation of the alcohol by sodium periodate utilizing ruthenium oxide as a catalyst gave an acid which was readily converted to cyclopropylurea **76-3** by Curtius rearrangement. The thymine derivative **76-4** was obtained by reacting the urea derivative **76-3** with β-methoxy-α-methylacryloyl chloride followed by deprotection of the MOM group with conc HCl in methanol. Treatment of the urea with β-methoxyacryloyl chloride gave the protected uracil derivative. Deprotection of the MOM group afforded the desired uridine analog **76-5**. The cytosine derivative **76-6** was obtained from the protected uracil derivative by treatment with chlorophenyl phoshorodichloridate and 3-nitro-1,2,4-triazole to obtain the 4-(3-nitro)triazolide. This was treated with ammonium hydroxide followed by the deprotection of the MOM group to give compound **76-6.**

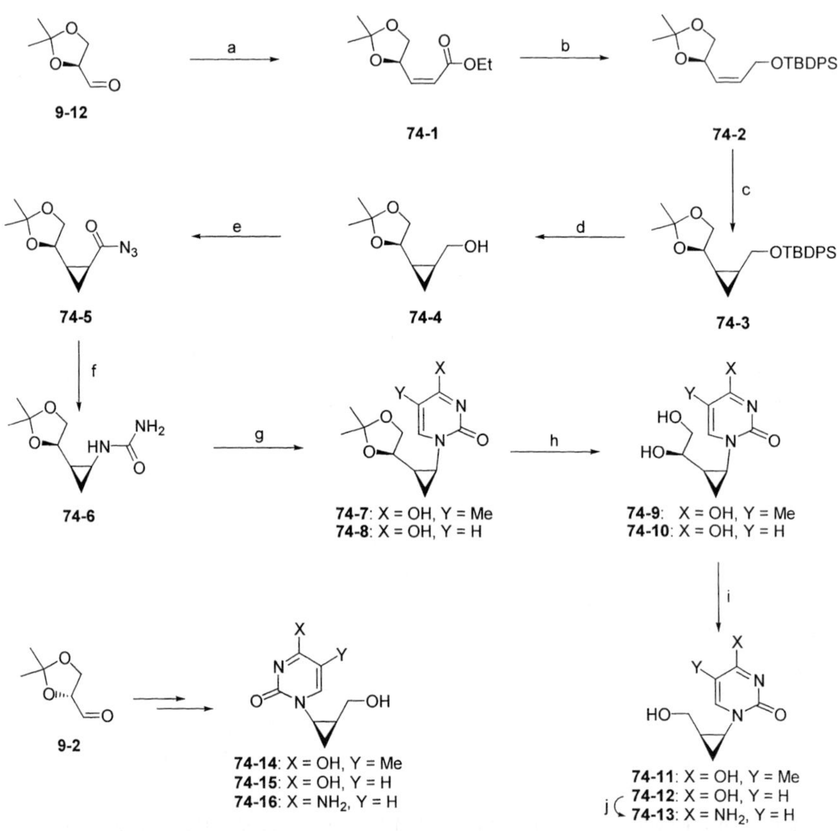

Reagents: a) Ph_3=CHCO$_2$Et, EtOAc; b) (1) DIBAL-H, -78 °C (2) TBDPSCl, Py.; c) Zn(Et)$_2$, CH_2Cl_2; d) TBAF, THF; e) (1) RuO_2/$NaIO_4$, (2) ClCO$_2$Et, Et$_3$N, (3) NaN$_3$; f) NH$_4$OH, 100 °C; g) CH$_3$OCH=C(CH$_3$)COCl or CH$_3$OCH=CHCOCl, Py, then NH$_4$OH, MeOH; h) 80% AcOH; i) NaIO$_4$, NaBH$_4$, MeOH; j) (1) 1,2,4-triazole, Cl$_2$P(O)OC$_6$H$_4$Cl, (2) NH$_4$OH.

Scheme 74

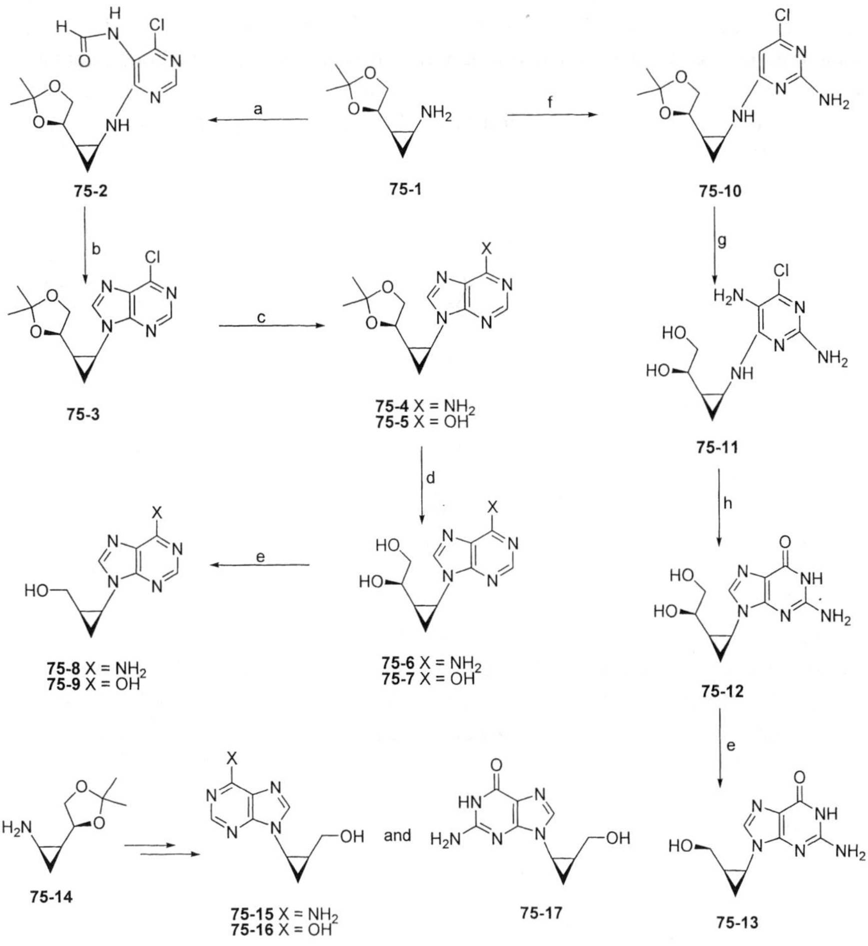

Reagents : a) 4,6-dichloro-5-formamidopyrimidine, Et_3N; b) diethoxymethylacetate, 120°C, NH_4OH; c) NH_3, 90°C, $HSCH_2CH_2OH$, MeONa; d) 80% HOAc; e) $NaIO_4$, $NaBH_4$, f) 4,6-dichloro-2-aminopyrimidine, Et_3N; g) HCl, *p*-$ClC_6H_4N_2Cl$, Zn/AcOH; h) $CH(OCH_3)_3$, HCl, aq.HCl.

Scheme 75

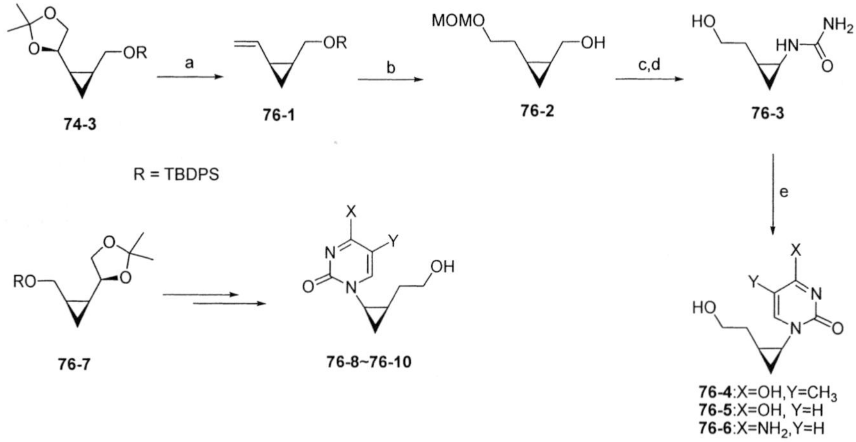

Reagents: a) 80% AcOH, then $NaIO_4$, then $Ph_3P=CH_2$; b) BH_3, then H_2O_2, then MOMCl, then TBAF; c) $RuO_2/NaIO_4$, then $ClCO_2Et$, Et_3N, then NaN_3; d) Tol, BnOH or Tol, NH_3, 100 °C; e) base construction.

Scheme 76

Recently, cyclopropane nucleosides with a methylene spacer between the cyclopropane ring and the heterocyclic moiety were also prepared[173] (Scheme 77). Mesylation of cyclopropyl methyl alcohol **74-4** with methanesulfonylchloride in the presence of triethylamine afforded the *O*-mesyl derivative **77-1**. This was not stable enough to be purified, thus it was treated directly with a base in the presence of K_2CO_3 and 18-crown-6 in DMF at 110°C to give the desired nucleosides in good yield. The final nucleosides were obtained by deprotection of the isopropylidene group and oxidative cleavage of the resulting diol followed by reduction of the resulting aldehyde.

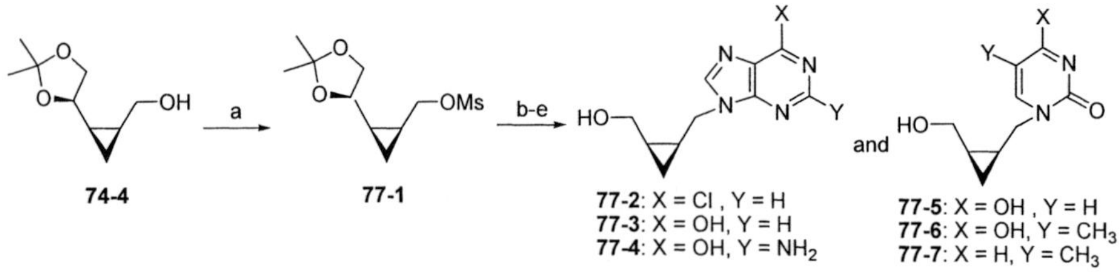

Reagents: a) MsCl, Et_3N; b) purine or pyrimidine base, K_2CO_3, 18-crown-6, 110°C; c) 80% AcOH; d) $NaIO_4$; e) $NaBH_4$.

Scheme 77

2.7. *C*-Nucleosides

In *C*-nucleosides, the sugar moiety is attached to a carbon atom of the heterocyclic moiety, rather than to a nitrogen. This causes significant changes in the geometry of the molecule, particularly in the spatial relationship between 5'-hydroxyl group and N-1, which may influence the biological feature of the molecule. Furthermore, the C-C bond is much resistant to chemical as well as enzymatic degradation than the anomeric bond, which makes *C*-nucleosides metabolically stable compared to their natural analogs. Natural *C*-nucleosides are shown in Figure 3; they are characterized by antibiotic, antiviral or antitumor properties.

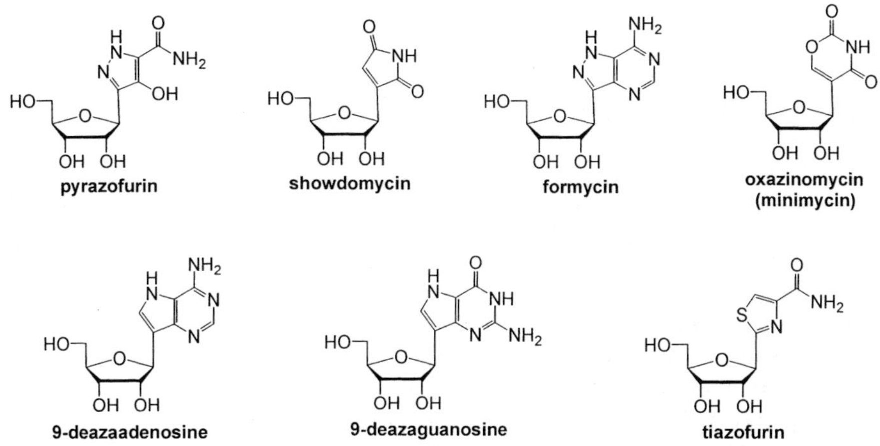

Figure 3

These natural antibiotics are all ribose-derivatives, and so are most synthetic analogs. It is beyond the scope of this chapter to explore syntheses of *C*-nucleoside analogs in which the sugar moiety maintains its identity (*i.e.* its stereo centers) in the final product, so we will consider only those examples where the starting carbohydrate is highly modified with respect to its chirality.

2', 3'-Dideoxy-D-*C*-nucleosides have been synthesized[174] starting from chiral γ-lactone **78-4**, prepared in four steps from protected D-mannitol **10-1** (Scheme 78).[175] Compound **78-4** was converted in two steps to protected lactol **78-6**, which was subject to the Horner-Emmons-Wadsworth type reaction to give intermediate **78-7**, on which the heterocyclic ring was built by a 4-step sequence *via* the methyl vinyl enolate **78-9**.

Chu and co-workers also reported the syntheses of L-4-amino-8-β-ribofuranosylpyrazolo[1,5-*a*]-1,3,5-triazine (L-APTR, **79-14**), the enantiomer of an antileukemic agent, and L-9-deazaadenosine **80-7** (Schemes 79 and 80).[176] Due to the limited availability of

L-ribose, the key intermediate **79-10** was prepared from L-xylose. This was converted to protected L-ribofuranose **79-4** in five steps, using a procedure previously reported by the same group.[177] After some deprotection/protection steps, the Horner-Emmons-Wadsworth type reaction similar to the one used in the synthesis of **78-8**, and formylation of the product **79-9** afforded enaminonitrile **79-10**. This was converted to L-APTR following the procedure developed for the synthesis of the D-isomer by Fox and co-workers.[178,179] Thus, **79-10** was converted to the unsaturated idrazine **79-11**, on which the pyrazolo[1,5-*a*]-1,3,5-triazinic system was built in two steps.

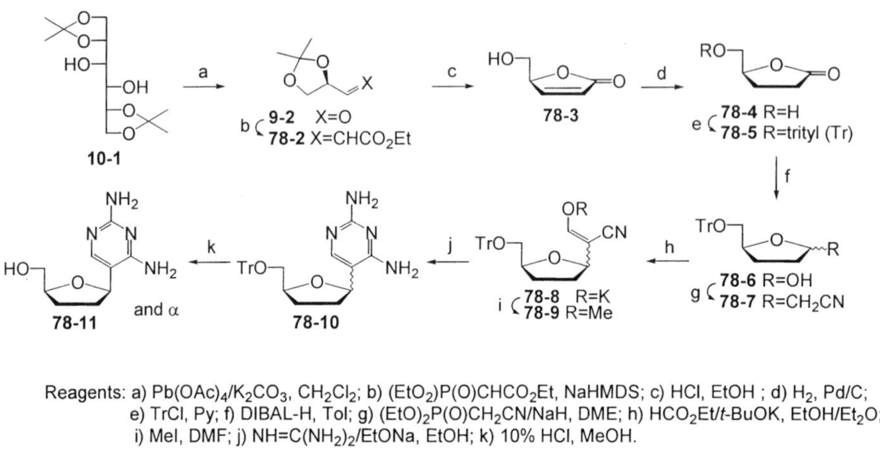

Scheme 78

Analogously, by using the procedure described by Lim and Klein[180] for the synthesis of the D-nucleoside, **79-10** was converted to β- and α-L-9-deazaadenosine (Scheme 80). In this case, the key intermediate was hydrolyzed to enolnitrile **80-1**, which was condensed with aminoacetonitrile to give, after protection, enaminodinitrile **80-3**. From this compound, the pyrrolo[3,2-*d*]pyrimidine system was built in three steps.

Hammerschmidt *et al.*[181] reported the synthesis of the D-apio analog of showdomycin from protected D-apio-β-D-furanose **34-1** (Scheme 34) as the starting material. This (Scheme 81) was converted in two steps to the protected nitrile **81-2**, whose *ribo* analog is a key synthon in many classical syntheses of ribofuranosyl-*C*-nucleosides. A number of hydrolytic and protective steps led to carboxylic acid **81-5** which, activated as the chloride, was converted to oxonitrile **81-7**. This was converted to β-cyanoacrylester **81-8** in Wittig conditions. Treatment of **81-8** with trifluoroacetic anhydride in trifluoroacetic acid allowed cyclization to protected nucleoside **81-9**, whose deprotection afforded the desired product **81-10**.

An alternative approach for the synthesis of **81-10** was based on a radical reaction (Scheme 82).[182] Here, the epimeric mixture **82-1** was fully protected in two steps to

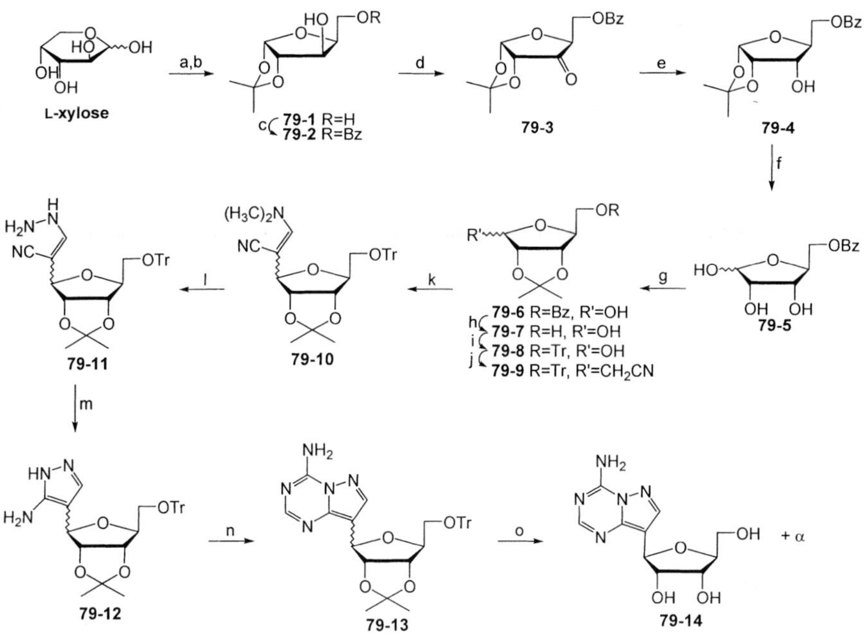

Reagents: a) acetone, $H_2SO_4/CuSO_4$; b) 0.2% HCl, H_2O; c) BzCl/Py, CH_2Cl_2; d) PDC/Ac_2O, CH_2Cl_2; e) $NaBH_4$, 1:2 EtOAc/EtOH; f) 85% HCO_2H; g) $CH_3C(OCH_3)_2CH_3/H_2SO_4$, acetone; h) NaOH, MeOH; i) TrCl, Py; j) $(EtO)_2P(O)CH_2CN$, DME; k) *t*-BuOCH[N(CH$_3$)$_2$]$_2$/DMF, CH_2Cl_2; l) $NH_2NH_2 \cdot HCl/NH_2NH_2$, MeOH/$H_2O$; m) CH_3CN (reflux); n) $NCNHCHOCH_3$, benzene; o) 10% HCl, MeOH.

Scheme 79

82-3 which, upon treatment with trimethylsilyl bromide/trimethylsilyl triflate afforded bromide **82-4** as the sole β-anomer. Treatment of **82-4** with *E*-methyl 3-cyanoacrylate in the presence of *tris*-trimetylsylilsilane/AIBN resulted in four diastereomer **82-5**, **82-6**. Following Kozikowski's procedure,[183] these were cyclized in oxidative fashion to succinimides **82-7**, which were oxidized to maleimide **81-9** and, hence, apio-showdomycin.

Chun and Chu[184] reported the synthesis of (-)-9-deazaaristeromycin, in which the structural features of two classes of compounds (carbocyclic and *C*-nucleosides) are combined. Intermediate **83-1** was synthesized analogously to its enantiomer **64-1** (Scheme 64).[149,150,151,152,153] Cyclopentanone **83-1** was condensed with ethyl cyanoacetate and, then, selective reduction of the double bond gave derivative **83-3** (Scheme 83). Different reducing agents led to different ratios of α/β anomers: trybutyltin hydride/AIBN in refluxing benzene afforded a 7/1 ratio. The carboxylic functionality of **83-3** was reduced to enolnitrile **83-4**, which was reacted with aminoacetonitrile to give dinitrile **83-5**. Protection and cyclization by treatment with DBN followed by deprotection afforded pyrrole derivative **83-6,** from which the purine ring was built by reaction with guanidine. Deprotection gave the target molecule **83-8**.

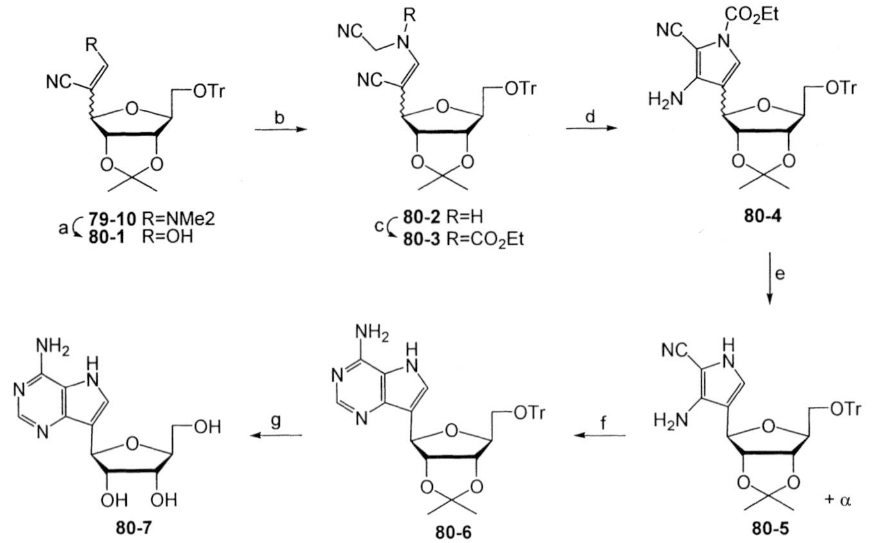

Reagents: a) CF_3CO_2H, $CHCl_3$; b) $NCCH_2NH_2 \cdot HCl/NaOAc \cdot 3H_2O$, $H_2O/MeOH$; c) $ClCO_2Et/DBN$, CH_2Cl_2; d) DBN; e) Na_2CO_3, MeOH; f) $NH_2CH{=}NH \cdot AcOH$, EtOH; g) 12% HCl, MeOH.

Scheme 80

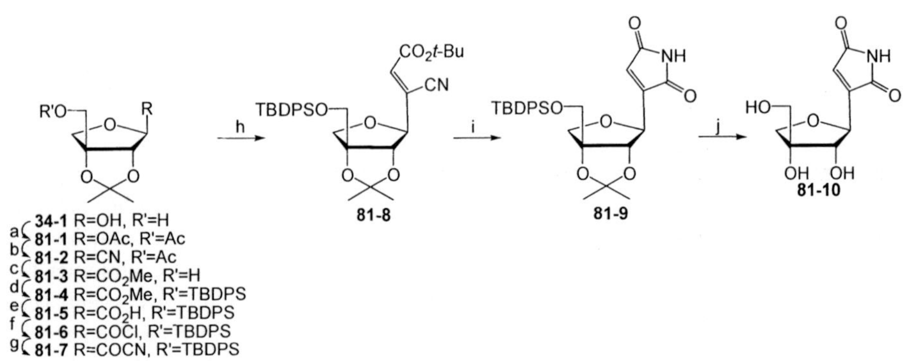

Reagents: a) Ac_2O, Py; b) TMSCN/TMSOTf, then chromatographic separation of β-isomer; c) MeONa, MeOH; d) TBDPSCl/imidazole/ DMAP, DMF; e) LiI, Py, then H^+; f) $SOCl_2$, Et_2O/DMF; g) TMSCN, CH_2Cl_2; h) $Ph_3PCH{=}CO_2t\text{-}Bu$; i) $(CF_3CO)_2O$, CF_3CO_2H; j) CF_3CO_2H, H_2O.

Scheme 81

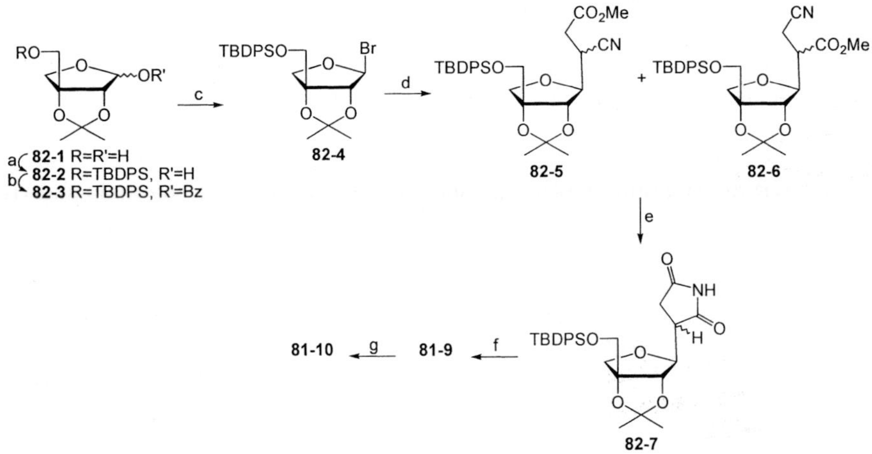

Reagents: a) TBDPSCl/imidazole/DMAP, DMF; b) BzCl, Py; c) TMSBr/TMSOTf; d) $(TMS)_3SiH$/AIBN, *(E)*-NCCH=CHCO2Me, benzene; e) H_2O_2, Na_2CO_3, H_2O/acetone; f) PhSeCl/(cC_6H_{11})*i*-PrNLi, THF, then $NaIO_4$, H_2O/ MeOH; g) 4:1 CF_3CO_2H/H_2O.

Scheme 82

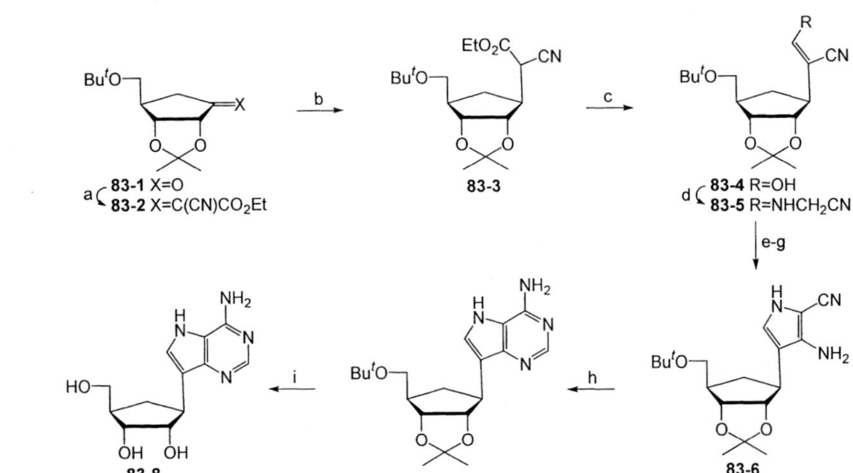

Reagents: a) $NCCH_2CO_2Et$/*t*-BuOK, EtOH; b) Bu_3SnH/AIBN, benzene; c) DIBAL-H, Et_2O; d) $H_2NCH_2CN{\cdot}H_2SO_4$, MeOH; e) $ClCO_2Et$/DBN, CH_2Cl_2; f) DBN; g) Na_2CO_3, MeOH; h) $HN_2C{=}NH{\cdot}AcOH$; i) CF_3CO_2H, H_2O.

Scheme 83

2.8. Fluorinated nucleosides

Among "classic" nucleoside analogs, various substitutions have been introduced on the sugar moiety in the attempt of obtaining more potent or metabolically stable derivatives. In some cases, an electron-withdrawing substituent such as fluorine has conferred favorable pharmacological properties. This approach has been rewarded by the syntheses of interesting molecules, such as the effective and non-toxic anti-HBV agent L-FMAU.[185,186]

As before, we will only consider those syntheses where the carbohydrate starting material has been considerably elaborated. One such example is the synthesis of 2'-deoxy-2',2"-difluoro-β-L-ribofuranosyl nucleosides from L-2,3-*O*-isopropylidene glyceraldehyde **9-12**, synthesized in two steps from L-gulonic acid-γ-lactone (Scheme 10).[187,188] The Reformatzkii reaction on compound **9-12** allowed the conversion to the 1:4 diastereomeric mixture of alcohols **84-1** and **84-2**, (Scheme 84), which could be separated by silica gel chromatography. Deprotection of the latter was followed by spontaneous lactonization to give **84-3** in 93% yield. Full protection of **84-3**, followed by reduction to epimeric lactols **84-5** and mesylation, afforded the key intermediates **84-6**. Despite the deactivating effect of the strongly electron-withdrawing difluoro substituent at C-2, **84-6** could be condensed with persilylated thymine or N^4-acetylcytosine to give the corresponding protected nucleosides and their α-epimers. These were separated, and each component was deprotected to the final product. Adenine derivative **84-12** and its α-epimer were obtained by Mitsunobu reaction of lactols **84-5** with 6-chloropurine, followed by separation of epimers, deprotection and ammonolysis. The β-adenine analog **84-12** showed moderately potent anti-HIV activity.

A similar method was used for the synthesis of the 3'-deoxy analogs.[189] In this case, deoxygenation of the mixture **84-1/84-2** was accomplished *via* reduction of dithiocarbonic *S*-methyl ester **85-1** with tributyltin hydride and 2,2'-bisisobutyronitrile (Scheme 85). Compound **85-2**, deoxy analog of **84-1/84-2**, was converted to the corresponding nucleosides by a sequence of reactions similar to the ones used in the synthesis of the oxygenated analogs.

Chu and co-workers reported the synthesis of 2',3'-dideoxy-2'-fluoro-β-L-*threo*-pentofuranosyl starting from L-xylose.[190] The starting sugar was converted to ketone **79-3** in four steps (Scheme 79). Ketone **79-3** was conveniently deoxygenated *via* the tosyl hydrazone **86-1**, which was subject to Wolff-Kishner conditions to obtain the intermediate **86-3** (Scheme 86). This was deprotected to 5'-*O*-benzoyl-3'-deoxy-L-ribose **86-4**, DAST-fluorination of which gave the difluoro derivative **86-5**. Treatment of **86-5** with hydrobromic acid gave the brominated intermediate **86-6**. Bromosugar **86-6** was coupled to persilylated cytosine or with the sodium salt of 6-chloropurine to give protected nucleosides **86-7** and **86-9**. These were elaborated in the usual way to afford the final products **86-8** and **86-9**.

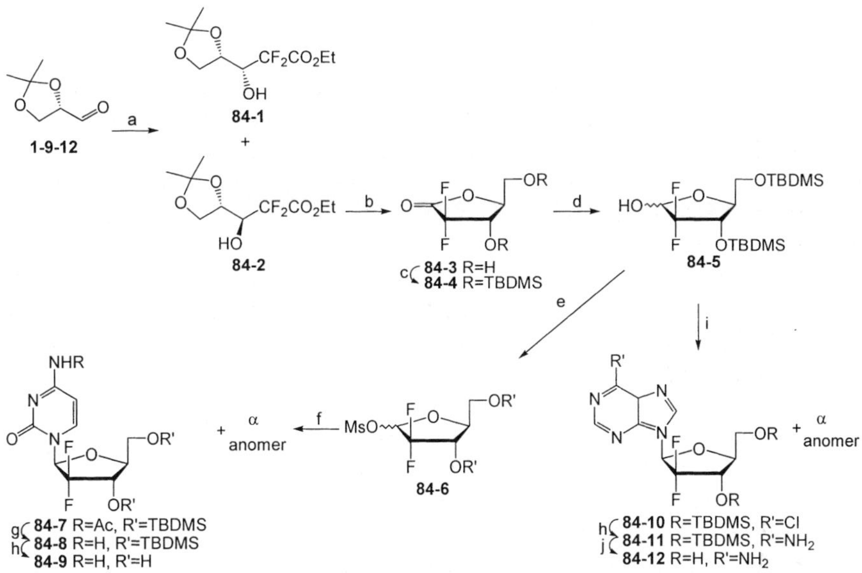

Reagents: a) $BrCF_2CO_2Et/Zn$, THF/Et_2O; b) Dowex 50(H^+); c) TBDMSCl/imidazole, DMF; d) DIBAL-H, Tol; e) $MsCl/Et_3N$, CH_2Cl_2; f) silylated N^4-acetylcytosine/NaI, CH_3CN; g) NaOMe, MeOH; h) TBAF, THF; i) 6-Cl-purine/DEAD/PPh_3, THF; j) NH_3, MeOH.

Scheme 84

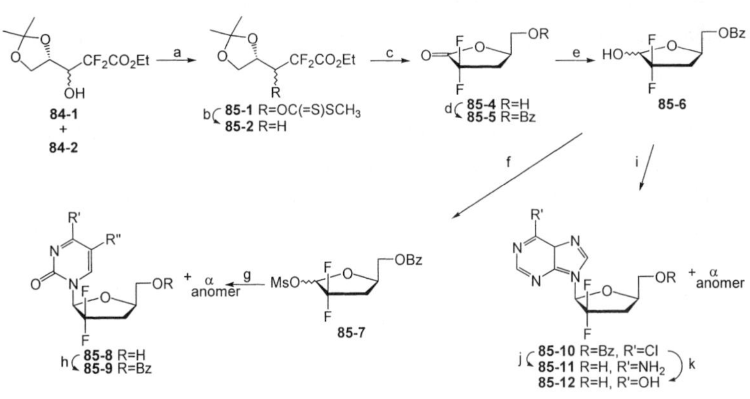

Reagents: a) $CS_2/NaH/CH_3I$, DMF; b) Bu_3SnH/AIBN, toluene; c) 5% HCl, EtOH, then benzene, Δ, Dean-Stark; d) BzCl/2,6-lutidine, CH_2Cl_2; e) Li(t-BuO)$_3$AlH, THF; f) $MsCl/Et_3N$, CH_2Cl_2; g) silylated base, NaI, MeCN, or silylated base, Me_3SiOTf, DCE; h) NH_3 or $MeNH_2$, MeOH; i) DEAD, Ph_3P, THF; j) NH_3/MeOH; k) $HSCH_2CH_2OH$, NaOMe, MeOH.

Scheme 85

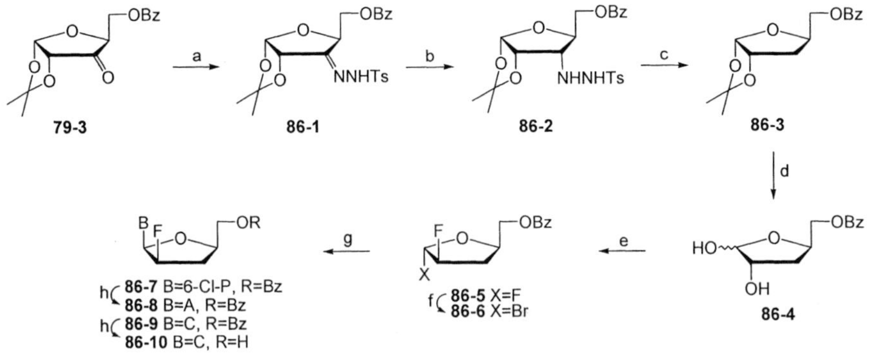

Reagents: a) $TsNHNH_2$, EtOH; b) $NaBH_3CN$; c) NaOAc·$3H_2O$/EtOH; d) 80% AcOH; e) DAST/DMAP, CH_2Cl_2; f) 45% HBr/AcOH; g) 6-Cl-Purine/NaH, CH_3CN, or silylated cytosine/TMSOTf, CH_3CN; h) NH_3, MeOH.

Scheme 86

Intermediate **79-4**, the product of stereoselective reduction of **79-3**, was used as the starting material in the synthesis of 2'-deoxy-2'-fluoro-β-L-arabinofuranosyl pyrimidine[177] and purine[191] nucleosides. A series of protection/deprotection steps afforded the conversion of **79-4** to **87-2**, whose treatment with HCl gas in methanol caused deacetylation with concomitant rearrangement of the benzoyl group to the anomeric position (Scheme 87). The resulting free hydroxyl group in **87-3** was activated as imidazolyl sulfonate, which was substituted by fluorine by treatment with potassium hydrofluoride and 48% hydrofluoric acid solution in 2,3-butanediol. Bromination of **87-5** and coupling of the intermediate **87-6** with silylated pyrimidines in 1,2-dichloroethane under reflux conditions gave mainly the protected β-isomers. The same reaction, conducted in acetonitrile, proceeded faster and in higher yield, but gave a 3:1 mixture of β and α isomers, the separation of which proved to be difficult. Debenzoylation of these compounds afforded pyrimidine derivatives. Purine nucleosides were synthesized by coupling **87-6** with the sodium salts of the corresponding bases, followed by derivatization/deprotection.

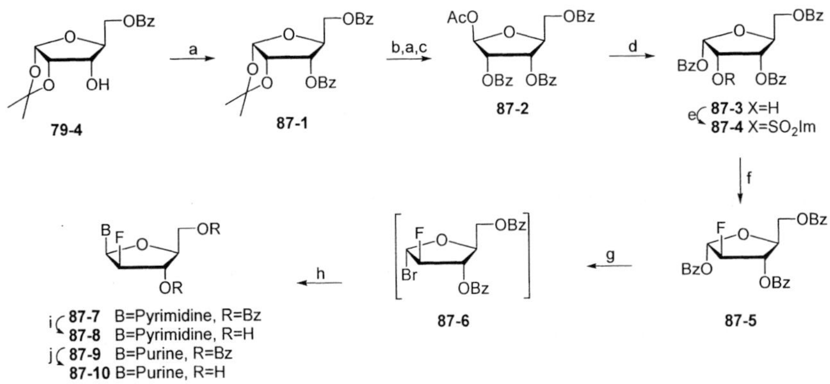

Reagents: a) BzCl, Py; b) 1% HCl, MeOH; c) conc H_2SO_4, Ac_2O/AcOH; d) HCl(g)/AcCl, CH_2Cl_2, then CH_3CN, H_2O; e) SO_2Cl_2, DMF/CH_2Cl_2, then imidazole; f) KHF_2/48% HF/H_2O, 2,3-butanediol; g) HBr/AcOH, CH_2Cl_2; h) sylilated pyrimidine, DCE, or Na-Purine, CH_3CN; i) NH_3, MeOH; j) derivatization of purine moiety and/or debenzoylation.

Scheme 87

Chu and co-workers have reported the synthesis of unsaturated 2'-fluoro-L-nucleosides from protected L-glyceraldehyde **9-12**.[107,192,193] In this case, fluorine is introduced as part of a reagent and not by a substitution reaction, thus avoiding complicated and low-yielding steps due the use of very reactive fluorinating agents. Thus, lactone **13-2**, obtained from protected L-glyceraldehyde **9-12** (Scheme 88), was reduced to epimeric lactols **88-1**. Acetylation of **88-1** afforded acetates **88-8**, which were condensed with silylated bases under Vorbrüggen conditions to give protected nucleoside analogs **88-3**. These were deprotected or chemically elaborated to give a complete series of derivatives **88-4**. Among these compounds, the adenine, cytosine and 5-F-cytosine derivatives show interesting anti-HIV-1 activity[107,192,193] and the cytidine and 5-fluorocytidine analogs are potent anti-HBV agents.[194] In the same way, the synthesis of D-analogs from protected D-glyceraldehyde **9-2** has been recently reported.[195,196]

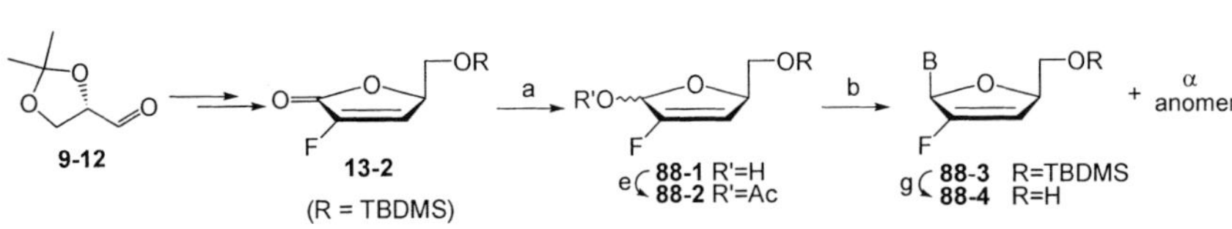

Reagents: CH_2Cl_2; a) DIBAL-H, CH_2Cl_2; b) Ac_2O/Py, CH_2Cl_2; c) silylated base/TMSOTf, DCE or CH_3CN; d) base modification and/or deprotection.

Scheme 88

Chu and co-workers have recently described the synthesis of unsaturated L- and D-3'-fluoro nucleosides as well as 2',3'-dideoxy-3',3'-difluoro nucleosides from L-xylose and protected D-glyceraldehyde, respectively. In the synthesis of L-analogs (Scheme 89),[197,198] L-xylose was converted to protected L-ribose **79-4** in 5 steps (Scheme 79). Methanolysis of this intermediate gave only the β-methyl glycoside, which was benzylated to alcohol **89-1**. Deoxygenation and debenzylation of the latter provided another alcohol, **89-3**, which was oxidized to the corresponding ketone **89-4**. Treatment with DAST afforded the difluorinated methyl glycoside that was converted to acetate key intermediate **89-6**. Condensation with persilylated N^4-benzoylcytosine followed by deprotection produced the 3',3'-difluorocytidine analog **89-9**, together with its α-isomer **89-10**. The unsaturated derivatives **89-11** and **89-12** were obtained by treatment of each isomer with sodium methoxide in DMF. Compound **89-11** showed potent anti-HIV activity.[197]

For the synthesis of D-analogs,[199] protected D-glyceraldehyde **9-2** was reacted with (1,3-dioxolan-2-ylmethyl)magnesium bromide to give epimeric alcohols **90-1**, which were oxidized to the ketone **90-2** (Scheme 90). Treatment with DAST gave difluoride **90-3**, which was deprotected and monobenzoylated to **90-5**. Acidic hydrolysis of the ketal functionality was followed by spontaneous cyclization to lactol, which was acetylated *in situ* to key intermediate **90-6**, enantiomer of **89-6**. The key intermediate was then converted to the D-3'-difluorinated and 3'-fluoro unsaturated nucleosides in the same way described for the L-analogs.

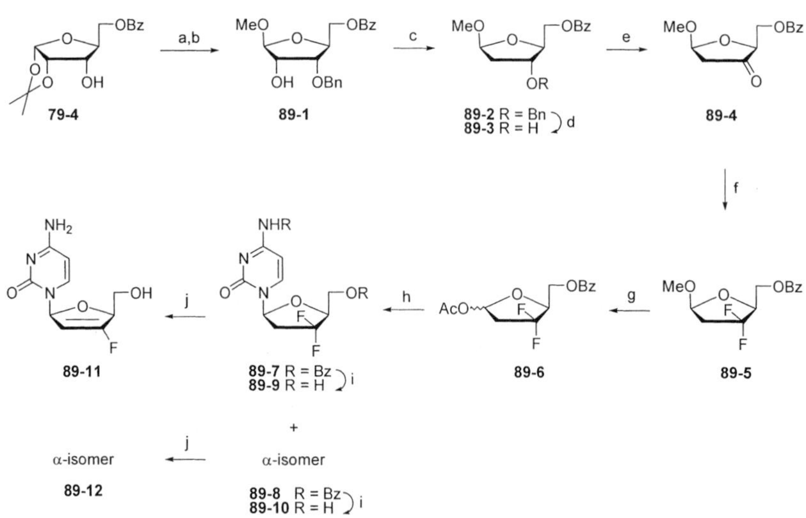

Reagents: a) NaH, THF, 0 °C to rt, 1h, then BnBr, TBAI; b) 1:2 (4.0 M HCl/dioxane)/MeOH; c) PhOC(S)Cl, DMAP, Tol, then Bu_3SnH, AIBN; d) H_2 (55 psi), 10% Pd/C, EtOH; e) CrO_3, Ac_2O, Py, CH_2Cl_2; f) DAST, CH_2Cl_2; g) conc. H_2SO_4, Ac_2O, AcOH; h) silylated N^4-BzCy, TMSOTf, CH_3CN; i) sat. NH_3/MeOH; j) MeONa, DMF.

Scheme 89

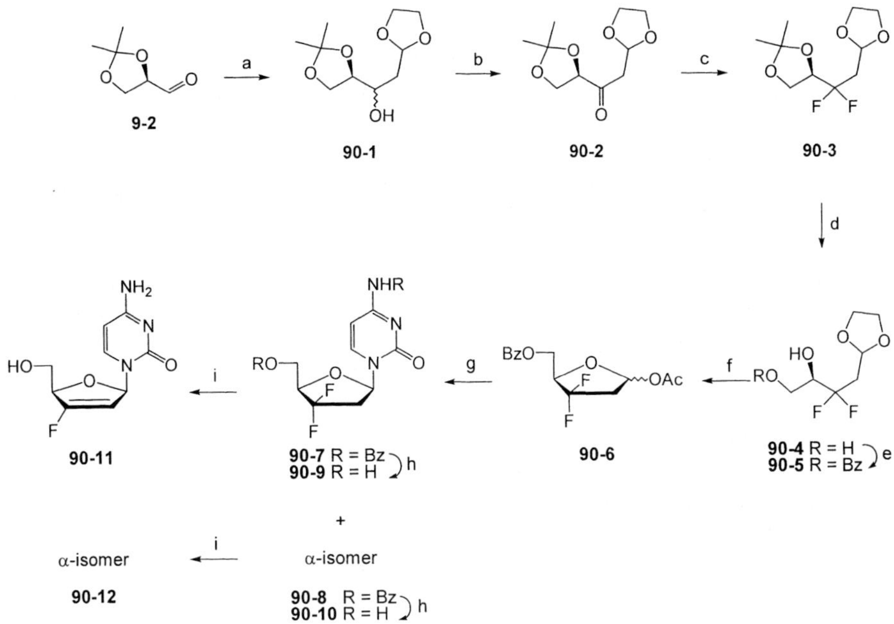

Reagents: a) (1,3-dioxolan-2-ylmethyl)MgBr, THF; b) DMSO, $(ClCO)_2$, TEA, CH_2Cl_2; c) DAST, CH_2Cl_2; d) 1:1 5% HCl/dioxane; e) BzCl, Py; f) conc. H_2SO_4, AcOH, Ac_2O; g) persilylated N^4-BzCy, TMSOTf, CH_3CN; h) NH_3/MeOH; i) NaOMe, DMF.

Scheme 90

2.9. Acyclonucleosides

Acyclonucleosides can be considered as derived from classical nucleosides by "removing" one or more bonds from the cyclic moiety. This interesting class of compounds has been subject of a number of reviews.[200,201,202,203]

Because of their structural flexibility, many acyclonucleosides possess biological properties despite their lack of chirality. Of course, a number of optically active derivatives are also known, and in most cases their syntheses are accomplished starting from sugars by cleavage of one or more bonds. As in the case of *C*-nucleosides, we will only consider those examples where sugar precursors are modified in a stereocontrolled way.

An early example of this kind is the synthesis of eritadenine **91-5** (Scheme 91), a natural acidic nucleoside endowed with hypocholesterolemic properties, obtained by condensation of the sodium salt of adenine with 2(*R*),3(*R*)-cyclohexylidenedioxybutyrolactone **91-3** followed by acidic hydrolysis.[204] Compound **91-3** was obtained by protection of D-erythronolactone **91-2**, prepared in three steps from D-glucose.[205]

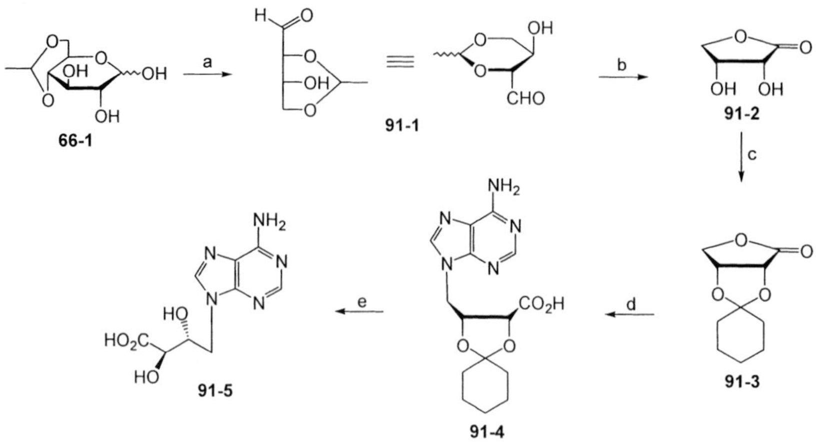

Reagents: a) $NaIO_4/NaOH$, H_2O; b) $CH_3CH_2CO_3H$, EtOAc, then H_2O; c) cyclohexanone/TsOH, benzene; d) [adenine/NaH], DMF; e) 10% HCl

Scheme 91

The adenosine deaminase inhibitor *erythro*-(2*S*,3*R*)-3-(adenin-9-yl)-2-nonanol (EHNA) **92-12**, its enantiomer **92-17** and their *threo* isomers were synthesized by Baker and Hawkins[206] starting from protected L- and D-rhamnitol **92-2** and **92-16** (Scheme 92), readily prepared from L-rhamnose[207] and D-mannose[208] in two and three steps, respectively. Thus, L-isomer **92-2** was benzylated and deprotected to give the benzyl tetrol **92-4**, the oxidative decomposition of which afforded aldehyde **92-5**. A Grignard reaction was used to attach a *n*-hexyl moiety, giving *threo* and *erythro* derivatives **92-6** and **92-7**, in a 1:3 ratio and without any detectable racemization. Each diastereomer, separated chromatographycally, was mesylated and condensed with the sodium salt of adenine to give, after debenzylation, compounds **92-12** and **92-13**. The same sequence of reactions was applied to the synthesis of **92-17** and **92-18** from **92-16**.

MacCoss *et al.*[209] synthesized all four possible diastereomers of 9-(1,3,4-trihydroxy-2-butoxymethyl)guanine (Scheme 93). Protected L-xylose **93-1** was converted to the open-chain dithioacetal **93-2**, which was benzoylated at the 4-position and, subsequently, deprotected to unmask the aldehyde group. Aldehyde **93-4** was then decarbonylated by Wilkinson's catalyst and debenzoylated to alcohol **93-6**, which was chloromethylated to give the key intermediate **93-7**. Condensation with persilylated 2-amino-6-chloropurine, followed by hydrolysis of the chloride functionality and debenzylation, afforded the guanine acyclonucleoside **93-10**. The other three isomers were synthesized in the same way starting from D-xylose, L-arabinose and D-arabinose, respectively.

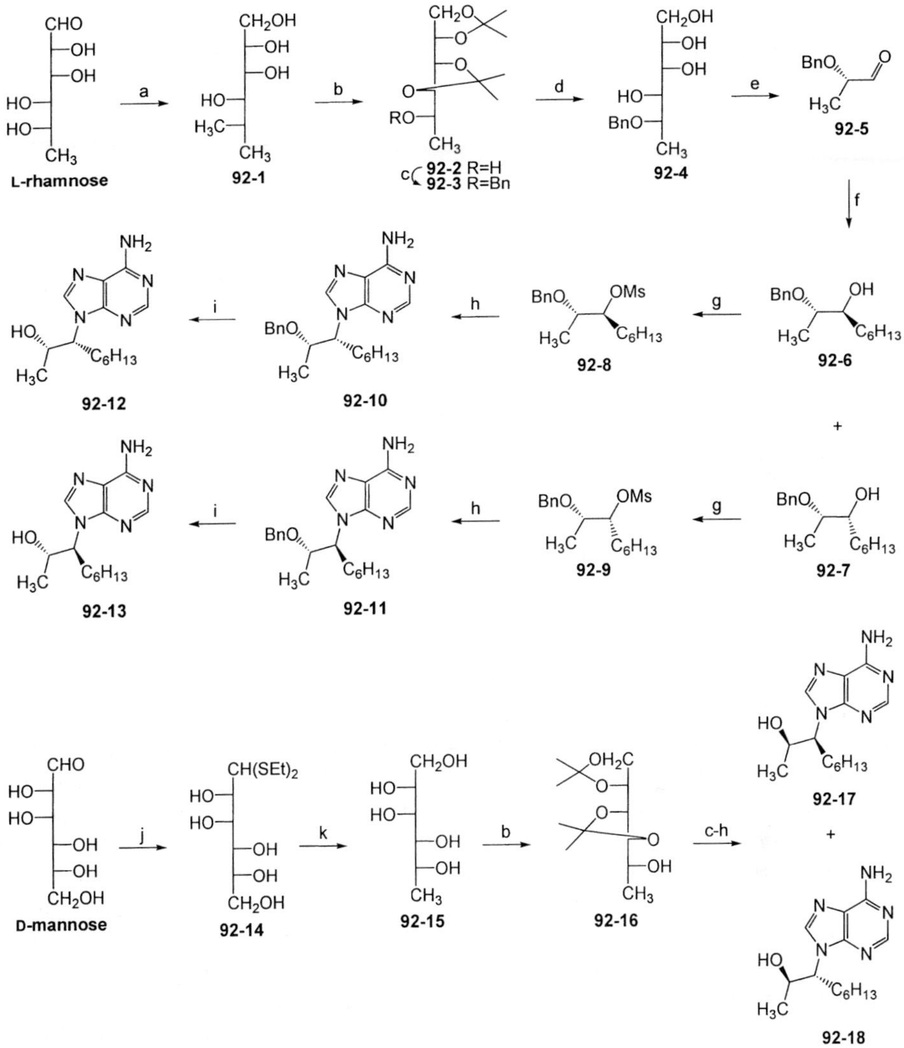

Reagents: a) $NaBH_4$; b) acetone, $CuSO_4/H_2SO_4$; c) BnCl/NaH, DMF; d) 9:1 CF_3CO_2H/H_2O; e) $NaIO_4$, NaOH/ H_2O; f) $C_6H_{13}MgBr$, THF; g) MsCl, Py; h) [adenine/NaH]/DMF; i) H_2, 10% Pd/C, HCl/*i*-PrOH ; j) EtSH, HCl; k) Raney nickel, 70% EtOH.

Scheme 92

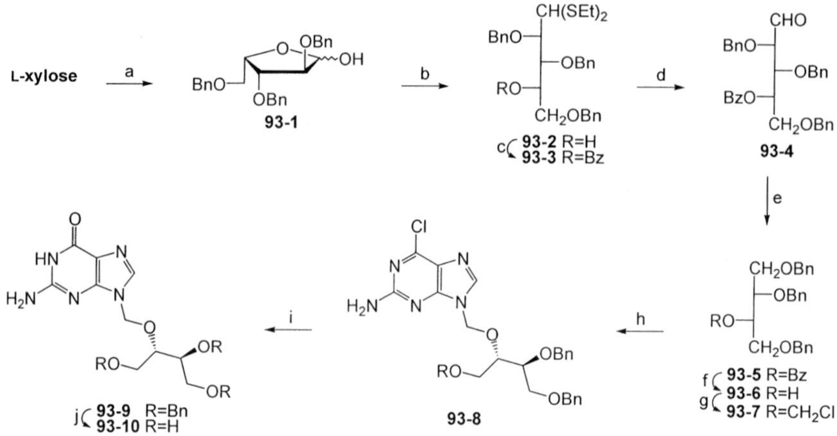

Reagents: a) BnBr, NaH; b) EtSH/HCl/$MgSO_4$; c) BzCl, Py; d) $HgCl_2$/ $CdCO_3$, acetone/H_2O; e) $(Ph_3P)_3RhCl$, CH_3CN; f) MeONa, MeOH; g) CH_2O/HCl (gas),CH_2Cl_2; h) silylated 2-NH_2-6-Cl-purine/$Hg(CN)_2$, PhH; i) 20% aq. Et_4NOH/25%aq. Me_3N, glyme; j) H_2/20%$Pd(OH)_2$/C, cyclohexene/EtOH.

Scheme 93

In an interesting stereocontrolled homologation of chiral α-hydroxyaldehydes using silylazoles, Dondoni and co-workers[210] described the addition of 2-trimethylsilylthiazole to 2,3-isopropylidene D-glyceraldehyde **9-2** to afford, after deprotection, the protected diastereomeric thiazole acyclonucleoside **94-2** and **94-3**, in 93% yield with a diastereomeric ratio >95:5 (Scheme 94). Partial deprotection of **94-2** gave **94-4**.

Aldehyde **9-2** was used by Ashton *et al.*[211] in the synthesis of (*S*)-9-[(2,3-dihydroxy-1-propoxy)methyl]guanine **95-10** (Scheme 95). Periodate oxidation of diisopropylidene mannitol afforded a mixture of **9-2** and its methyl hemiacetal **95-1**. The crude mixture was reduced to 1,2-*O*-isopropylidene-L-glycerol **95-2** by sodium borohydride in Wickberg conditions. A series of protection-deprotection steps afforded chloride **95-8**, which was condensed with persilylated guanine to give, after debenzylation, **95-10**. The *R* enantiomer **95-17** was obtained in the same way *via* 1,2-di-*O*-benzyl-L-glycerol **95-13**, prepared by oxidative cleavage of 1,2,5,6-tetra-*O*-benzyl-D-mannitol **95-11**. Intermediate **95-13** was purified as the trityl derivative **95-14**.

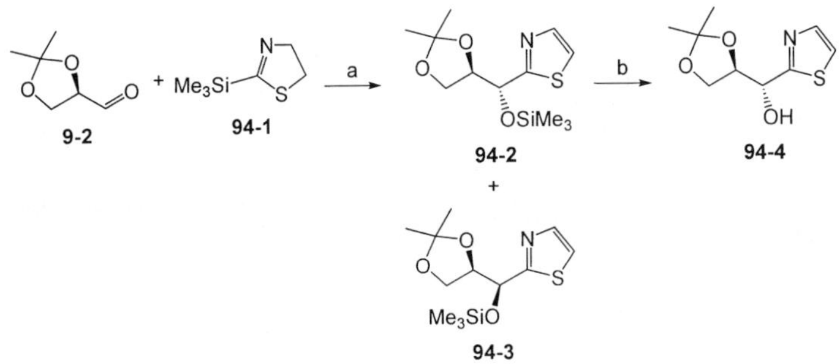

Reagents: a) CH_2Cl_2 ; b) THF, $H_2O.F^-$.

Scheme 94

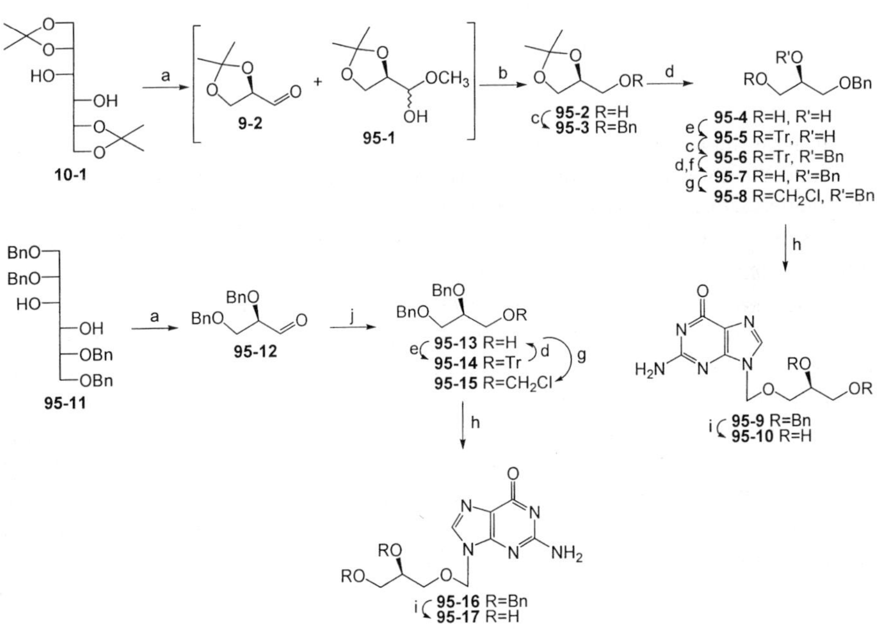

Reagents: a) $NaIO_4$, MeOH; b) $NaBH_4$, 0.2M Na_2HPO_4; c) NaH/BnCl, DMSO; d) AcOH, H_2O; e) TrCl, Py; f) MeONa, MeOH; g) $(CH_2O)_x$/HCl, CH_2Cl_2; h) silylated guanine, xylene, then *n*-PrOH, AcOH; i) H_2, 20% $Pd(OH)_2$/C, TsOH, MeOH; j) $NaBH_4$, MeOH.

Scheme 95

Chu and co-workers[212] combined acyclonucleoside and *C*-nucleoside chemistry in the synthesis of *C*-acyclonucleosides (Scheme 96). Isopropylidene D-glyceraldehyde **9-2** was reacted with (cyanomethylene)triphenylphosphorane in Wittig conditions to yield the *E*/*Z* mixture of α,β-unsaturated nitriles **96-1**, which were regioselectively reduced to **96-2** with sodium borohydride. Nitrile **96-2** was converted to the α-formyl derivative **96-3**, which was condensed with thiosemicarbazide to afford thiosemicarbazone **96-4**. The two latter compounds were reacted without purification. Compound **96-4** was cyclized to pyrazole **96-5** under basic conditions. Hydrolysis of the latter afforded the versatile intermediate **96-6** which was reacted with *N*-cyanoformimidate to give, after deprotection, the adenosine analog **96-8**. Reaction of **96-6** with *N*-(ethoxycarbonyl)-*S*-methylisothiourea, followed by basic hydrolysis and deprotection gave the guanosine analog **96-10**. The 6-mercaptopurine analog **96-11** was synthesized from **96-5** by reaction with triethyl orthoformate, followed by deprotection. Finally, in order to synthesize the inosine analog **96-15**, **96-6** was reacted with ethoxycarbonyl isothiocyanate to give a mixture of **96-12** and **96-13**; the latter was converted into the former by refluxing in ethanol. **96-12** was readily cyclized in basic conditions to give **96-14**, which was deprotected, desulfurized, and methylated to give the methlythio analog **96-16**.

A similar approach had been used by Buchanan *et al.*[213] in the synthesis of pyrrolo- and thieno-[3,2-*d*]pyrimidine *C*-nucleosides (Scheme 97). α,β-Unsaturated nitrile **96-1** was obtained from protected mannitol **10-1** in a "one-pot" process by treatment with aqueous sodium periodate and reaction of the resulting aldehyde **9-2** with diethyl cyanomethylphosphonate and potassium carbonate in Horner-Wadsworth-Emmons fashion. The advantage of this procedure was a better yield and optical stability throughout the whole process. Saturation of **96-1** was accomplished by catalytic hydrogenation; the derived nitrile **96-2** was formylated in its α position to **96-3**, which was directly converted to the enaminodinitrile **97-1** by treatment with aminoacetonitrile. Compound **97-1** was protected at its secondary amino functionality and cyclized "one pot" to the enaminonitrilpyrrole **97-3**, which was deprotected and reacted with formamidine acetate to give, after deisopropylidation, the adenosine analog **97-6**. A thiophene system could be obtained from **96-2**, *via* intermediate **97-7**, by *in situ* mesylation, followed by treatment with acetylthioacetonitrile and sodium carbonate. The resulting **97-8** was converted to the thienopyrimidine derivative **97-9** by the same procedure used in the synthesis of **97-6**.

Lin *et al.*[214] reported the synthesis of (*R*)- and (*S*)-1-{[2-hydroxy-1-(aminomethyl)ethoxy]-methyl}-5-benzyluracil (AHPBU) **98-7** and **98-16** (Scheme 98), starting from 1-*O*-benzyl-D-glycerol **95-4**, prepared from D-mannitol according to Howe and Malkin.[215] In the synthesis of (*R*)-(AHPBU), alcohol **95-4** was converted to the azido derivative **98-2** *via* tosylate **98-1**. Compound **98-2** was then chloromethylated and coupled with persilylated 5-benzyluracil. Reduction of the azido group followed by debenzylation gave the desired product. Interestingly, the free amine **98-5** produced by the first hydrogenation was not reactive enough to give the debenzylated derivative under the reaction conditions. Its hydrochloric salt **98-6**, however, could be debenzylated using the same catalyst and higher hydrogen pressure. In order to synthesize (*S*)-(AHPBU), azidoalcohol **98-11**, enantiomer of **98-2**, was needed. This was elegantly synthesized by full tosylation of diol **95-4**, displacement of the primary tosyl group by sodium benzoate, hydrolysis of the benzoate with concomitant cyclization to epoxide **98-10,** and nucleophilic opening of the epoxide by lithium azide. Compound **98-11** was converted to **98-16** by the same route used in the synthesis of the (*R*)- isomer.

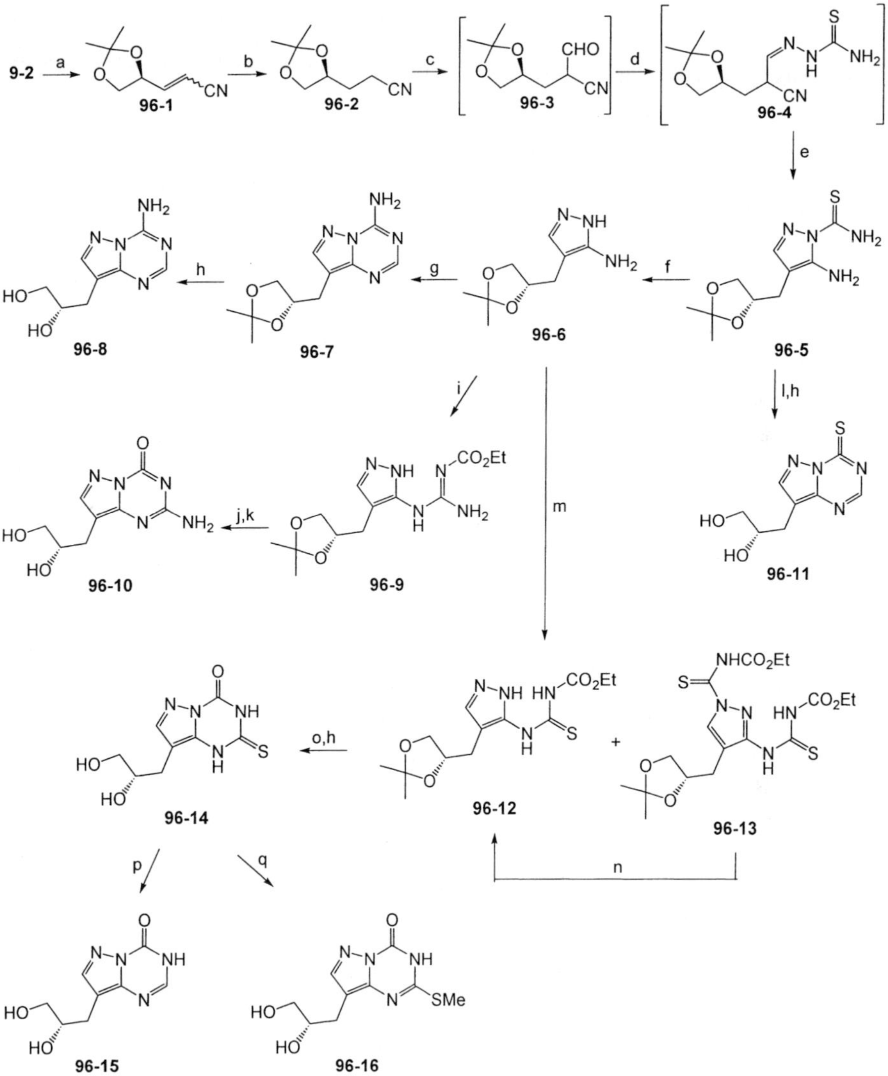

Reagents: a) $(Ph_3)P{=}CHCN$, MeCN; b) $NaBH_4$, EtOH; c) HCO_2Et/NaH, *t*-BuOH; d) $H_2NNHCSNH_2$, 20% AcOH, MeOH; e) EtONa, EtOH; f) NaOH, MeOH; g) NC-N=CHOEt, PhH; h) 80% CF_3CO_2H; i) $EtO_2CN{=}C(NH_2)SMe$, Et_2O; j) NaOH, DMF; k) 75% CF_3CO_2H; l) $HC(OEt)_3$; m) $S{=}C{=}NCO_2Et$, MeCN; n) EtOH, Δ; o) 2N NaOH; p) Raney Nickel, 3% NH_4OH; q) MeI, NaOMe/MeOH.

Scheme 96

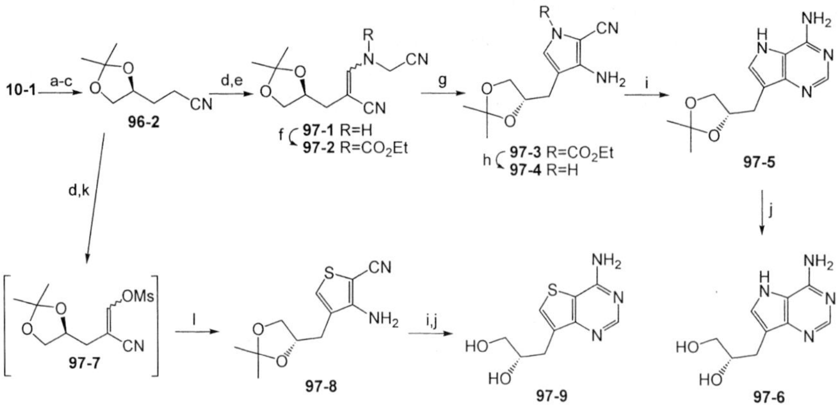

Reagents: a) $NaIO_4/NaHCO_3$, H_2O; b) $(EtO)_2P(O)CHCN$, K_2CO_3/H_2O; c) H_2, 5% Pd/C, MeOH; d) HCO_2Et/NaH, $EtOH/Et_2O$; e) $H_2NCH_2CN \cdot HCl/AcONa$, $MeOH/H_2O$; f) $ClCO_2Et/DBN$, CH_2Cl_2; g) DBN, CH_2Cl_2; h) Na_2CO_3, MeOH; i) $HC(=NH)NH_2 \cdot AcOH$, EtOH; j) 80% AcOH; k) $MsCl/Et_3N$, $CHCl_3$; l) $CH_3C(O)SCH_2CN/Na_2CO_3$, EtOH.

Scheme 97

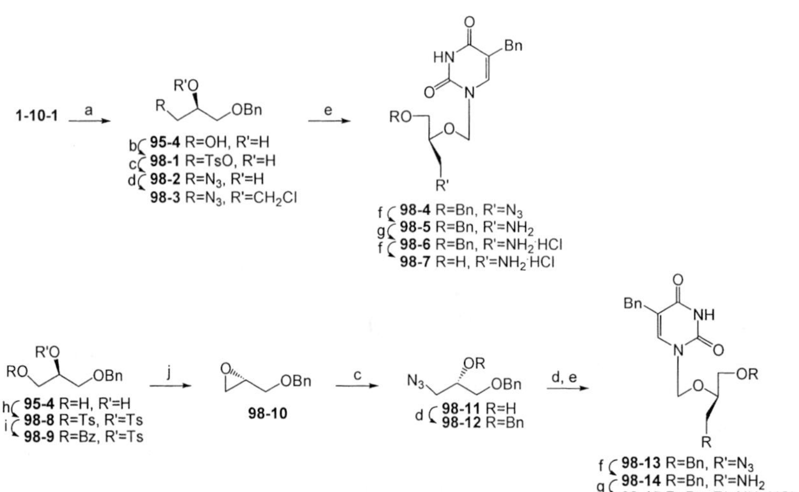

Reagents: a) Howe and Malkin conditions; b) TsCl/Py, CH_2Cl_2; c) LiN_3, DMF; d) $(CH_2O)_n$/ HCl, DCE; e) silylated 5-benzyluracil, Tol; f) H_2, 10% Pd/C, EtOH; g) HCl/EtOH; h) TsCl, Py; i) NaOBz, DMF; j) NaOMe, CH_2Cl_2.

Scheme 98

Acyclic fluorinated nucleosides were synthesized by Herdewijn *et al.*[216] starting from 2-deoxy-2-fluoro-D-arabinose **99-1** (Scheme 99). This was dibenzoylated, then condensed with silylated bases to give, after debenzoylation, nucleosides **99-4** and **99-6**. Each epimer was converted to the acyclic derivatives **99-7** and **99-8** by oxidative ring opening with sodium periodate and reduction of the intermediate dialdehyde by sodium borohydride.

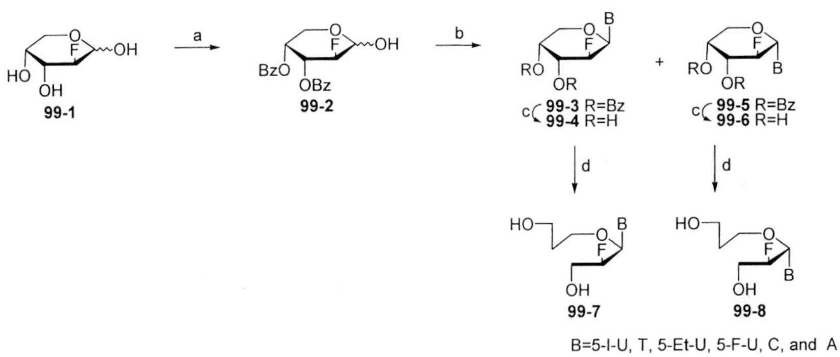

Reagents: a) BzCl, Py; b) silylated base/TMSOTf, DCE; c) NH_3, MeOH; d) $NaIO_4$, H_2O/dioxane, then $NaBH_4$, EtOH.

Scheme 99

2.10. Miscellaneous nucleosides

Oxetanocin (**100-13**, Scheme 100) is a microbial nucleoside endowed with antibacterial, antitumor and antiviral activity.[217,218] Due to its unique structure, characterized by the presence of an oxetane ring, it has been the object of numerous synthetic studies.[219]

Wilson *et al.*[220] proposed an interesting stereoselective one-carbon extension at the C-3 position of diacetone-D-glucose **100-1** (Scheme 100). Thus, **100-1** was converted to the trimethylsilyl alcohol **100-2** in two steps. Peterson fragmentation of **100-2** and subsequent hydroboration gave the C-3 branched derivative **100-3**. Selective deprotection of the exocyclic isopropylidene group, followed by oxidative cleavage of the resulting diol, borohydride reduction and benzylation gave the fully protected ribose homologue **100-4**. Deprotection and oxidation of the latter afforded lactone **100-5**, which was triflated to compound **100-6**. Methanolysis of this triflate lactone was followed by spontaneous intramolecular cyclization to the oxetane ester **100-7**. Hydrolysis of the latter, followed by treatment with oxalyl chloride and reaction with the sodium salt of 2-mercaptopyridine *N*-oxide, afforded the thiohydroxamic ester **100-9** that was subject to a radical decarboxylative chlorination to the unstable epimeric chlorides **100-11**, which were trapped *in situ* as the corresponding thiophenyl glycosides **100-10**. Regeneration of **100-11** was accomplished by treatment with chlorine, and their *in situ* condensation with N^6-benzoyladenine followed by deprotection afforded oxetanocin **100-13** and its α-epimer epioxetanocin.

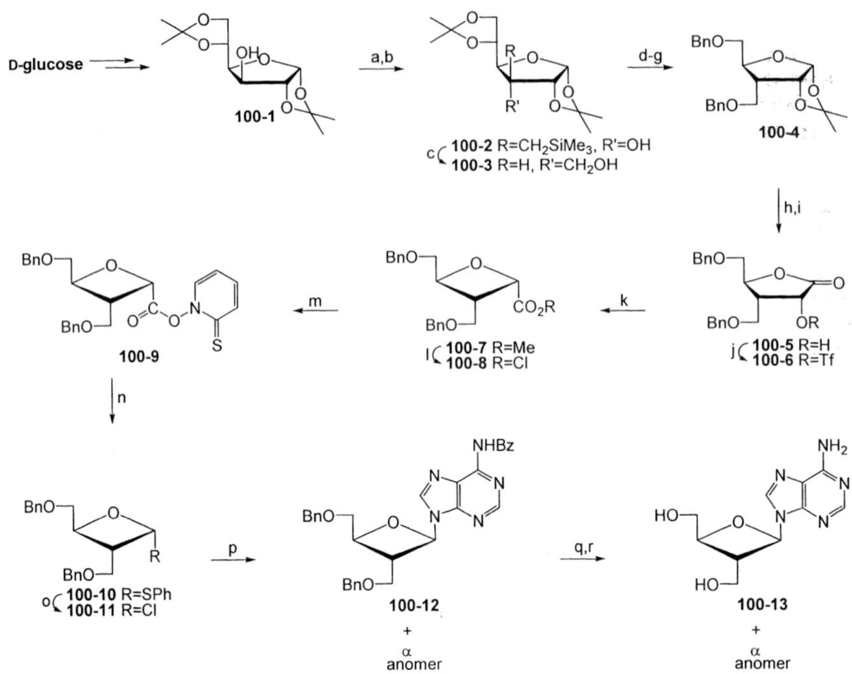

Reagents: a) PCC/4Å mol. sieves, CH_2Cl_2; b) Me_3SiCH_2MgCl; c) NaH, THF, then BH_3-DMS, then aq. NaOH; d) aq. AcOH; e) $NaIO_4$, aq. EtOH; f) $NaBH_4$, EtOH; g) BnBr, DMF; h) aq. TFA; i) $Br_2/BaBz_2$, aq. dioxane; j) Tf_2O/Py, CH_2Cl_2; k) K_2CO_3, MeOH; l) hydrolysis, then $(COCl)_2$; m) Na-2-mercaptopyridine *N*-oxide; n) decarboxylative chlorination , then PhSH/18-crown-6, K_2CO_3 ; o) Cl_2, $CHCl_3$; p) N^6-BzAd, K_2CO_3; q) NaOH, MeOH; r) $H_2/Pd(OH)_2$, cyclohexene.

Scheme 100

3'-*O*-methyloxetanocin has been prepared by Saksena *et al.*[221] (Scheme 101). Lactone **101-1** was synthesized following the same route described in the synthesis of **100-6**. Ring contraction was accomplished by a different method *via* a mesylate instead of a triflate with improved yield. The acids **101-2** were converted to the thiopyridyl glycosides **101-5** in a three steps-one-pot reaction including activation of the acids as isobutyryl mixed anhydrides, condensation with 2-mercaptopyridine *N*-oxide, and photolytic decarboxylative rearrangements of the resulting thiohydroxamic esters **101-4**. Thiopyridyl glycosides **101-5** were coupled with N^6-benzoyladenine in the presence of bromine and under strictly anhydrous conditions. It is notable that the use of bromine in the combination with the thiopyridyl activating group worked as an effective coupling condition, whereas the use of a Lewis acid would destroy the ring system. Debenzoylation of coupling product **101-6** provided the benzyl derivative **101-7**, which could be separated from its α-isomer. Debenzylation of each epimer afforded the final product **101-8** and its α-isomer.

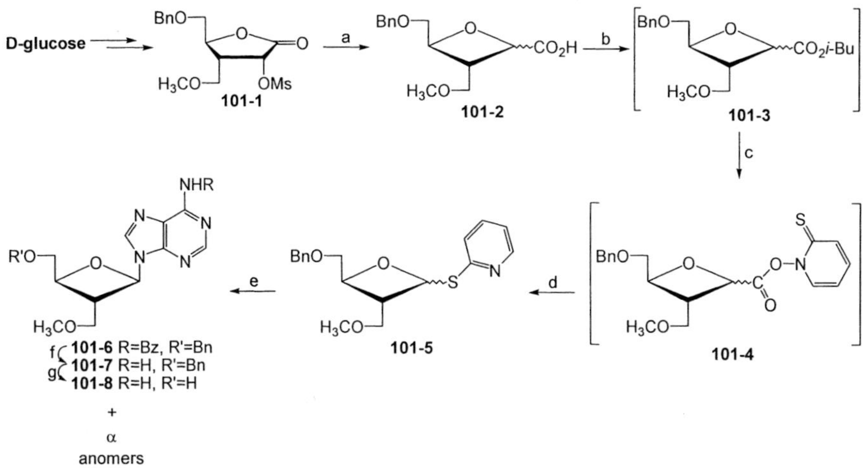

Reagents: a) NaOH, H_2O/MeOH; b) *i*-BuOC(O)Cl/*N*-methylmorpholine, THF; c) 2-mercaptopyridine *N*-oxide/TEA, THF; d) hν; e) N^6-benzoyladenine/Br_2/4Å mol. sieves, DMF; f) NaOMe, MeOH; g) H_2/Pd black, EtOH.

Scheme 101

A similar approach has recently allowed the synthesis of L-oxetanocin.[222,223] The Wittig olefination of ketone **102-1**, enantiomer of **31-4**, derived from L-xylose in five steps, gave intermediate **102-2** (Scheme 102). Hydroboration, benzylation of the resulting alcohol and hydrolysis of the acetonide protecting group afforded lactol **102-5**, which was oxidized to lactone **102-6** by bromine in barium carbonate buffer. Lactone **102-6** was mesylated to **102-7**, which was subject to saponification with concomitant ring contraction to give the oxetanecarboxylic acids **102-8** as an epimeric mixture. This was converted to thiopyridil oxetanes **102-9**, which were coupled with protected adenine to give, after two deprotection steps, L-oxetanocin **102-12** and its α-epimer. The final debenzylation could be accomplished by hydrogenation over palladium black, but failed when other catalysts were used.

Lactones **78-4** and **78-3** have proved very versatile intermediates, having been used in the synthesis of a wide variety of nucleoside analogs (Schemes 103 and 104). For the syntheses of 2',3'-dideoxy- and 2',3'-didehydro-2',3'-dideoxynucleosides,[224,225] compound **103-1**, obtained by protection of **78-4**, was enolized and blocked as silyl enol ether **103-2** which, without isolation, was treated with phenylselenium bromide to give epimeric selenides **103-3** and **103-4** (Scheme 103). The former could be converted to the latter by base-catalyzed epimerization. Reduction of **103-4** followed by acetylation of the resulting alcohol **103-5** afforded the key intermediate **103-6**, which was coupled with nucleobases to give β-nucleosides **103-7** with high stereoselectivity due to the orientating effect of the phenylselenyl group. These nucleosides could be either hydrogenated to saturated compounds **103-8** or oxidatively deselenated to unsaturated derivatives **103-10**.

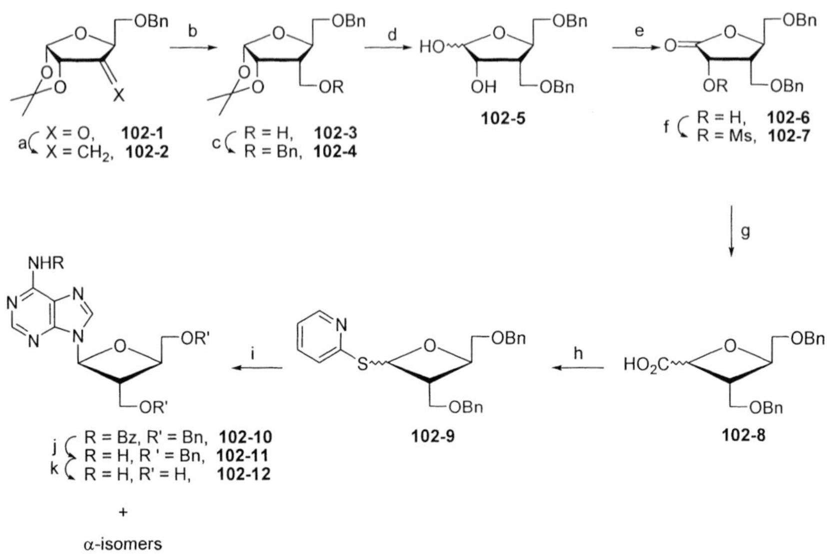

Reagents: a) $Ph_3P{=}CH_2$, THF; b) BH_3.DMS, THF, then 30% H_2O_2, NaOH; c) NaH, THF, then TBAI, BnBr; d) 10% HCl, 1,4-dioxane; e) Br_2, $BaCO_3$, 3:1 H_2O/1,4-dioxane; f) MsCl, Py, CH_2Cl_2; g) 1N NaOH, MeOH; h) *i*-BuOC(O)Cl, *N*-methylmorpholine, THF, then 2-mercaptopyridine-*N*-oxide, Et_3N, then hν; i) N^6-benzoyladenine, Br_2, 4 Å mol. sieves, DMF; j) MeONa, MeOH; k) H_2, Pd black, EtOH.

Scheme 102

In a synthesis of AZT and related compounds,[226] lactone **104-1**, product of the silylation of **78-3**, was reacted with azide ion in Michael reaction to give the azido lactone **104-2** (Scheme 104). Reduction and acetylation afforded the key intermediate **104-4**, which was used in the syntheses of the series of analogs **104-6**.

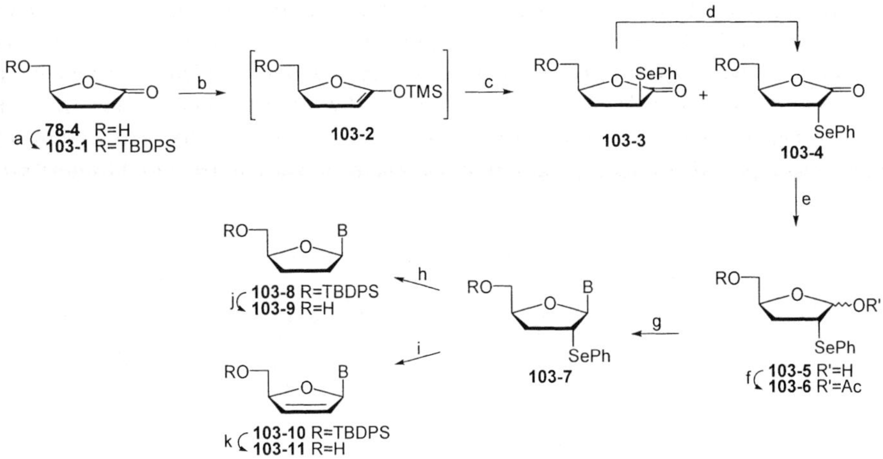

Reagents: a) TBDPSCl/imidazole, DMF; b) LiHMDS, THF, then TMSCl; c) PhSeBr; d) DBU or Et_2NH; e) DIBAL-H, Tol; f) Ac_2O/DMAP, Py; g) Silylated base, TMSOTf or $SnCl_4$, DCE; h) Bu_3SnH/Et_3B, benzene; i) 30% H_2O_2/Py, CH_2Cl_2; j) base modification and/or TBAF, THF.

Scheme 103

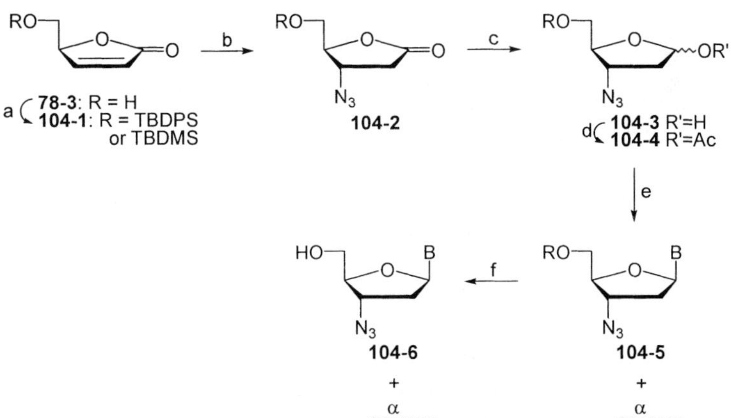

Reagents: a) TBDPSCl or TBDMSCl/imidazole, DMF; b) NaN_3, AcOH/H_2O/THF; c) DIBAL-H, Hexanes/CH_2Cl_2; d) Ac_2O, Py; e) silylated base/$SnCl_4$, TMSOTf, or $BF_3 \cdot Et_2O$, MeCN, DCE, or 1,4-dioxane; f) base modification and/or TBAF, THF.

Scheme 104

Treatment of intermediate **103-2** with *N*-bromosuccinimide gave bromide **105-1** and **105-2** in 1:4 epimeric ratio (Scheme 105).[227] Basic hydrolysis of these bromides produced furanoic acids **105-3**, products of a silyl migration from the *O*-5 to the *O*-4 position. Decarboxylative oxidation of acids **105-3** gave the acetate intermediates **105-4**, which were used to synthesize the novel class of [4-(hydroxy)tetrahydrofuran-2-yl]nucleosides. Of these, the thymidine analog was found to be a good inhibitor of uridine phosphorylase.

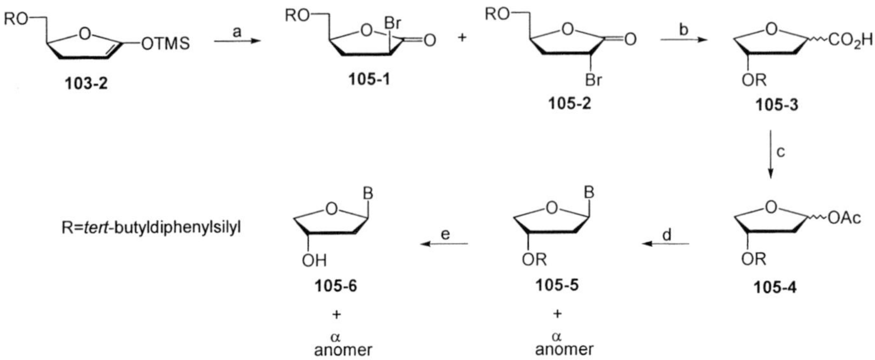

Reagents: a) NBS, THF; b) 1 N NaOH, MeOH; c) $Pb(OAc)_4$/Py, THF; d) Silylated thymine/TMSOTf, DCE or sylilated 6-Cl-purine/TMSOTf, CH_2Cl_2; e) TBAF, THF or TBAF, THF, then NH_3, MeOH.

Scheme 105

In the synthesis of 2'-*C*-α-methyl-2',3'-dideoxynucleosides,[228] lactone **103-1** was methylated to give epimers **106-1** and **106-2** in 1:10 ratio (Scheme 106). Reduction of **106-2**, acetylation and coupling with nucleobases afforded the target compounds.

Protected D-xylose **24-1** was used in the synthesis of the reversed aminoacyl nucleoside **107-8** (Scheme 107).[229] Protection of **24-1** as a carbonate, followed by oxidation and treatment of the resulting ketone **107-2** with hydroxylamine afforded oxime **107-3**. Treatment with lithium aluminum hydride reduced the carbonate functionality and, by complexing the oxime from the "up" side, promoted its stereoselective reduction to amine **107-4**. This was condensed with benzyloxycarbonyl-L-phenylalanine and tosylated to give intermediate **107-6**, which could be condensed with the sodium salt of adenine. Deprotection of the isopropylidene and benzyloxycarbonyl groups afforded the target molecule **107-8**, "reversed" analog of the broad-spectrum antibiotic and antitumor compound puromycin.

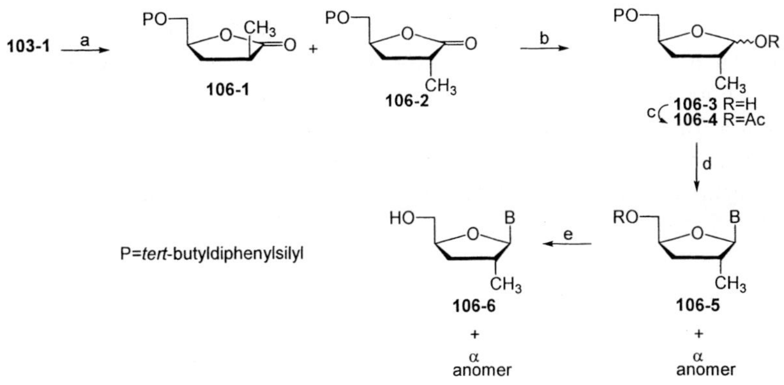

Reagents: a) LDA, THF, then CH_3I; b) DIBAL-H, CH_2Cl_2; c) Ac_2O/Py/DMAP, CH_2Cl_2; d) Silylated base, TMSOTf, DCE; e) base modification and/or TBAF, THF.

Scheme 106

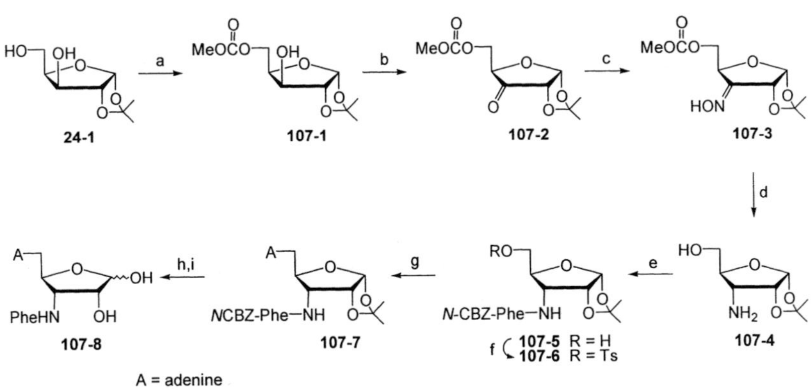

Reagents: a) MeOCOCl, Py; b) H_3PO_4, DCC, DMSO; c) $NH_2OH{\cdot}HCl$, Py; d) LAH, THF; e) *N*-CBZ-Phe, EEDQ; f) TsCl, Py; g) adenine, NaH; h) HCl, H_2O; i) H_2, 10% Pd/C.

Scheme 107

The bicyclic carbohydrate isosorbide **108-1** (Scheme 108) has been used as starting material for the synthesis of bicyclic nucleosides **108-4** and **108-11**, structurally related to the natural nucleosides griseolic acids, endowed with inhibitory activity against cyclic nucleotide phosphodiesterase.[230,231] In the synthesis of *nor*-derivatives **108-4**, isosorbide

108-1 was selectively benzoylated, then triflated to compound **108-2**. Reductive azidation of the latter afforded the key amine **108-3**, upon which different bases were built by conventional methods.[230]

Isosorbide **108-1** was also converted to the protected secondary alcohol **108-7** in several steps. This was dehydrated to unsaturated compound **108-8**, which was condensed with heterocyclic bases in the presence of NIS to give β-protected nucleosides **108-10**. The β-orienting effect of iodine is shown in transition state **108-9** (Scheme 108). Deiodination and deprotection afforded the final nucleosides **108-11**.[231]

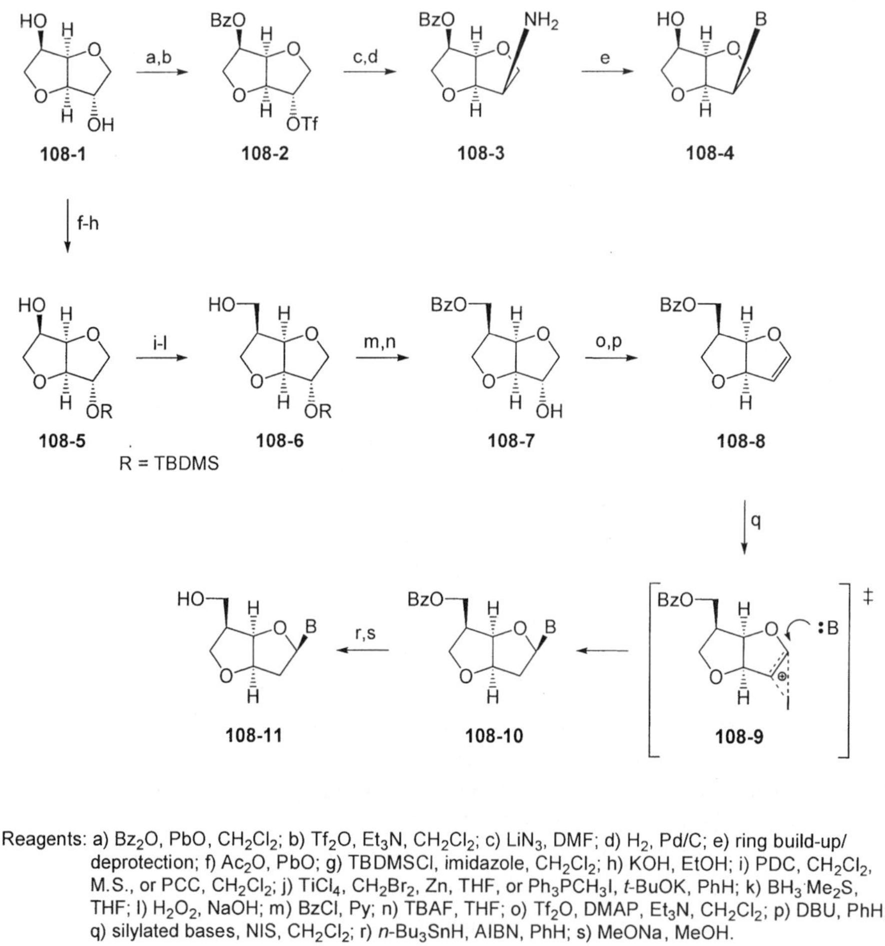

Reagents: a) Bz_2O, PbO, CH_2Cl_2; b) Tf_2O, Et_3N, CH_2Cl_2; c) LiN_3, DMF; d) H_2, Pd/C; e) ring build-up/deprotection; f) Ac_2O, PbO; g) TBDMSCl, imidazole, CH_2Cl_2; h) KOH, EtOH; i) PDC, CH_2Cl_2, M.S., or PCC, CH_2Cl_2; j) $TiCl_4$, CH_2Br_2, Zn, THF, or Ph_3PCH_3I, *t*-BuOK, PhH; k) $BH_3 \cdot Me_2S$, THF; l) H_2O_2, NaOH; m) BzCl, Py; n) TBAF, THF; o) Tf_2O, DMAP, Et_3N, CH_2Cl_2; p) DBU, PhH; q) silylated bases, NIS, CH_2Cl_2; r) *n*-Bu_3SnH, AIBN, PhH; s) MeONa, MeOH.

Scheme 108

2.11. References

1. (a) Gumina, G.; Song, G. -Y.; Chu, C. K. L-Nucleosides as chemotherapeutic agents. *FEMS Microbiol. Lett.* **2001**, *202*, 9-15. (b) Gumina, G.; Song, G. -Y.; Chu, C. K. Advances in antiviral agents for hepatitis B virus. *Antiviral Chem. Chemother.* **2001**, *12*, 89-112. (c) Cooperwood, J. S.; Gumina, G.; Boudinot, F. D.; Chu, C. K. Nucleoside and nucleotide prodrugs. In *Recent Advances in Nucleosides: Chemistry and Chemotherapy*. C. K. Chu Ed. **2002**, 91-147. (d) Kinchington, D. Recent advances in antiviral therapy. *J. Clin. Pathol.* **1999**, *52*, 89-94. (e) Mansour, T. S.; Storer, R. Antiviral nucleosides. *Curr. Pharm. Design* **1997**, *3*, 227-264. (f) De Clercq, E. HIV inhibitors targeted at the reverse transcriptase. *AIDS Res. Hum. Retroviruses* **1992**, *8*, 119-134. (g) Schinazi, R. F.; Mead, J. R.; Feorino, P. M. Insight into HIV chemotherapy. *AIDS Res. Hum. Retroviruses* **1992**, *8*, 963-990. (h) De Clercq, E. Antiviral activity spectrum and target of action of different classes of nucleoside analogs. *Nucleosides Nucleotides* **1994**, *13*, 1271-1295.
2. Reist, E. J.; Goodman, L.; Lee, W.W.; Baker, B. R.; Benitez, A. Potential anticancer agents. 76. Synthesis of purine nucleosides of β-D-arabinofuranose. *J. Org. Chem.* **1962**, *27*, 3274-3279.
3. Parks, R.; Stoeckler, J.; Cambor, C.; Savarese, T.; Crabtree, G.; Chu, S. Molecular actions and targets for cancer chemotherapeutic agents. Sartorelli, A., Lazo, T., Bertino, J. eds., Academic Press, New York, **1981**, 229.
4. Paulsen, H.; Todt, K. *Adv. Carbohydr. Chem.* **1968**, *23*, 115-232.
5. Dyson, M. R.; Coe, P. L.; Walker, R. T. An improved synthesis of benzyl 3,5-di-O-benzyl-2-deoxy-1,4-dithio-D-eryhtro-pentofuranoside, an intermediate in the synthesis of 4'-thionucleosides. *Carbohyd. Res.* **1991**, *216*, 237-248.
6. Bredenkamp, M. W.; Holzapfel, C. W.; Swanepoel, A. D. Stannylene directed selective acylation of some open-chain-arabinose derivatives. *Tetrahedron Lett.* **1990**, *31*, 2759-2762.
7. Bellon, L.; Barascut, J. L.; Imbach, J. L. Efficient synthesis of 4-thio-D-ribofuranose and some 4'-thioribonucleosides. *Nucleosides Nucleotides* **1992**, *11*, 1467-1479.
8. Classon, B.; Garegg, P. J.; Samuelsson, B.; Liu, Z. C. Novel route to 4-thiofuranosides - synthesis of methyl 4-thio-α-D-talofuranosides. *J. Carbohydr. Chem.* **1987**, *6*, 593-597.
9. Huang, B.; Hui, Y. A facile synthesis of 4'-thio-2'-deoxypyrimidine nucleosides and preliminary studies on their properties. *Nucleosides Nucleotides* **1993**, *12*, 139-147.
10. Tiwari, K. N.; Montgomery, J. A.; Secrist, J. A., III. The synthesis and biological-activity of 1-(2-deoxy-4-thio-α-L-*threo*-pentofuranosyl)thymine. *Nucleosides Nucleotides* **1993**, *12*, 841-846.
11. Dyson, M. R.; Coe, P. L.; Walker, R. T. The synthesis and antiviral properties of E-5-(2-bromovinyl)-4'-thio-2'-deoxyuridine. *J. Chem. Soc., Chem. Commun.* **1991**, 741-742.
12. Dyson, M. R.; Coe, P. L.; Walker, R. T. The synthesis and antiviral activity of some 4'-thio-2'-deoxy nucleoside analogs. *J. Med. Chem.* **1991**, *34*, 2782-2786.
13. Secrist, J. A.; Tiwari, K. N.; Riodan, J. M.; Montgomery, J. A. Synthesis and biological activity of 2'-deoxy-4'-thio pyrimidine nucleosides. *J. Med. Chem.* **1991**, *34*, 2361-2366.
14. Fu, Y. L.; Bobek, M. An alternative synthesis of anomeric methyl 2-deoxy-4-thio-D-*erythro*-pentofuranosides. *J. Org. Chem.* **1976**, *41*, 3831-3834.
15. Leydier, C.; Bellon, L.; Barascut, J. L.; Deydier, J.; Georges, M.; Helene, P.; Moulay, A. E.; Imbach, J. L. A new synthesis of some 4'-thio-D-ribonucleosides and preliminary enzymatic evaluation. *Nucleosides Nucleotides* **1994**, *13*, 2035-2050.
16. Reist, E. J.; Gueffroy, D. E.; Goodman, L. Synthesis of 4-thio-D- + -L-ribofuranose + corresponding adenine nucleosides. *J. Am. Chem. Soc.* **1964**, *86*, 5658-5663.

17. Bobek, M.; Whistler, R. L.; Bloch, A. Synthesis and biological activity of 4'-thio analogs of the antibiotic toyocamycin. *J. Med. Chem.* **1972**, *15*, 168-171.

18. Bobek, M.; Bloch, A.; Parthasaarthy, R.; Whistler, R. L. Synthesis and biological activity of 5-fluoro-4'-thiouridine and some related nucleosides. *J. Med. Chem.* **1975**, *18*, 784-787.

19. Tber, B., Fahmi; N., Ronco; G., Villa; P., Ewing; D. F., Mackenzie, G. An alternative strategy for the synthesis of 3'-azido-2',3'-dideoxy-4'-thionucleosides starting from D-xylose. *Carbohyd. Res.* **1995**, *267*, 203-215.

20. Wong, M. Y. H.; Gray, G. R. 2-deoxypentoses - stereoselective reduction of ketene dithioacetals. *J. Am. Chem. Soc.* **1978**, *100*, 3548-3553.

21. Aitsir, H.; Ewing, D. F.; Fahmi, N.; Goethals, G.; Mackenzie, G.; Ronco, G.; Tber, B.; Villa, P. Synthesis of 2-deoxy-4-thio-D-ribofuranose and its 3-azido analog from L-arabinose intermediates in the synthesis of 4'-thiodeoxynucleosides. *Nucleosides Nucleotides*, **1995**, *14*, 359-362.

22. Jandu, K. S.; Selwood, D. L. 4'-Thionucleosides *via in situ* pyranose-furanose rearrangement- A short synthesis of the antiherpes agent 2'-deoxy-5-ethyl-4'-thiouridine *via* direct coupling of a silylated pyrimidine base with a 4-thiopyranose sugar. *J. Org. Chem.* **1995**, *60*, 5170-5173.

23. Tsuda, Y.; Kanemitsu, K.; Kakimoto, K.; Kikuchi, T. Utilization of sugars in organic synthesis. 19. Trialkyltin radical used to catalyze the O,S-rearrangement of cyclic thiocarbonate - A new entry to thio-sugars. *Chem. Pharm. Bull.* **1987**, *35*, 2148-2150.

24. Rassu, G.; Spanu, P.; Pinna, L.; Zanardi, F.; Casiraghi, G. 2-(*tert*-Butyldimethylsiloxy) thiophene: Application to total synthesis of both enantiomer of 2',3'-dideoxy-4'-thiocytidine. *Tetrahedron Lett.* **1995**, *36*, 1941-1944.

25. Baer, E. L-α-Glycerophosphoric acid - barium salt. *Biochem. Prep.* **1952**, *2*, 31-38.

26. Eibl, H. An improved method for the preparation of 1,2-isopropylidene-*SN*-glycerol. *Phys. Lipids* **1981**, *28*, 1-5.

27. Hubschwerlen, C. A convenient synthesis of L-(*S*)-glyceraldehyde acetonide from L-ascorbic acid. *Synthesis* **1986**, 962-964.

28. Young, R. J.; Shaw-Ponter, S.; Thomson, J. B.; Miller, J. A.; Cumming, J. G.; Pugh, A. W.; Rider, P. Synthesis and antiviral evaluation of enantiomeric 2',3'-dideoxy- and 2',3'-dideohydro-2',3'-dideoxy-4'-thionucleosides. *Bioorg. Med. Chem. Lett.* **1995**, *5*, 2599-2604.

29. Pederson, R. L.; Liu, K. K. C.; Rutan, J., F.; Chen, L.; Wong, C., -H. Enzymes in organic synthesis - synthesis of highly enantiomerically pure 1,2-epoxy aldehyde, epoxy alcohols, thirane, aziridine, and glyceraldehyde-3-phosphate. *J. Org. Chem.* **1990**, *55*, 4897-4901.

30. Taguchi, Y.; Suhara, Y. The synthesis of dihydro-2(3H)-thiophenone derivative. *Bull. Chem. Soc. Jpn.* **1986**, *59*, 2321-2324.

31. Kawakami, H.; Ebata, T.; Koseki, K.; Matsushita, H.; Naoi, Y.; Itoh, K. Stereoselectivity in the coupling reaction between 2-phenylthio-2,3-dideoxyribose and silylated pyrimidine bases. *Chem. Lett.* **1990**, 1459-1462.

32. Choi, Y.; Choo, H.; Chong, Y.; Lee, S.; Olgen, S.; Schinazi, R. F.; Chu, C. K. Synthesis and potent anti-HIV activity of L-2',3'-didehydro-2',3'-dideoxy-2'-fluoro-4'-thiocytidine. *Org. Lett.* **2002**, *4*, 305-307.

33. Choi, Y.; Choo, H.; Chong, Y.; Lee, S.; Schinazi, R. F.; Chu, C. K. Synthesis and antiviral activities of L-2',3'-didehydro-2',3'-dideoxy-2'-fluoro-4'-thionucleosides. *Abstr. Papers Am. Chem. Soc.* **2001**, CARB 029.

34. Choo, H.; Chong, Y.; Lee, S.; Choi, Y.; Schinazi, R. F.; Chu, C. K. Synthesis and anti-HIV activity of D-2',3'-unsaturated-2'-fluoro-4'-thionucleosides. *Abstr. Papers Am. Chem. Soc.* **2002**, MEDI 244.

35. Chu, C. K.; Choi, Y.; Choo, H.; Chong, Y.; Lee, S.; Zhu, W.; Schinazi, R. F. Synthesis and anti-HIV activity

of D- and L-2',3'didehydro-2',3'-dideoxy-2'-fluoro-4'-thiocytidines. *Antiviral Res.* **2002**, *53*, A36-A36.

36. Van Draanen, N. A.; Freeman, G. A.; Short, S. A.; Harvey, R.; Jansen, R.; Szczech, G.; Kozalka, G. W. Synthesis and antiviral activity of 2'-deoxy-4'-thiopurine nucleosides. *J. Med. Chem.* **1996**, *39*, 538-542.
37. Rahim, S.G.; Tricedi, N.; Bogunovic-Batchelor, M. V.; Hardy, G. W.; Mills, G.; Selway, J. W. T.; Snowden, W.; Littler, E.; Coe, P. L.; Basnak, I.; Whale, R. F.; Walker, R. T. Synthesis and anti-herpes virus activity of 2'-deoxy-4'-thiopyrimidine nucleosides. *J. Med. Chem.* **1996**, *39*, 789-795.
38. Birk, C.; Voss, J.; Wirsching, J. Preparation, structural elucidation and reactions of benzyl 2-deoxy-3,5-di-O-methyl-1,4-dithio-L-*threo*-pentofuranoside and synthesis of the corresponding 2'-deoxy-4'-thionucleosides. *Carbohydr. Res.* **1997**, *304*, 239-247.
39. Yoshimura, Y.; Kitano, K.; Yamada, K.; Satoh, H.; Watanabe, M.; Miura, S.; Sakata, S.; Sasaki, T.; Matsuda, A. A novel synthesis of 2'-modified 2'-deoxy-4'-thiocytidines from D-glucose. *J. Org. Chem.* **1997**, *62*, 3140-3152.
40. Yoshimura, Y.; Kitano, K.; Satoh, H.; Watanabe, M.; Miura, S.; Sakata, S.; Sasaki, T.; Matsuda, A. A novel synthesis of new antineoplastic 2'-deoxy-2'-substituted-4'-thiocytidines. *J. Org. Chem.* **1996**, *61*, 822-823.
41. Abbruzzese, J. L.; Grunewald, R.; Weeks, E. A.; Gravel, D.; Adams. T.; Nowak, B.; Mineishi, S.; Tarassoff, P.; Satterlee, W.; Raber, M. N.; Plinkett, W. A phase-1 clinical, plasma, and cellular pharmacology study of gemcitabine. *J. Clin. Oncol.* **1991**, *9*, 491-498.
42. Jeong, L. S.; Nicklaus, M. C.; George, C.; Marquez, V. E. Unanticipated retention of configuration in the DAST fluorination of deoxy-4'-thiopyrimidine nucleosides with up-hydroxyl groups. *Tetrahedron Lett.* **1994**, *35*, 7569-7572.
43. Hughes, N. A.; Kuhajda, K. M.; Miljkovic, D. A. New syntheses of methyl 5-thio-β-D- arabinopyranoside and (+)-biotin. *Carbohydr. Res.* **1994**, *257*, 299-304.
44. Machida, H.; Ashida, N.; Miura, S.; Endo, M.; Yamada, K.; Kitano, K.; Yoshimura, Y.; Sakata, S.; Ijichi, O.; Eizuru, Y. Anti-herpesvirus activity profile of 4'-thioarabinofuranosyl purine and uracil nucleosides and activity of 1-β-D-2'-fluoro-4'-thioarabinofuranosyl guanine and 2,6-diaminopurine against clinical isolates of human cytomegalovirus. *Antiviral Res.* **1998**, *39*, 129-137.
45. An, S. H.; Bobek, M. Methyl 3,4-di-O-acetyl-2-deoxy-2(R)-[1H,3H-2,4-dioxo-1-pyrimidinyl]-2-fluoro-β-D-arabinopyranoside - novel nucleosides analogs *via* condensation of gem-2,2-difluoromethyl glycosides with silylated heterocyclic bases. *Tetrahedron Lett.* **1986**, *27*, 3219-3222.
46. Yoshimura, Y.; Kitano, K.; Watanabe, M.; Satoh, H.; Sakata, S.; Miura, S.; Ashida, N.; Machida, H.; Matsuda, A. Synthesis and biological activities of 2'-modified 4'-thionucleosides. *Nucleosides Nucleotides* **1997**, *16*, 1103-1106.
47. Yoshimura, Y.; Watanabe, M.; Satoh, H.; Ashida, N.; Ijichi, K.; Sakata, S.; Machida, H.; Matsuda, A. A facile, alternative synthesis of 4'-thioarabinonucleosides and their biological activities. *J. Med. Chem.* **1997**, *40*, 2177-2183.
48. Haraguchi, K.; Nishikawa, A.; Sasakura, E.; Tanaka, H.; Nakamura, K. T.; Miyasaka, T. Electrophilic addition to 4-thio furanoid glycal: a highly stereoselective entry to 2'-deoxy-4'-thio pyrimidine nucleosides. *Tetrahedron Lett.* **1998**, *39*, 3713-3716.
49. DeValette, F.; Barascut, J. L.; Imbach, J. L. Synthesis and antiviral evaluation of 2'-deoxy- 4'-thio-L-nucleosides and their phosphotriester derivatives bearing S-acyl-2-thioethyl bioreversible phosphate-protecting groups. *Nucleosides Nucleotides* **1998**, *17*, 2289-2310.
50. Bellon, L.; Leydier, C.; Barascut, J. L.; Imbach, J. L. A reinvestigation of the synthesis of 4'-thio-pentofuranosides for further nucleosidic synthesis. *Nucleosides Nucleotides* **1993**, *12*, 847-852.
51. Debernardo, S.; Tengi, J. P.; Sasso, G. J.; Weigele, M. Clavalanine (RO 22-5417), a new clavam antibiotic from sreptomyces-clavuligerus.4. A stereorational synthesis. *J. Org. Chem.* **1985**, *50*, 3457-3462.

52. Rasmussen, J. R.; Slinger, C. J.; Kordish, R. J.; Newman-Evans, D. D. Synthesis of deoxy sugars - deoxygenation by treatment with N,N'-thiocarbonyldiimidazole-tri-normal-butyl stannane. *J. Org. Chem.* **1981**, *46*, 4843-4846.
53. Barton, D. H. R.; Jang, D. O.; Jaszberenyi, J. C. The invention of radical reactions. 29. Radical monodeoxygenation and dideoxygenation with silanes. *Tetrahedron* **1993**, *49*, 2793-2804.
54. Vorbrüggen, H.; Bennua, B. Nucleoside syntheses. 19. New simplified nucleoside synthesis. *Tetrahedron Lett.* **1978**, *15*, 1339-1342.
55. Secrist, J. A.; Tiwari, K. N.; Shorynacy-Fowler, A. T.; Messini, L.; Riordan, J.; Montgomery, J.A. Synthesis and biological activity of certain 4'-thio-D-arabinofuranosylpurines. *J. Med. Chem.* **1998**, *41*, 3865-3871.
56. Jeong, L. S.; Moon, H. Y.; Choi, Y. J.; Chun, M. W.; Kim, H. O. A short and efficient synthesis of L-thioarabino derivative: A versatile synthon for the synthesis of L-2'-deoxy-2',3'-disubstituted-4'-thionucleosides. *J. Org. Chem.* **1998**, *63*, 4821-4825.
57. Blumberg, A.; Fuccello, A.; Van Es, T. Intramolecular cyclization of pentose and hexose dithio-acetals. *Carbohydr. Res.* **1979**, *70*, 217-232.
58. Jeong, L. S.; Moon, H. Y.; Yoo, S. J.; Lee, S. N.; Chun, M. W.; Lim, Y. H. Ring contraction and rearrangement of 4-thiofuranose derivatives observed during DAST fluorination in the synthesis of L-2'-"UP"-fluoro-4'-thiothymidine. *Tetrahedron Lett.* **1998**, *39*, 5201-5204.
59. Tiwari, K. N.; Montgomery, J. A.; Secrist III, J. A. The synthesis and biological activity of 1-(2-deoxy-4-thio-α-L-*threo*-pentofuranosyl)thymine. *Nucleosides Nucleotides* **1993**, *12*, 841-846.
60. Montgomery, J. A.; Thomas, H. J. Isonucleosides. 2. Purine and pyrimidine derivatives of 1,4-anhydro-2-deoxy-D-arabinitol. *J. Org. Chem.* **1978**, *43*, 541-544.
61. Kjølberg, O. A new method of degrading glucose to xylose and galactose to arabinose. *Acta Chem. Scan.* **1960**, *14*, 1118-1123.
62. Bobek, M.; An, S. -H.; Skrincosky, D.; De Clerq, E.; Bernacki, R. J. 2'-Fluorinated isonucleosides. 1. Synthesis and biological activity of some methyl 2'-deoxy-2'-fluoro-2'-pyrimidinyl-D-arabinopyranosides. *J. Med. Chem.* **1989**, *32*, 799-807.
63. Huryn, D. M.; Sluboski, B. C.; Tam, S. Y.; Todaro, L. J.; Weigele, M. Synthesis of iso-ddA, member of a novel class of anti-HIV agents. *Tetrahedron Lett.* **1989**, *30*, 6259-6262.
64. Nair, V.; Nuesca, Z. M. Isodideoxynucleosides: A conceptually new class of nucleoside antiviral agents. *J. Am. Chem. Soc.* **1992**, *114*, 7951-7953.
65. Zintek, L. B.; Jeon, G. S.; Nair, V. The synthesis of (*S*) and (*R*) enantiomers of novel hydroxymethylated isodideoxynucleosides. *F* **1994**, *37*, 1853-1864.
66. Tino, J. A.; Clark, J. M.; Field, A. K.; Jacobs, G. A.; Lis, K. A.; Michalik, T. L.; McGeever-Rubin; B., Slusarchyk, W. A.; Spergel, S. H.; Sundeen, J. E.; Tuomari, A. V.; Weaver, E. R.; Young, M. G.; Zahler, R. Synthesis and antiviral activity of novel isonucleoside analogs. *J. Med. Chem.* **1993**, *36*, 1221-1229.
67. Kira, T.; Kakefuda, A.; Shuto, S.; Matsuda, A.; Baba, M.; Shigeta, S. Anti-herpes activities of isonucleoside analogs with variable bases at 2' position. *Nucleosides Nucleotides* **1995**, *14*, 571-574.
68. (a) Levene, P. A.; Raymond, A. L. Derivative of monoacetone xylose. J. Biol. Chem. **1933**, 102, 317-330; (b) Rosenthal, A.; Baker, D. A. New routes to branched-chain amino sugars by application of modified Wittig reaction to ketoses. *Tetrahedron Lett.* **1969**, 397-400.
69. Rosenthal, A.; Sprinzl, M. Branched-chain sugars *via* the hydroboration-oxidation of unsaturated sugar derivatives. *Carbohydrate Res.* **1971**, *16*, 337-342.
70. Franchetti, P.; Capellacci, L.; Grifantini, M.; Messini, L.; Sheikha, G. A.; Loi, A. G.; Tramontano, E.; DeMontis, A.; Spiga, M. G.; LaColla, P. Synthesis and evaluation of the anti-HIV activity of aza and

deaza analogs of isoddA and their phosphates as prodrugs. *J. Med. Chem.* **1994**, *37*, 3534-3541.

71. Varela, O.; De Fina, G. M.; DeLederkremer, R. M. The reaction of 2-hydroxyglycal esters with alcohols in the presence of N-iodosuccinimide stereoselective synthesis of α-anomers of alkyl 3-deoxyhex-2-enopyranosides and 3,4-dideoxyhex-3-enopyranoside-2-uloses. *Carbohydr. Res.* **1987**, *167*, 187-196.
72. Lemieux, R.U. Newer developments in the conformational analysis of carbohydrates. *Pure Appl. Chem.* **1971**, *27*, 527-547.
73. Hanessian, S.; Banoub, J. Chemistry of glycosidic linkage - efficient synthesis of 1,2-trans-disaccharides. *Carbohydr. Res.* **1977**, *53*, C13-C16.
74. Herscovici, J.; Bessodes, M.; Antonakis, K. Studies on sodium-borohydride reduction of unsaturated keto nucleosides – novel route to deoxy nucleosides. *J. Org. Chem.* **1976**, *41*, 3827-3830.
75. Hammerschmidt, F.; Ohler, E.; Polster, J. P.; Zbiral, E.; Balzarini, J.; DeClercq, E. A Convenient route to D-apio-β-D-furanosyl- and 2'-deoxyapio-β-D-furanosyl nucleosides. *Liebigs Ann.* **1995**, 551-558.
76. Hammerschmidt, F., Ohler, E., Polster, J. P., Zbiral, E., Balzarini, J., DeClercq, E. A Convenient route to 3'-deoxy-α-L- and 3'-deoxy-β-D-apionucleosides. *Liebigs Ann.* **1995**, 559-565.
77. Purdy, D. F.; Zintek, L. B.; Nair, V. Synthesis of isonucleosides related to AZT and AZU. *Nucleosides Nucleotides*, **1994**, *13*, 109-126.
78. Barker, R.; Fletcher, H. G. 2,3,5-Tri-O-benzyl-D-ribosyl and -L-arabinosyl bromides. *J. Org. Chem.* **1961**, *26*, 4605-4609.
79. Bennek, J. A.; Gray, G. R. An efficient synthesis of anhydroalditols and ally *C*-glycosides. *J. Org. Chem.* **1987**, *52*, 892-897.
80. Heath, P.; Mann, J.; Walsh, E. B.; Wadsworth, A. H. The preparation of dioxaprostacyclin analogs from D-(-)-ribose. *J. Chem. Soc., Perkin Trans. 1* **1983**, 2675-2679.
81. Nicotra, F.; Panza, L.; Russo, G.; Zuchelli, L. Chemoenzymatic approaches to carbohydrate-derivative anaolgs of platelet-activating-factor. *J. Org. Chem.* **1992**, *57*, 2154-2158.
82. Montgomery, J. A.; Thomas, H. J. Isonucleosides 2. Purine and pyrimidine-derivatives of 1,4-anhydro-2-deoxy-D-arabinitol. *J. Org. Chem.* **1978**, *43*, 541-544.
83. Shaw, G.; Warrener, R.N. Purines, pyrimidines, and glyoxalines. 8. New syntheses of uracils and thymines. *J. Chem. Soc.* **1958**, 157-161.
84. Shealy, Y. F.; O'Dell, C. A. Acid-catalyzed cyclization of alkoxyacryloloureas to 2,4(1H,3H) pyrimidinediones. *J. Heterocycl.. Chem.* **1976**, *13*, 1041-1047.
85. Sochacka, E.; Nawrot, B.; Pankiewicz, K. W.; Watanabe, K. A. Synthesis of 1-methyl-5-(3-azido-2,3-dideoxy-β-D-*erythro*-pentofuranosyl)uracil and 1-methyl-5-(3-azido-2,3-dideoxy-2-fluoro-β-D-arabinofuranosyl)uracil - the C-nucleoside isostere of 3'-azido-3'-deoxythymidine and its 2'-up-fluoro analogs. *J. Med. Chem.* **1990**, *33*, 1995-1999.
86. Dunkel, M.; Pfleiderer, W. Nucleosides 52. Synthesis and properties of quinazoleine-3'-azidonucleosides. *Nucleosides Nucleotides* **1993**, *12*, 125-137.
87. Kakefuda, A.; Shuto, S.; Nagahata, T.; Seki, J.; Sasaki, T.; Matuda, A. Nucleosides and nucleotides. 132. Synthesis and biological evaluations of ring-expanded oxetanocine analogs: purine and pyrimidine analogs of 1,4-anhydro-2-deoxy-D-arabitol and 1,4-anhydro-2-deoxy-3-hydroxymethyl-D-arabitol. *Tetrahedron* **1994**, *50*, 10167-10182.
88. Barker, D.C.; Horton, D.; Tindall Jr., C.G. Large-scale preparation of D-allose: observations on stereoselectivity of reduction of 1,2-5,6-di-*O*-isopropylidene-α-D-*ribo*-hexofuranos-3-ulose hydrate. *Carbohydrate Res.* **1972**, *24*, 192-197.
89. Acton, E. M.; Goerner, R. N.; Uh, H. S.; Ryan, K. J.; Henry, D. W. Improved anti-tumor effects in 3'-branched homologs of 2'-deoxythioguanosine - synthesis and evaluation of thioguanosine nucleosides

of 2,3-dideoxy-3-(hydroxymethyl)-D-*erythro*-pentofuranose. *J. Med. Chem.* **1979**, *22*, 518-525.

90. Ohrui, H.; Waga, T.; Ueno, A.; Okamoto, M.; Horie, H.; Meguro, H.; Kabuto, C. A new reaction of adenine with a 1,4-anhydroribitol having cis-orientated azide and O-tosylate: formation of a novel aziridinoadenine isonucleoside. *Biosci. Biotech. Biochem.* **1995**, *59*, 1801-1803.
91. Yu, H. W.; Zhang, L. R.; Zhou, J. C.; Ma, L. T.; Zhang, L. H. Studies on the synthesis and biological activities of 4'-(R)-hydroxy-5'-(S)-hydroxymethyl-tetrahydrofuranyl purines and pyrimidines. *Bioorg. Med. Chem.* **1996**, *4*, 609-614.
92. Prévost N.; Rouessac, F. Isonucleosides by Michael addition of pyrimidine bases on 2,6-disubstituted 2H-pyran-3(3H)-ones. *Tetrahedron Lett.* **1997**, *38*, 4215-4218.
93. Prévost, N.; Rouessac, F. A novel route to pyrimidine isodideoxynucleosides *via* Michel-type addition on unsaturated modified sugars. *Synthetic Commun.* **1997**, *27*, 2325-2335.
94. Fraser-Reid, B. Some progeny of 2,3-unsaturated sugars - they little resemble grandfather glucose: ten years later. *Acc. Chem. Res.* **1985**, *18*, 347-354.
95. Koreeda, M.; Houston, T. A.; Shull, B. K.; Klemke, E.; Tuinman, R. J. Iodine-catalyzed Ferrier reaction. 1. A mild and highly versatile glycosylation of hydroxyl and phenolic groups. *Synlett* **1995**, 90-92.
96. Marco, J. L. A new and simple synthesis of (+)- *threo*-ephedrine and (-)-*threo*-ephedrine. *J. Chem. Res.* **1988**, *S2*, 51-51.
97. Prasad, J. S.; Clive, D. L. J.; Da Silva, G. V. J. Synthetic studies related to compactin - use of tri-O-acetyl-D-glucal for preparation of chiral cyclohexenes. *J. Org. Chem.* **1986**, *51*, 2717-2721.
98. Card, P. J. Synthesis of benzannelated pyranosides. *J. Org. Chem.* **1982**, *47*, 2167-2173.
99. Talekar, R. R.; Wightman, R. H. Synthesis of some pyrrolo[2,3-d]pyrimidine and 1,2,3-triazole isonucleosides. *Tetrahedron*, **1997**, *53*, 3831-3842.
100. Moravcova, J.; Capkova, J.; Stanek, J. One-pot synthesis of 1,2-O-isopropylidene-α-D-xylofuranose. *Carbohydr. Res.* **1994**, *263*, 61-66.
101. Defaye, J.; Horton, D.; Muesser, M. Methanolysis of 1,2-*O*-isopropylidene-5-*O*-*p*-tolylsulfonyl-pentofuranoses - influence of steric factors on the formation of 2,5-anhydropentose dimethyl acetals. *Carbohydr. Res.* **1971**, *20*, 305-318.
102. Cristalli, G.; Franchetti, P.; Grifantini, M.; Vittori, S.; Lupidi, G.; Riva, F.; Bordoni, T.; Geroni, C.; Verini, M. A. Adenosine-deaminase inhibitors - synthesis and biological-activity of deaza anaolgs of *erythro*-9-(2-hydroxy-3-nonyl)adenine. *J. Med. Chem.* **1988**, *31*, 390-393.
103. Yang, Z. J.; Yu, H. W.; Min, J. M.; Ma, L. T.; Zhang, L. H. Stereoselective synthesis of 4-deoxy-4-nucleobase-2,5-anhydro-L-mannitol derivatives. *Tetrahedron Asymm.* **1997**, *8*, 2739-2747.
104. Yu, H. W.; Zhang, H. Y.; Yang, Z. J.; Min, J. M.; Ma, L. T.; Zhang, L. H. Studies on the syntheses and biological activities of isonucleosides. *Pure Appl. Chem.* **1998**, *70*, 435-438.
105. Hong, J. H.; Lee, K.; Choi, Y.; Chu, C. K. Enantiomeric synthesis of 3'-fluoro-apionucleosides using Claisen rearrangement. *Tetrahedron Lett.* **1998**, *39*, 3443-3446.
106. David, S.; Thieffry, A.; Veyrieres, A. A mild procedures for the regiospecific benzylation and allylation of polyhydroxy-compounds *via* their stannylene derivatives in non-polar solvents. *J. Chem. Soc., Perkin Trans. I* **1981**, 1796-1801.
107. Lee, K.; Choi, Y.; Hong, J. H.; Schinazi, R. F.; Chu, C. K. New classes of fluorinated L-nucleosides; synthesis and antiviral activity. *Nucleosides Nucleotides* **1999**, *18*, 537-540.
108. Hong, J. H.; Gao, M. Y.; Chu, C. K. Synthesis of novel 3'-C-methyl-4'-thio apionucleosides *via* highly enantioselective elaboration of quaternary carbon by [3,3]-sigmatropic rearrangement. *Tetrahedron Lett.* **1999**, *40*, 231-234.
109. Hong, J. H.; Gao, M. -Y.; Choi, Y.; Cheng, Y. -C; Schinazi, R. F. Chu, C. K. Synthesis of novel 3'-*C*-

methyl-apionucleosides: an asymmetric construction of a quaternary carbon by Claisen rearrangement. *Carbohydr. Res.* **2000**, *328*, 37-48.

110. Chu C. K.; Beach J. W.; Jeong L. S.; Choi B. G.; Comer F. I.; Alves A. J.; Schinazi R. F. Enantiomeric synthesis of (+)-BCH-189-[(+)-(2S,5R)-1-[2-(hydroxymethyl)-1,3-oxathiolan-5-yl]-cytosine] from D-Mannose and its anti-HIV activity. *J. Org. Chem.* **1991**, *56*, 6503-6505.
111. Jeong L. S.; Alves A. J.; Carrigan S. W.; Kim H. O.; Beach J. W.; Chu C. K. An efficient synthesis of enantiomerically pure (+)-(2S,5R)-1-[2-(hydroxymethyl)-1,3-oxathiolan-5-yl]-cytosine [(+)-BCH-189] from D-galactose. *Tetrahedron Lett.* **1992**, *33*, 595-598.
112. Jeong L. S.; Schinazi R. F.; Beach J. W.; Kim H. O.; Shanmuganathan K.; Nampalli S.; Chun M. W.; Chung W. K.; Choi B. G.; Chu C. K. Structure-activity relationships of β-D-(2S,5R)- and α-D-(2S,5S)-1,3-oxathiolanyl nucleosides as potential anti-HIV agents. *J. Med. Chem.* **1993**, *36*, 2627-2638.
113. Beach J. W.; Jeong L. S.; Alves A. J.; Pohl D.; Kim H. O.; Chang C. N.; Doong S. L.; Schinazi R. F.; Cheng Y. -C.; Chu C. K. Synthesis of enantiomerically pure (2'R,5'S)-(–)-1-[2-(hydroxymethyl)oxathiola n-5-yl]-cytosine as a potent antiviral agent against hepatitis B virus (HBV) and human immunodeficiency virus (HIV). *J. Org. Chem.* **1992**, *57*, 2217-2219.
114. Jeong L. S.; Shinazi R. F.; Beach J. W.; Kim H. O.; Nampalli S.; Shanmuganathan K.; Alves A. J.; McMillan A.; Chu C. K.; Mathis R. Asymmetric synthesis and biological evaluation of β-L-(2R,5S)- and α-L-(2R,5R)-1,3-oxatholanyl-pyrimidine nucleosides as potential anti-HIV agents. *J. Med. Chem.* **1993**, *36*, 181-195.
115. Kim H. O.; Ahn S. K.; Alves A. J.; Beach J. W.; Jeong L. S.; Choi B. G.; VanRoey P.; Schinazi R. F.; Chu C. K. Asymmetric synthesis of 1,3-dioxolane pyrimidine nucleosides and their anti-HIV activity. *J. Med. Chem.* **1992**, *35*, 1987-1995.
116. Kim H. O.; Schinazi R. F.; Nampalli S.; Shanmuganathan, K.; Cannon, D. L.; Alves A. J.; Jeong L. S.; Beach J. W.; Chu, C. K. 1,3-Dioxolanylpurine nucleosides (2*R*, 4*R*) and (2*R*, 4*S*) with selective anti-HIV activity in human lymphocytes. *J. Med. Chem.* **1993**, *36*, 30-37.
117. Kim H. O.; Shanmuganathan K.; Alves A. J.; Jeong L. S.; Beach J. W.; Schinazi R. F.; Chang C. N.; Cheng Y. -C.; Chu, C. K. Potent anti-HIV and anti-HBV activities of (-)-L-β-dioxolane-C- and (+)-L-β-dioxolane-T and their asymmetric syntheses. *Tetrahedron Lett.* **1992**, *33*, 6899-6902.
118. Kim H. O.; Schinazi R. F.; Shanmuganathan K.; Jeong L. S.; Beach J. W.; Nampalli S.; Cannon D.; Chu C. K. L-β-(2*S*,4*S*)-dioxolanyl and L-α-(2*S*,4*R*)-dioxolanyl nucleosides as potential anti-HIV agents - asymmetric synthesis and structure-activity relationships. *J. Med. Chem.* **1993**, *36*, 519-528.
119. Evans C. A.; Dixit D. M.; Siddiqui A.; Jin H.; Allan Tse H. L.; Cimpoia A.; Bednarski K.; Breining T.; Mansour T. S. Divergent asymmetric synthesis of dioxolane nucleosides analogs. *Tetrahedron Asymm.* **1993**, *4*, 2319-2322.
120. Du J.; Qu F. C.; Lee D. W.; Newton M. G.; Chu C. K. 1,3-Dioxolane *C*-nucleosides – asymmetric synthesis of 4 stereoisomers of 2-[2-(hydroxymethyl)-1,3-dioxolan-5-yl)]-1,3-thiazole-4-carboxamide. *Tetrahedron Lett.* **1995**, *36*, 8167-8170.
121. Qu F.; Hong J. H.; Du J.; Newton M. G.; Chu C. K. Asymmetric synthesis of (2'R,4'R) and (2'S,4'S)-1,3-dioxolanyl triazole *C*-nucleosides. *Tetrahedron* **1999**, *55*, 9073-9088.
122. Shealy, Y. F.; Clayton, J. D. 9-[β-DL-2α,3α-dihydroxy-4β-(hydroxymethyl)cyclopentyl]adenine, the carbocyclic analog of adenosine. *J. Am. Chem. Soc.* **1966**, *88*, 3885-3887.
123. Kusaka, T.; Yamamoto, H.; Shibata, M.; Muroi, M.; Kishi, T.; Mizuno, K. Streptomyces citricolor nov. sp. And a new antibiotic, aristeromycin. *J. Antibiot.* **1968**, *21*, 255-263.
124. Hayashi, M.; Yaginuma, S.; Muto, N.; Tsujino, M. Structures of neplanocins, new antitumor antibiotics. *Nucleic Acids Synp. Ser.* **1980**, *8*, S65-67.

125. De Clerq, E. Antiviral and antimetabolic activities of neplanocins. *Antimicrob. Agents. Chemother.* **1985**, *28*, 84-89.
126. Borthwick, A. D.; Butt, S.; Biggadike, K.; Exall, A.M.; Roberts, S. M.; Youds, P. M.; Kirk, B. E.; Booth, B. R.; Cameron, J. M.; Cox, S. W.; Marr, C. L. P.; Shill, M. D. Synthesis and enzymatic resolution of carbocyclic 2'-*ara*-fluoro-guanosine - a potent new anti-herpetic agent. *J. Chem. Soc., Chem. Commun.* **1988**, 656-658.
127. Borthwick, A. D.; Kirk, B. E.; Biggadike, K.; Exall, A.M.; Butt, S.; Roberts, S. M.; Knight, D. J.; Coates, J. A. V.; Ryan, D. M. Fluorocarbocyclic nucleosides – synthesis and antiviral activity of 2'-fluorocarbocyclic and 6'-fluorocarbocyclic 2'-deoxyguanosines. *J. Med. Chem.* **1991**, *34*, 907-914.
128. De Clercq, E. Carbocyclic adenosine analogs as S-adenosylhomocysteine hydrolase inhibitors and antiviral agents: recent advances. *Nucleosides Nucleotides.* **1998**, *17*, 625-634.
129. Boehme, R. E.; Storer, R.; Williamson, C.; Clemens, I.R.; Cameron, J. M. Abstract of Papers, Interscience Conference on Antimicrobial Agents and Chemotherapy, Atlanta, October **1990**; Abstract 1088.
130. Bisacchi, G. S.; Chao, S. T.; Bachard, C.; Daris, J. P.; Innaimo, S.; Jacobs, G. A.; Kocy, O.; Lapointe, P.; Martel, A.; Merchant, Z.; Slusarchyk, W. A.; Sundeen, J. E.; Youg, M. G.; Colonno, R.; Zahler, R. BMS-200475, a novel carbocyclic 2'-deoxyadenosine analog with potent and selective anti-hepatitis B virus activity *in vitro*. *Bioorg. Med. Chem. Lett.* **1997**, *7*, 127-132.
131. Katagiri, N.; Nomura, M.; Sato, H.; Kaneko, C.; Yusa, K.; Tsuruo, T. Synthesis of nucleosides and related compounds -synthesis and anti-HIV activity of 9-[*C*-4,T-5-bis (hydroxymethyl)cyclopent-2-en-*R*-1-yl]-9*H*-adenine. *J. Med. Chem.* **1992**, *35*, 1882-1886.
132. Vince, R.; Hua, M. Synthesis and anti-HIV activity of carbocyclic 2',3'-didehydro-2',3'-dideoxy 2,6-disubstituted purine nucleosides. *J. Med. Chem.* **1990**, *33*, 17-21.
133. Daluge S. M.; Good S. S.; Faletto M. B.; Miller W. H.; St Clair M. H.; Boone L. R.; Tisdale M.; Parry N. R.; Reardon J. E.; Dornsife R. E.; Averett D. R.; Krenitsky T. A. 1592U89, A novel carbocyclic nucleoside analog with potent, selective anti-human immunodeficiency virus activity. *Antimicrob. Agents Chemother.* **1997**, *41*, 1082-1093.
134. Dobkin J. F. Abacavir enters the clinic. *Infect. Med.* **1999**, *16*, 7-7.
135. Borthwick, A. D.; Biggadike, K. Synthesis of chiral carbocyclic nucleosides. *Tetrahedron.* **1992**, *48*, 571-623.
136. Agrofoglio, L.; Suhas, E.; Farese, A.; Condom, R.; Challand, S. R.; Earl, R. A.; Guedj, R. Synthesis of carbocyclic nucleosides. *Tetrahedron* **1994**, *50*, 10611, 10670.
137. Crimmins, M. T. New developments in the enantioselective synthesis of cyclopentyl carbocyclic nucleosides. *Tetrahedron* **1998**, *54*, 9229-9272.
138. Marquez, V.E.; Lim, M. -I.; Tseng, C. K. -H.; Markovac, A.; Priest, M. A.; Khan, M. S.; Kaskar, B. Total synthesis of (-)-neplanocin A. *J. Org. Chem.* **1988**, *53*, 5709-5714.
139. Lim, M. -I.; Marquez, V. E. Total synthesis of (-)-neplanocin A. *Tetrahedron Lett.* **1983**, *24*, 5559-5562.
140. Lim, M. -I.; Marquez, V. E. A synthetic approach towards neplanocin A: preparation of the optically active cyclopentene moiety from D-ribose. *Tetrahedron Lett.* **1983**, *24*, 4051-4054.
141. Altmann, K. -H.; Kesselring, R.; Francotte, E.; Rihs, G. 4',6'-Methano carbocyclic thymidine: a conformationally constrained building block for oligonucleotides. *Tetrahedron Lett.* **1994**, *35*, 2331-2334.
142. Yoshikawa, M.; Okaichi, Y.; Cha, B. C.; Kitagawa, I. Synthesis of (-)-aristeromycin from D-glucose. *Tetrahedron* **1990**, *46*, 7459-7470.
143. Yoshikawa, M.; Okaichi, Y.; Cha, B. C.; Kitagawa, I. Synthesis of (-)-aristeromycin from D-glucose. *Chem. Pharm. Bull.* 1989, *37*, 2555-2557.
144. Wolfe, M. S.; Anderson, B. L.; Borcherding, D. R.; Borchardt, R. T. Enantiospecific syntheses of aristeromycin and neplanocin A. *J. Org. Chem.* **1990**, *55*, 4712-4717.

145. Wolfe, M. S.; Borcherding, D. R.; Borchardt, R. T. A 9-step enantiospecific synthesis of (-)-aristeromycin from D-ribonic acid γ-lactone. *Tetrahedron Lett.* **1989**, *30*, 1453-1456.
146. Borcherding, D. R.; Scholtz, S. A.; Borchardt, R. T. Synthesis of analogs of neplanocin A: utilization of optically active dihydroxycyclopentenones derived from carbohydrates. *J. Org. Chem.* **1987**, *52*, 5457-5461.
147. Bélanger, P.; Prasit, P. Carbocycles from carbohydrates: a simple route to an enantiomerically pure prostaglandin intermediate. *Tetrahedron Lett.* **1988**, *29*, 5521-5524.
148. Ali, S. M.; Ramesh, K.; Borchardt, R.T. Efficient enantioselective syntheses of carbocyclic nucleoside and prostaglandin synthons. *Tetrahedron Lett.* **1990**, *31*, 1509-1512.
149. Wang, P.; Agrofoglio, L.; Newton, M. G.; Chu, C. K. Asymmetric synthesis of L-cyclopentyl carbocyclic nucleosides. *Tetrahedron Lett.* **1997**, *38*, 4207-4210.
150. Wang, P.; Schinazi, R. F.; Chu, C. K. Asymmetric synthesis and anti-HIV activity of L-carbocyclic 2',3'-didehydro-2',3'-dideoxyadenosine. *Bioorg. Med. Chem. Lett.* **1998**, *8*, 1585-1588.
151. Wang, P. Agrofoglio, L. A.; Newton, M. G.; Chu, C. K. Chiral synthesis of carbocyclic analogs of L-ribofuranosides. *J. Org. Chem.* **1999**, *64*, 4173-4178.
152. Wang, P. Gullen, B.; Newton, M. G.; Cheng, Y. -C.; Schinazi, R. F.; Chu, C. K. Asymmetric synthesis and antiviral activities of L-carbocyclic 2',3'-didehydro-2',3'-dideoxy and 2',3'-dideoxy nucleosides. *J. Med. Chem.* **1999**, *42*, 3390-3399.
153. Song, G. Y.; Paul, V.; Choo, H.; Morrey, J.; Sidwell, R. W.; Schinazi, R. F.; Chu, C. K. Enantiomeric synthesis of D- and L-cyclopentenyl nucleosides and their antiviral activity against HIV and West Nile virus. *J. Med. Chem.* **2001**, *44*, 3985-3993.
154. Tadano, K. -I.; Hoshino, M.; Ogawa, S.; Suami, T. Transformation of D-erythrose to some pseudo-aldopentofuranoses. Syntheses of (1S,2R,3S,4S)-, (1R,2R,3S,4S)-, and 1R,2S,3R,4R)-2,3,4-trihydroxy-1-(hydroxymethyl)cyclopentanes and (1R,2S,3R,4R)-2,3-dihydroxy-4-(hydroxymethyl)-1-cyclopentanamine. *J. Org. Chem.* **1988**, *53*, 1427-1432.
155. Tadano, K. -I.; Maeda, H.; Hoshino, M.; Iimura, Y.; Suami, T. A novel transformation of four aldoses to some optically pure pseudohexopyranoses and a pseudopentofuranose, carbocyclic analogs of hexopyranoses and pentofuranose. Synthesis of derivatives of (1S,2S,3R,4S,5S)-, (1S,2S,3R,4R,5S)-, (1R,2R,3R,4R,5S)-, (1S,2S,3R,4S,5R)-2,3,4,5-tetrahydroxy-1-(hydroxymethyl)cyclohexanes and (1S,2S,3S,4S)-2,3,4-trihydroxy-1-(hydroxymethyl)cyclopentane. *J. Org. Chem.* **1987**, *52*, 1946-1956.
156. Tadano, K.-I.; Hoshino, M.; Ogawa, S.; Suami, T. An alternative synthesis of (1R,2S,3R,4R)-2,3-dihydroxy-4-hydroxymethyl-1-cyclopentanamine, a synthetic intermediate of (-)-aristeromycin. *Tetrahedron Lett.* **1987**, *28*, 2741-2744.
157. Tadano, K. -I.; Maeda, H.; Hoshino, M.; Iimura, Y.; Suami, T. A new transformation of aldose derived synthons to pseudo-hexopyranose or pseudo-pentofuranose derivatives. *Chem. Lett.* **1986**, 1081-1084.
158. Ballou, C. E.; Fischer, H. O. L.; MacDonald, D. L. The synthesis and properties of D-erythrose 4-phosphate. *J. Am. Chem. Soc.* **1955**, *77*, 5967-5970.
159. Vanhessche, K.; González Bello, C.; Vandewalle. Total synthesis of (-)-neplanocin A from L-ribulose. *Synlett.* **1991**, 921-922.
160. Vanhessche, K.; Van der Eycken, E.; Vandewalle, M. L-Ribulose: a novel chiral pool compound. *Tetrahedron Lett.* **1990**, *31*, 2337-2340.
161. Park, K. H.; Rapoport, H. Enantioselective synthesis of (1R,4S)-1-amino-4-(hydroxymethyl)-2-cyclopentene, a precursor for carbocyclic nucleoside synthesis. *J. Org. Chem.* **1994**, *59*, 394-399.
162. Gerspacher, M.; Rapoport, H. 2-Amino-2-deoxyhexoses as chiral educts for hydroxylated indolizidines. Synthesis of (+)-castanospermine and (+)-epicastanospermine. *J. Org. Chem.* **1991**, *56*, 3700-3706.
163. Yoshikawa, M.; Yokokawa, Y.; Inoue, Y.; Yamaguchi, S.; Murakami, N. Facile syntheses of pseudo-α-

D-arabinofuranose, and two pseudo-D-arabinofuranosylnucleosides, (+)-cyclaradine and (+)-1-pseudo-β-D-arabinofuranosyluracil, from D-arabinose. *Tetrahedron*, **1994**, *50*, 9961-9974.

164. Ohira, S.; Sawamoto, T.; Yamato, M. Synthesis of (-)-neplanocin A *via* C-H insertion of alkylidenecarbene. *Tetrahedron Lett.* **1995**, *36*, 1537-1538.
165. Chong, Y.; Gumina, G.; Chu, C. K. A divergent synthesis of D- and L-carbocyclic 4'-fluoro-2',3'-dideoxynucleosides as potential antiviral agents. *Tetrahedron Asymm.* **2000**, *11*, 4853-4875.
166. Gumina, G.; Chong, Y.; Choi, Y.; Chu, C. K. Stereoselective synthesis of carbocyclic L-4'-fluoro-2',3'-dideoxyadenosine. *Org. Lett.* **2000**, *2*, 1229-1231.
167. Lee, M G.; Du, J. F.; Chun, M. W.; Chu, C. K. Enantiomeric synthesis of L-(or 1'R, 2'S)-carbocyclic cyclopropyl nucleosides. *J. Org. Chem.* **1997**, *62*, 1991-1995.
168. Zhao, Y.; Yang, T. -F.; Lee, M. G.; Lee, D. -W.; Newton, M. G.; Chu, C. K. Asymmetric synthesis of (1'S, 2'R)-cyclopropyl carbocyclic nucleosides. *J. Org. Chem.* **1995**, *60*, 5236-5242.
169. Lee, M. G.; Lee, D.; Zhao, Y.; Newton, M.G.; Chu, C. K. Enantioselective synthesis of cyclopropyl carbocyclic L-nucleosides. *Tetrahedron Lett.* **1995**, *36*, 3499-3502.
170. Yang, T. -F.; Kim, H.; Kotra, L.P.; Chu, C. K. Design and synthesis of 2'-hydroxyethylcyclopropyl carbocyclic nucleosides. *Tetrahedron Lett.* **1996**, *37*, 8849-8852.
171. Zhao, Y. F.; Lee, M. G.; Yang, T. -F.; Chun, B.K.; Du, J. F.; Schinazi, R. F.; Chu, C. K. Asymmetric synthesis of cyclopropyl carbocyclic nucleosides. *Nucleosides Nucleotides* **1995**, *14*, 303-305.
172. Zhao, Y.; Yang, T. -F.; Lee, M. G.; Chun, B. K.; Du, J.; Schinazi, R. F.; Lee, D. -W.; Newton, M. G.; Chu, C. K. Synthesis of enantiomerically pure cyclopropyl carbocyclic nucleosides. *Tetrahedron Lett.* **1994**, *35*, 5405-5408.
173. Pierra, C.; Olgen, S.; Cavalcanti, S. C. H.; Cheng, Y. -C.; Schinazi, R. F.; Chu, C. K. Synthesis and antiviral activities of enantiomeric 1-[2-(hydroxymethyl)-cyclopropyl]methyl nucleosides. *Nucleosides Nucleotides* **2000**, *19*, 253-268.
174. Xiang, Y.; Du, J.; Chu, C. K. Syntheses of 2',3'-dideoxy-D-*C*-nucleosides from γ-lactone. *Nucleosides Nucleotides* **1996**, *15*, 1821-1834.
175. Beach, J. W.; Kim, H. O.; Jeong, L. S.; Nampalli, S.; Islam, Q.; Ahn, S. K.; Babu, J. R.; Chu, C. K. A highly stereoselective synthesis of anti-HIV 2',3'-dideoxynucleosides and 2',3'-didehydro-2',3'-dideoxynucleosides. *J. Org. Chem.* **1992**, *57*, 3887-3894.
176. Liang, C.; Ma, T.; Cooperwood, J. S.; Du, J.; Chu, C. K. Synthesis of L-ribofuranosyl C-nucleosides. *Carbohydrate Research* **1997**, *303*, 33-38.
177. Ma, T.; Pai, S. B.; Zhu, Y. L.; Lin, J. S.; Shanmuganathan, K.; Du, J.; Wang, C.; Kim, H.; Newton, M. G.; Cheng, Y. -C.; Chu, C. K. Structure-activity relationships of 1-(2-deoxy-2-fluoro-β-L-arabinofuranosyl)pyrimidine nucleosides as anti-hepatitis B virus agents. *J. Med. Chem.* **1996**, *39*, 2835-2843.
178. Tam, S. Y. -K.; Klein, R. S.; Wempen, I.; Fox, J. J. Nucleosides. 112. Synthesis of some new pyrazolo[1,5-*a*]-1,3,5-triazines and their *C*-nucleosides. *J. Org. Chem.* **1979**, *44*, 4547-4553.
179. Tam, S. Y. -K.; Hwang, J. S.; De Las Heras, F. G.; Klein, R. S.; Fox, J. J. Nucleosides. 105. Synthesis of 8-(β-D-ribofuranosyl)pyrazolo [1,5-*a*]-1,3,5-triazine isosteres of adenosine and inosine. *J. Heterocycl. Chem.* **1976**, *13*, 1305-1308.
180. Lim, M. -I.; Klein, R. S. Synthesis of 9-deazaadenosine - a new cytotoxic *C*-nucleoside isostere of adenosine. *Tetrahedron Lett.* **1981**, *22*, 25-28.
181. Hammerschmidt, F.; Peric, B.; Öhler, E. Synthesis of D-apio-β-D-furanosyl maleimide, an analog of showdomycin with transposed hydroxymethyl group. *Monatsh. Chem.* **1997**, *128*, 183-190.
182. Drescher, M.; Hammerschmidt, F. A second synthesis of D-apio-β-D-furanosyl maleimide as a structural analog of showdomycin. *Monatsh. Chem.* **1997**, *128*, 713-723.

183. Kozikowski A. P.; Ames, A. Total synthesis of the *C*-nucleoside DL-showdomycin by a Diels-Alder, retrograde Dieckmann strategy. *J. Am. Chem. Soc.* **1981**, *103*, 3923-3924.

184. Chun, B. K.; Chu, C. K. Asymmetric synthesis of carbocyclic *C*-nucleoside, (-)-9-deazaaristeromycin. *Tetrahedron Lett.* **1999**, *40*, 3309-3312.

185. Chu, C. K.; Ma, T.; Shanmuganathan, K.; Wang, C.; Xiang, Y.; Pai, S. B.; Yao, G. -Q.; Sommadossi, J. -P.; Cheng, Y. -C. Use of 2'-fluoro-5-methyl-β-L-arabinofuranosyluracil as a novel antiviral agent for hepatitis B virus and Empstein-Barr Virus. *Antimicrob. Agents Chemother.* **1995**, *39*, 979-981.

186. Chu, C. K.; Hong, J. H.; Choi, Y.; Du, J.; Lee, K.; Chun, B. K.; Boudinot, F. D.; Peek, S. F.; Korba, B. E.; Tennant, B. C.; Cheng, Y. -C. L-FMAU. *Drugs of the Future.* **1998**, *23*, 821-826.

187. Xiang, Y.; Kotra, L. P.; Chu, C. K.; Schinazi, R. F. Synthesis and anti-HIV activities of 2'-deoxy-2',2"-difluoro-β-L-ribofuranosyl-pyrimidine and -purine nucleosides. *Bioorg. Med. Chem. Lett.* **1995**, *5*, 743-748.

188. Kotra, L. P.; Xiang, Y.; Newton, M. G.; Schinazi, R. F.; Cheng, Y. -C.; Chu, C. K. Structure-activity relationships of 2'-deoxy-2',2'-difluoro-L-*erythro*-pentofuranosyl nucleosides. *J. Med. Chem.* **1997**, *40*, 3635-3644.

189. Kotra, L. P.; Newton, M. G.; Chu, C. K. Synthesis of 2,3-dideoxy-2,2-difluoro-L-glycero-pentofuranosyl nucleosides. *Carbohydr. Res.* **1998**, *306*, 69-80.

190. Xiang, Y.; Cavalcanti, S.; Chu, C. K.; Schinazi, R. F.; Pai, S. B.; Zhu, Y. -L.; Cheng, Y. -C. Anti-human immunodeficiency virus type-1 (HIV-1) and anti-hepatitis B virus (HBV) activities of (2,3-dideoxy-2-fluoro-β-L-*threo*-pentofuranosyl)nucleosides. *Bioorg. Med. Chem. Lett.* **1995**, *5*, 877-880.

191. Ma, T.; Lin, J. -S.; Newton, M. G.; Cheng, Y. -C.; Chu, C. K. Synthesis and anti-hepatitis B virus activity of 9-(2-deoxy-2-fluoro-β-L-arabinofuranosyl)purine nucleosides. *J. Med. Chem.* **1997**, *40*, 2750-2754.

192. Choi, Y.; Lee, K.; Hong, J. H.; Schinazi, R. F.; Chu, C. K. Synthesis and anti-HIV activity of L-2'-fluoro-2',3'-unsaturated purine nucleosides. *Tetrahedron Lett.* **1998**, *39*, 4437-4440.

193. Lee, K.; Choi, Y.; Gullen, E.; Schlueter-Wirts, S.; Schinazi, R. F.; Cheng, Y. -C.; Chu, C. K. Synthesis and anti-HIV and anti-HBV activities of 2'-fluoro-2',3'-unsaturated L-nucleosides. *J. Med. Chem.* **1999**, *42*, 1320-1328.

194. Pai, S. B.; Liotta, D. C.; Chu, C. K.; Schinazi, R. F. Inhibition of hepatitis B virus by novel 2'-fluoro-2',3'-unsaturated D- and L-nucleosides. *Abstracts of Papers*, 3rd International Conference on Therapies for Viral Hepatitis, Maui, USA; International Society for Antiviral Therapy: 1999; Abstract 125.

195. Lee, K; Choi, Y.; Schinazi, R. F.; Chu, C. K. Structure-activity relationships of 2'-fluoro-2',3'-unsaturated D-nucleosides as anti-HIV-1 agents. *218th American Chemical Society National Meeting* **1999**, August 22-26, New Orleans, LA, USA.

196. Lee, K.; Choi, Y.; Gumina, G.;Zhou, W.; Schinazi, R. F.; Chu, C. K. Structure-activity relationships of 2'-fluoro-2',3'-unsaturated D-nucleosides as anti-HIV agents. *J. Med.Chem.* **2002**, *45*, 1313-1320.

197. Gumina, G.; Schinazi, R. F.; Chu, C. K. Synthesis and potent anti-HIV activity of L-3'-fluoro unsaturated cytidine. *Org. Lett.* **2001**, *3*, 4177-4180.

198. Gumina, G.; Zhou, W.; Schinazi, R. F.; Chu, C. K. Synthesis of L-3'-fluoro unsaturated nucleosides as potential antiviral agents. *222nd American Chemical Society National Meeting* ***2001***, August 26-30, Chicago, IL, USA.

199. Zhou, W.; Gumina, G.; Schinazi, R. F.; Chu, C. K. Synthesis of 3'-fluoro-2',3'-unsaturated D-nucleosides. *222nd American Chemical Society National Meeting* **2001**, August 26-30, Chicago, IL, USA.

200. El Ashry, E. S. H.; El Kilany, Y. Acyclonucleosides: Part 1. Seco-nucleosides. *Adv. Heterocycl. Chem.* **1997**, *67*, 391-438.

201. El Ashry, E. S. H.; El Kilany, Y. Acyclonucleosides: Part 2. Diseco-nucleosides. *Adv. Heterocycl. Chem.* **1997**, *68*, 1-88.

202. El Ashry, E. S. H.; El Kilany, Y. Acyclonucleosides: Part 3. Tri- tetra-, and pentaseco-nucleosides. *Adv. Heterocycl. Chem.* **1998**, *69* 129-215.

203. Chu, C. K.; Cutler, S. Chemistry and antiviral activities of acyclonucleosides. *J. Heterocyclic Chem.* **1986**, *23*, 289-319.

204. Okumura, K.; Oine, T.; Yamada, Y.; Tomie, M.; Adachi, T.; Nagura, T.; Kawazu, M.; Mizoguchi, T.; Inoue, I. Synthetic studies on eritadenine. I. Reactions of some purines with the 2,3-O-protected dihydroxybutyrolactone. *J. Org. Chem.* **1971**, *36*, 1573-1579.

205. Barker, R.; MacDonald, D. L. Some oxidation and reduction products of 2,4-O-ethylidene-D-*erythrose*. *J. Am. Chem. Soc.* **1960**, *82*, 2301-2303.

206. Baker, D. C.; Hawkins, L. D. Synthesis of inhibitors of adenosine deaminase. A total synthesis of *erythro*-3-(adenin-9-yl)-2-nonanol and its isomers from chiral precursors. *J. Org. Chem.* **1982**, *47*, 2179-2184.

207. Bukhari, M. A.; Foster, A. B.; Lehmann, J.; Webber, J. M. Aspects of stereochemistry. Part X. Isopropylidene derivatives of L-rhamnitol. *J. Chem. Soc.* **1963**, 2287-2890.

208. Richtmyer, N. K.; Hudson, C. S. A direct proof of the equivalence of carbon atoms 1 and 6 in D-mannitol and a new method of correlating certain higher-carbon sugars and alcohols. *J. Am. Chem. Soc.* **1950**, *72*, 3880-3884.

209. MacCoss; M. Chen, A.; Tolman, R. L. Syntheses of all four possible diastereomers of the acyclonucleoside 9-(1,3,4-trihydroxy-2-butoxymethyl)guanine from carbohydrate precursors. *Tetrahedron Lett.* **1985**, *26*, 4287-4290.

210. Dondoni, A.; Fogagnolo, M.; Medici, A.; Pedrini, P. Diastereoselectivity in the 1,2-addition of silylazoles to chiral aldehydes. Stereocontrolled homologation of α-hydroxyaldehydes. *Tetrahedron Lett.* **1985**, *26*, 5477-5480.

211. Ashton, W. T.; Canning, L. F.; Reynolds, G. F.; Tolman, R. L.; Karkas, J. D.; Liou, R.; Davies, M. -E. M.; DeWitt, C. M.; Perry, H. C.; Field, A. K. Synthesis and antiherpetic activity of (S)-, (R)-, and (±)-9-[(2,3-dihydroxy-1-propoxy)methyl]guanine, linear isomers of 2'-nor-2'-deoxyguanosine. *J. Med. Chem.* **1985**, *28*, 926-933.

212. Ullas, G. V.; Chu, C. K.; Ahn, M. K.; Kosugi, Y. Synthesis of *C*-nucleoside analog of (S)-9-(2,3-dihydroxypropyl)adenine and related acyclonucleosides. *J. Org. Chem.* **1998**, *53*, 2413-2418.

213. Buchanan, J. G.; Craven, D. A.; Wightman, R. H.; Harnden, M. R. *C*-Nucleoside studies. Part 21. Synthesis of some hydroxylalkylated pyrrolo- and thieno-[3,2-d]pyrimidines related to known antiviral acyclonucleosides. *J. Chem. Soc. Perkin Trans. 1* **1991**, 195-202.

214. Lin, T. -S.; Mu, S. -P.; Liu, M. -C.; Mancini, W. R. Synthesis of (R)- and (S)-1-{[2-hydroxy-1-(aminome thyl)ethoxy]methyl}-5-benzyluracil, potent inhibitors of uridine phosphorylase. *Nucleosides Nucleotides* **1990**, *9*, 559-568.

215. Howe, R. J.; Malkin, T. J. An X-ray and thermal examination of the glycerides. Part XI. The 1:2-diglycerides, and futher observations on 1:3-diglycerides. *J. Chem. Soc.* **1951**, 2663-2667.

216. Herdewijn, P.; Van Aerschot, A.; Busson, R.; Claes, P.; De Clerq, E. Synthesis of 2'-deoxy-2'-fluoro-D-arabinopyranosyl nucleosides and their 3',4'-seco analogs. *Nucleosides Nucleotides* **1991**, *10*, 1525-1549.

217. Shimada, N.; Hasegawa, S.; Harada, T.; Tomisawa, T.; Fujii, A.; Takita, T. Oxetanocin, a novel nucleoside from bacteria. *J. Antibiotics* **1986**, *39*, 1623-1625.

218. Hoshino, H.; Shimizu, N.; Shimada, N.; Takita, T.; Takeuchi, T. Inhibition of infectivity of human immunodeficiency virus by oxetanocin. *J. Antibiotics* **1987**, *40*, 1077-1078.

219. Ichikawa, E.; Kato, K. Synthesis of oxetanocin A and related unusual nucleosides with bis(hydroxymethyl)-branched sugars. *Synthesis* **2002**, 1-28.

220. Wilson, F. X.; Fleet, G. W. J.; Vogt, K.; Wang, Y.; Witty, D. R.; Choi, S.; Storer, R.; Myers, P. L.; Wallis,

C. J. Synthesis of oxetanocin. *Tetrahedron Lett.* **1990**, *31*, 6931-6934.

221. Saksena, A. K.; Ganguly, A. K.; Girihavallabhan, V. M.; Pike, R. E.; Chen, Y. -T.; Puar, M. S. Ring contraction reactions of 2-O-methanesulfonates of α-hydroxy-γ-lactones in aqueous medium to oxetane-2-carboxylic acids: a convenient synthesis of 3'-*O*-methyloxetanocin and a formal synthesis of oxetanocin. *Tetrahedron Lett.* **1992**, *33*, 7721-7724.
222. Gumina, G.; Chu, C. K. Total synthesis of L-oxetanocin. *Abstr. Papers Am. Chem. Soc.* **2002**, CARB 002.
223. Gumina, G.; Chu, C. K. Synthesis of L-Oxetanocin. *Org. Lett.* **2002**, *4*, 1147-1149.
224. Chu, C. K.; Beach, J. W.; Babu, J. R.; Jeong, L. S.; Jeong, H. K.; Ahn, S. K.; Islam, Q.; Lee, S. J.; Chen, Y. Stereoselective synthesis of 2',3'-dideoxy and 2',3'-didehydro-2',3'-dideoxy-nucleosides. *Nucleosides Nucleotides* **1991**, *10*, 423-426.
225. Beach, J. W.; Kim, H. O.; Jeong, L. S.; Nampalli, S.; Islam, Q.; Ahn, S. K.; Babu, J. R.; Chu, C. K. A highly stereoselective synthesis of anti-HIV 2',3'-dideoxy- and 2',3'-didehydro-2',3'-dideoxynucleosides. *J. Org. Chem.* **1992**, *57*, 3887-3894.
226. Jeong, L. S.; Beach, J. W.; Chu, C. K. Stereoselective synthesis of 3-azido-2,3-dideoxy-D-ribose derivatives and its utilization for the synthesis of anti-HIV nucleosides. *J. Heterocycl. Chem.* **1993**, *30*, 1445-1452.
227. Yang, T. -F., Kotra, L. P.; Teng, Q.; Naguib, F. N. M.; Sommadossi, J. -P., El Kouni, M.; Chu, C. K. *Tetrahedron Lett.* **1995**, *36*, 983-986.
228. Giri, I.; Bolon, P. J.; Chu, C. K. Synthesis of 2'-*C*-α-methyl-2',3'-dideoxynucleosides. *Nucleosides Nucleotides* **1996**, *15*, 183-204.
229. Nair, V.; Emanuel, D. J. Synthetic design, stereochemistry, and enzymatic activity of a reversed aminoacyl nucleoside: an analog of puromycin. *J. Am. Chem. Soc.* **1977**, *99*, 1571-1576.
230. Pickering, L.; Nair, V. Novel bicyclic nucleoside analogs related to natural griseolic acids. *Nucleosides Nucleotides* **1997**, *16*, 1435-1438.
231. Chao, Q.; Zhang, J.; Pickering, L.; Jahnke, T. S.; Nair, V. Concise and stereospecific synthesis of novel bicyclic dideoxynucleosides as potential antiviral agents. *Tetrahedron* **1998**, *54*, 3113-3124.

CHAPTER 3

OXATHIOLANE AND DIOXOLANE NUCLEOSIDES: SYNTHESIS AND ANTIVIRAL ACTIVITY

GIUSEPPE GUMINA, JOHN S. COOPERWOOD and CHUNG K. CHU

3.1. Introduction

Before the emergence of AIDS (acquired immunodeficiency syndrome), antiviral chemotherapy had met with limited success and only a few antiviral drugs had been available. However, AIDS has created an urgent need for the development of new antiviral agents due the global nature of the epidemic. The study of HIV replication has offered several viral targets, including the HIV reverse transcriptase (RT) and protease. Currently seven nucleoside/nucleotide RT inhibitors (NRTIs), three non-nucleoside RT inhibitors (NNRTIs), and four protease inhibitors have been approved by the FDA. Since the discovery of 3'-azido-3'-deoxythymidine (AZT, zidovudine or retrovir)[1] as an effective anti-HIV agent, 2',3'-dideoxynucleosides, acting as inhibitors of the viral reverse transcriptase, have demonstrated the biological importance of nucleosides lacking a 3'-hydroxy functionality. As mentioned, seven nucleosides [AZT, 2',3'-dideoxyinosine (didanosine, ddI),[2] 2',3'-dideoxycytidine (zalcitabine, ddC), 2',3'-didehydro-3'-deoxythymidine (stavudine, D4T),[3,4] 2',3'-dideoxy-3'-thiacytidine (3TC, lamivudine),[5] 1592U89 succinate (abacavir),[6,7] and tenofovir disoproxil[8]] have been approved by the FDA for the treatment of HIV infection, along with NNRTIs and protease inhibitors. Among NRTIs, 3TC has unique chemical and biological properties, having a sulfur atom in place of the C-3' and possessing the unnatural L-sugar configuration. It is currently included in over 80% of the combinations used in AIDS therapy and is the drug of choice for the treatment of hepatitis B virus (HBV) infection. Intensive efforts have been directed towards the synthesis of 3TC analogs during the past ten years. For this reason, this chapter will focus on the syntheses and biological activities of oxathiolane and dioxolane nucleosides.

3.2. Oxathiolane nucleosides

3.2.1. Synthesis

The first synthesis of an oxathiolane nucleoside, as a racemic mixture, was reported by Belleau *et al.*[9] in 1989 (Scheme 1). The oxathiolane **1-3**, key intermediate for the synthesis of nucleosides, was obtained as a 1:1 mixture of anomers from the conden-

Antiviral Nucleosides: Chiral Synthesis and Chemotherapy, Ed. by C.K. Chu. 191 — 258

sation of benzoyloxyacetaldehyde **1-1** with 2-mercaptodimethylacetal **1-2**. The oxathiolane derivative was then coupled with silylated cytosine to give a 1:1 mixture of *cis* and *trans* protected cytosine nucleosides **1-4**. After separation by chromatography followed by deprotection, (±)-BCH-189 was obtained as a racemic mixture **1-5**.

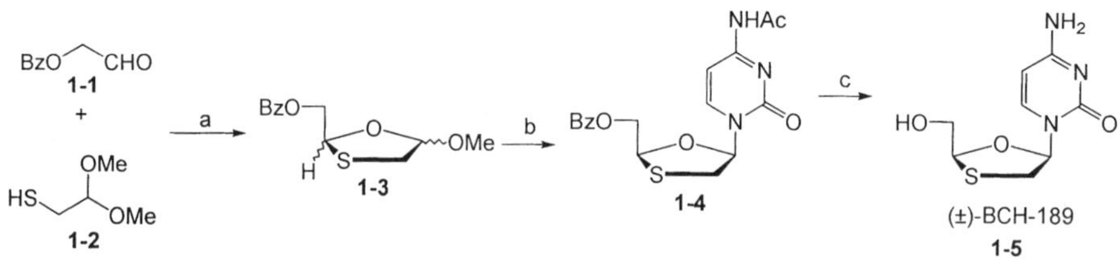

Reagents: a) *p*-TsOH, toluene; b) Silylated N^4-acetylcytosine, DCE, $TiCl_4$; c) NH_3/MeOH.

Scheme 1

Liotta and co-workers[10] reported the synthesis of 1,3-oxathiolane **1-5** using a highly stereoselective base-coupling reaction which operated *via* the *in situ* chelation of a complex between the oxathiolane intermediates **2-4** and an appropriate Lewis acid (Scheme 2). The key intermediates **2-4** were synthesized from protected glycolic aldehyde **2-2**, which was obtained from alkene **2-1** by ozonolysis. Aldehyde **2-2** was reacted with 2-mercaptoacetic acid in toluene under refluxing conditions to obtain the 1,3-oxathiolane lactone **2-3**, which was reduced by diisobutylaluminum hydride (DIBAL-H) in toluene at -78 °C or by lithium tri-*tert* -butoxyaluminum hydride in THF at 0 °C followed by acetylation with acetic anhydride to obtain **2-4** as a 2:1 mixture of anomers. Condensation of the anomeric mixture **2-4** with silylated cytosine in the presence of stannic chloride (2 equiv, CH_2Cl_2) at room temperature gave exclusively the β-anomer **2-5**, probably due to *in situ* chelation. The level of selectivity was determined by HPLC to be >300:1 in favor of the β-isomer.

In efforts to synthesize the optically pure isomer of BCH-189, Chu *et al.*[11] reported the synthesis of enantiomerically pure (+)-(2'*S*,5'*R*)-1-[2-(hydroxymethyl)-1,3-oxathiolan-5-yl]-cytosine [(+)-BCH-189] from D-mannose *via* 1,6-thioanhydro-D-mannose **3-5** (Scheme 3). 2,3,4-Triacetyl-1,6-thioanhydro-D-mannose **3-4** was synthesized from D-mannose by tosylation of the primary alcohol followed by acetyl protection of the 1,2,3,4-hydroxyl groups, and bromination of the anomeric acetyl group to give bromosugar **3-3**. Cyclization with 3 equivalents of potassium *O*-ethylxanthate gave 1,6-thioanhydro-β-mannose derivative **3-4**. Without further purification, **3-4** was treated with ammonia in methanol to give triol **3-5**. The *cis*-2,3-vicinal hydroxyl groups of **3-5** were selectively protected with an isopropylidene group; the following benzoylation gave fully protected compound **3-6**. The isopropylidene group of **3-6** was then selectively removed using 2% aqueous sulfuric acid in dioxane at 70 °C to obtain diol **3-7**,

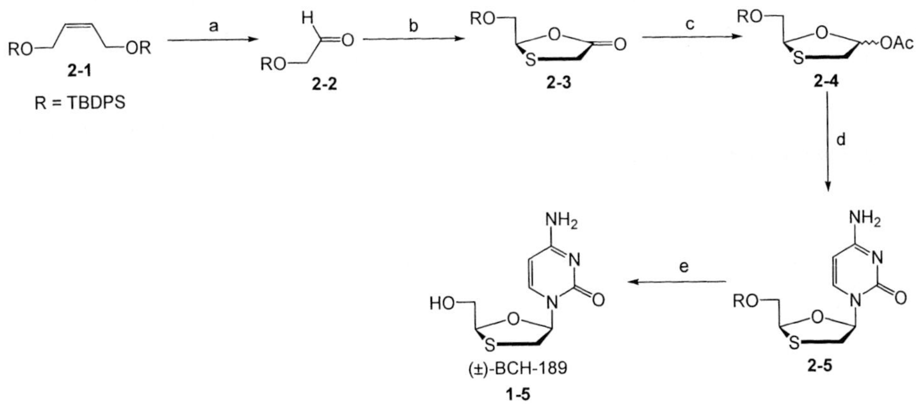

Reagents: a) O_3/DMS; b) 2-Mercaptoacetic acid; c) DIBAL-H, then Ac_2O; d) silylated base, $SnCl_4$; e) TBAF/THF.

Scheme 2

which was subjected to oxidative cleavage using lead tetraacetate followed by reduction with sodium borohydride to give compound **3-8**, where the benzoyl group had migrated to diol's primary hydroxyl group. The primary 5'-hydroxyl group of compound **3-8** was then selectively protected using *tert*-butyldiphenylsilyl chloride (TBDPSCl); the following ammonolysis yielded **3-9**. Compound **3-9** was subjected to oxidative cleavage by lead tetraacetate and oxidation of the resulting aldehyde to the carboxylic acid using sodium chlorite to afford the acid derivative **3-10** as a mixture of *endo* and *exo* sulfoxide. Crude **3-10** was esterified using dimethyl sulfate to give methyl ester **3-11**, which was reduced to sulfide **3-12** in 80% yield using dichloroborane and dimethylsulfide in THF. Compound **3-12** was, then, hydrolyzed by LiOH to give the corresponding carboxylic acid, which underwent oxidative decarboxylation by lead tetraacetate/pyridine in ethyl acetate to afford the key intermediate **3-13**. Crude glycosyl donor **3-13** was condensed with silylated N^4-acetylcytosine in 1,2-dichloroethane in the presence of trimethylsilyl trifluoromethansufonate (TMSOTf) as a Lewis acid to give a β:α mixture (2:1) of **3-14** and **3-15**. After separation of anomeric mixtures by silica gel column chromatography, compounds **3-14** and **3-15** were deacetylated by ammonolysis in methanol and desilylated with tetrabutylammonium fluoride (TBAF) to give β-(+)-BCH-189 **3-16** and α isomer **3-17**, respectively.

In a related work, Chu and co-workers reported a more efficient methodology for the synthesis of (+)-BCH-189 from 1,6-thioanhydro-D-galactose **4-1** (Scheme 4).[12] Oxidative cleavage of *cis* diol **4-1** was accomplished using $NaIO_4$. The resulting unstable aldehyde was reduced with sodium borohydride to give a vicinal diol, which was protected by 2,2-dimethoxypropane to afford the 1,3-oxathiolane derivative **4-2**. Benzoylation of the primary hydroxyl group gave compound **4-3** in high yield. Selective deprotection of the isopropylidene group followed by oxidative cleavage of the resulting diol with $NaIO_4$ and reduction with sodium borohydride gave compound **4-4**. The hydroxyl group of this compound was protected by TBDPSCl, and the benzoyl group was removed by ammonolysis to give silylated compound **4-5**. Treatment of **4-5** with pyridinium dichromate (PDC) in DMF afforded the acid derivative **4-6**, which was converted to the key intermediate **3-13** by oxidative decarboxylation.

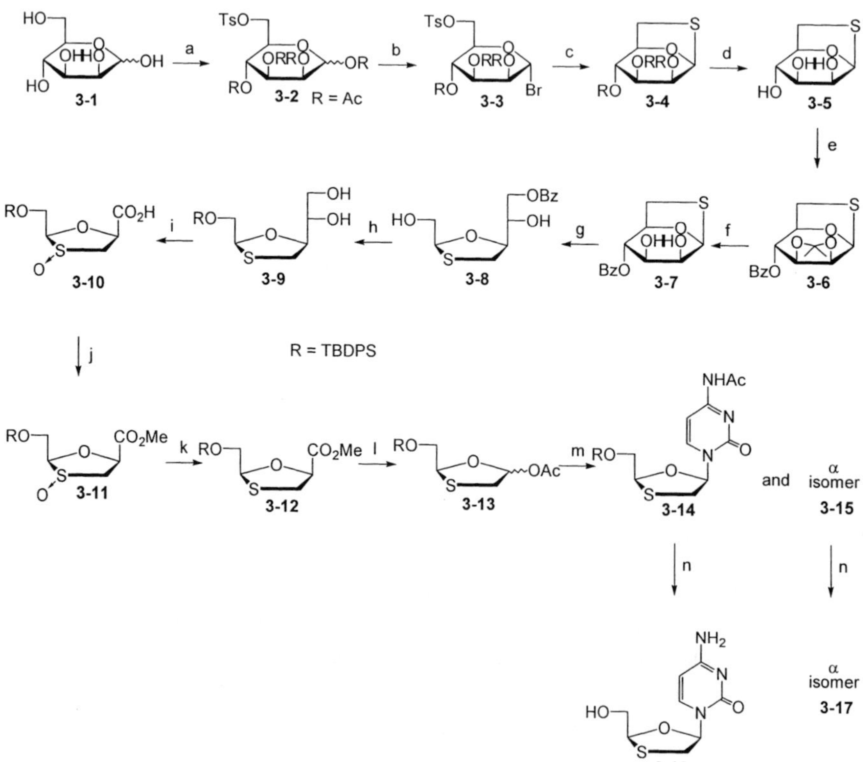

Reagents: a) TsCl, then Ac_2O; b) HBr/CH_3CO_2H; c) potassium *O*-ethylxanthate; d) NH_3/MeOH; e) DMP, H^+, then BzCl; f) 2% aq H_2SO_4; g) $Pb(OAc)_4$, then $NaBH_4$; h) TBDPSCl, then NH_3/MeOH; i) $Pb(OAc)_4$, then $NaClO_2$; j) $(CH_3)_2SO_4$, K_2CO_3, acetone; k) $BHCl_2$, THF, 0-5 °C; l) LiOH, then $Pb(OAc)_4$, Py; m) Silylated N^4-acetylcytosine, TMSOTf; n) NH_3/MeOH, then TBAF.

Scheme 3

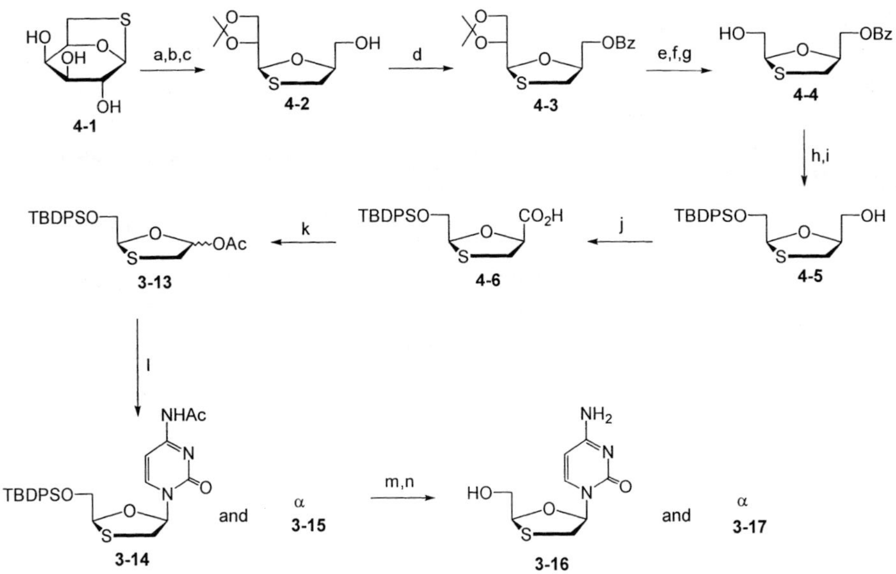

Reagents: a) $NaIO_4$, $MeOH/H_2O$; b) $NaBH_4$; c) $CH_3C(OMe)_2CH_3$, *p*-TsOH; d) BzCl, Py; e) 10% HCl, MeOH; f) $NaIO_4$; g) $NaBH_4$; h) TBDPSCl, DMF, imidazole; i) NaOMe, MeOH; j) PDC, DMF; k) $Pb(OAc)_4$, Py; l) silylated N^4-acetylcytosine, DCE, TMSOTf; m) NH_3, MeOH; n) TBAF, THF.

Scheme 4

Condensation of **3-13** with silylated N^4-acetylcytosine in dichloroethane in the presence of TMSOTf gave **3-14** and **3-15** as an α/β mixture. After separation and deprotection, the desired nucleosides, **3-16** and **3-17**, respectively, were obtained. Jeong *et al.* synthesized various optically pure β-D- and α-D-1,3-oxathiolane-pyrimidine and -purine nucleosides with natural nucleosides configuration (Scheme 5, 6) to study the structure-activity relationships as anti-HIV-1 agents (Table 1). The key intermediate **3-13** synthesized from D-mannitol was condensed with various pyrimidine (**5-23~5-44**) and purine (**6-3~6-22**) bases.[13]

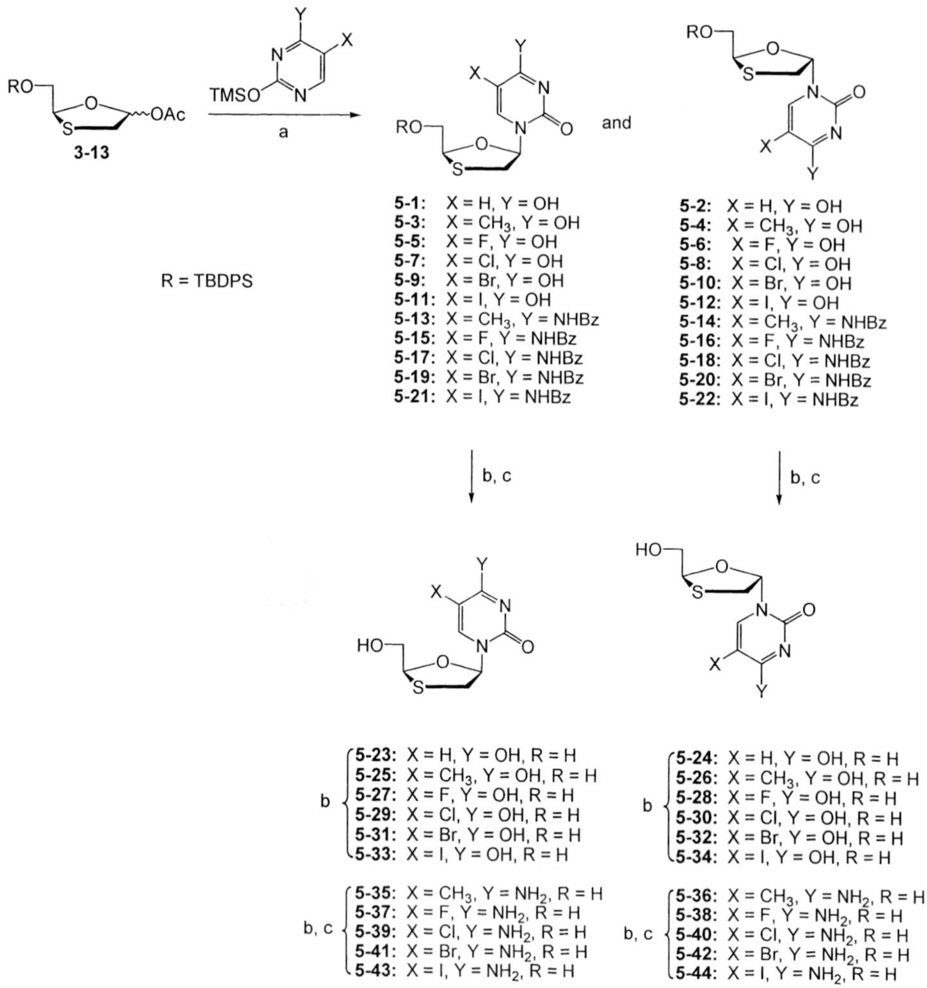

Reagents: a) TMSOTf, DCE, rt; b) TBAF, THF, rt; c) NH_3, MeOH, rt.

Scheme 5

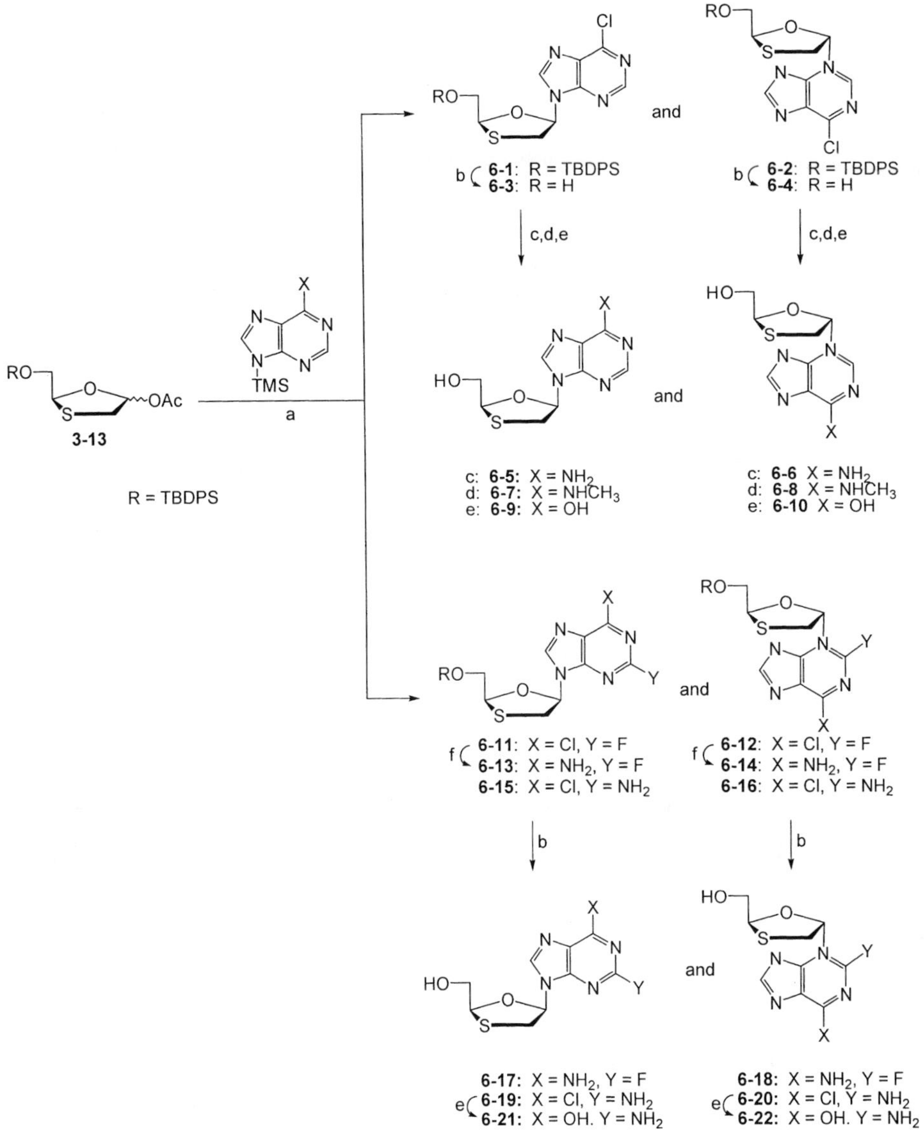

Reagents: a) TMSOTf, DCE, -20 °C to rt; b) TBAF, THF, rt; c) NH_3, MeOH, 80-90 °C; d) CH_3NH_2, MeOH, 80-90 °C; e) NaOMe, $HSCH_2CH_2OH$, MeOH, reflux. f) NH_3, DME, rt.

Scheme 6

Table 1. AntiL-HIVL-1 activity (EC_{50}) and inhibitory (IC_{50}) concentration of DL-oxathiolanylL-pyrimidine and purine nucleosides in human peripheral blood mononuclear (PBM) and in Vero cells

Compd	Anomer	EC_{50} (μM) anti-HIV-1 (PBM)	IC_{50} (μM) cytotoxicity (PBM)	IC_{50} (μM) cytotoxicity (Vero)
3-16	(+)-β	0.21	> 100	> 100
3-17	(–)-α	> 100	> 100	> 100
5-23	(+)-β	94.7	> 100	> 100
5-24	(–)-α	> 100	> 100	> 100
5-25	(+)-β	11.6	> 100	> 100
5-26	(–)-α	> 100	> 100	> 100
5-27	(+)-β	6.3	> 100	> 100
5-28	(–)-α	2.3	> 100	> 100
5-29	(+)-β	30.4	> 100	> 100
5-30	(–)-α	> 100	> 100	> 100
5-31	(+)-β	> 100	> 100	> 100
5-32	(–)-α	> 100	> 100	> 100
5-33	(+)-β	29.3	> 100	> 100
5-34	(–)-α	112.1	> 100	> 100
5-35	(+)-β	0.172	> 100	> 100
5-36	(–)-α	> 100	> 100	> 100
5-37	(+)-β	0.38	> 100	> 100
5-38	(–)-α	77.5	> 100	> 100
5-39	(+)-β	0.28	> 100	> 100
5-40	(–)-α	> 100	> 100	31.15
5-41	(+)-β	0.012	> 100	> 100
5-42	(–)-α	> 100	> 100	> 100
5-43	(+)-β	1.9	> 100	> 100
5-44	(–)-α	0.28	> 100	> 100
6-3	(–)-β	0.48	> 100	46.5

Table 1. Continued

Compd	Anomer	EC_{50} (μM) anti-HIV-1 (PBM)	IC_{50} (μM) cytotoxicity (PBM)	IC_{50} (μM) cytotoxicity (Vero)
6-4	(–)-α	2.7	> 100	86.1
6-5	(–)-β	0.28	> 100	> 100
6-7	(–)-β	1.1	> 100	> 100
6-9	(–)-β	0.05	> 100	> 100
6-6	(–)-α	11.9	> 100	> 100
6-8	(–)-α	11.7	> 100	> 100
6-10	(–)-α	> 100	> 100	> 100
6-17	(–)-β	2.8	> 100	5.6
6-18	(–)-α	61.9	> 100	45.4
6-19	(–)-β	0.52	> 100	> 100
6-20	(+)-α	5.6	> 100	> 100
6-21	(–)-β	0.83	> 100	> 100
6-22	(+)-α	7.62	> 100	63.6
AZT		0.004	> 100	> 100

By using a similar strategy for the synthesis of (+)-BCH-189 from 1,6-thioanhydro-D-galactose **4-1**, Chu and co-workers developed a synthetic methodology to access (-)-(2'*R*,5'*S*)-1-[2-(hydroxymethyl)-1,3-oxathiolan-5-yl]-cytosine [(-)-BCH-189, 3TC] (**7-8**) from 1,6-thioanhydro-L-gulose (**7-1**) (Scheme 7).[14] Conversion of **7-1** to **7-2** and protection of the primary hydroxyl group, followed by deprotection of the isopropylidene were accomplished in five steps. In contrast to the synthesis of the (+)-2'*S*,5'*R*-isomer, in which several steps were required to obtain the key intermediate, direct cleavage of diol **7-3** by lead tetraacetate at room temperature, oxidation with a mild oxidizing agent (PDC), and oxidative decarboxylation by lead tetraacetate gave **7-5** in only three steps. Condensation of **7-5** with silylated N^4-acetylcytosine in DCE using TMSOTf as a catalyst gave a 2:1 mixture (β:α) of **7-6** and **7-7**. Chromatographic separation and deprotection afforded the 2'*R*,5'*S*-(-)-isomer **7-8** (3TC) and the 2'*R*,5'*R*-(+)-isomer **7-9**. It was found that the use of stannic chloride instead of TMSOTf as a Lewis acid during the condensation gave a racemic mixture, probably by the opening and closing of the oxathiolane ring under the reaction conditions.

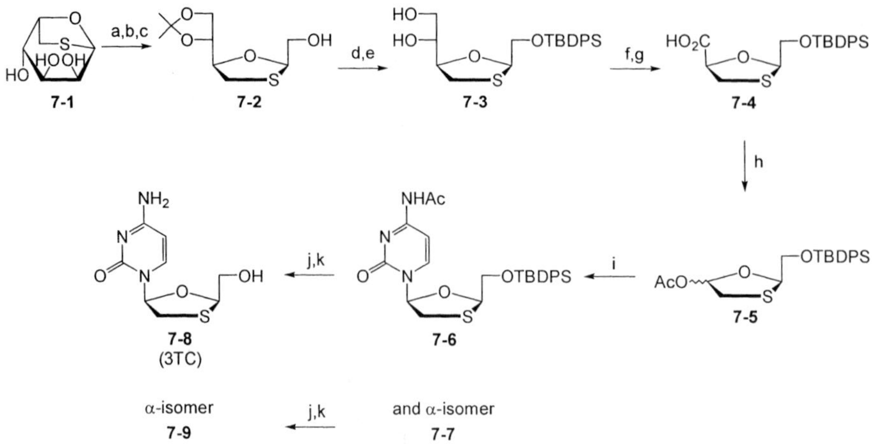

Reagents: a) $NaIO_4$, MeOH/H_2O; b) $NaBH_4$, MeOH; c) $CH_3C(OMe)_2CH_3$, *p*-TsOH; d) TBDPSCl, DMF, imidazole; e) *p*-TsOH, MeOH; f) $Pb(OAc)_4$, EtOAc; g) PDC, DMF; h) $Pb(OAc)_4$, THF; i) silylated N^4-acetylcytosine, DCE, TMSOTf; j) NH_3, MeOH; k) TBAF, THF.

Scheme 7

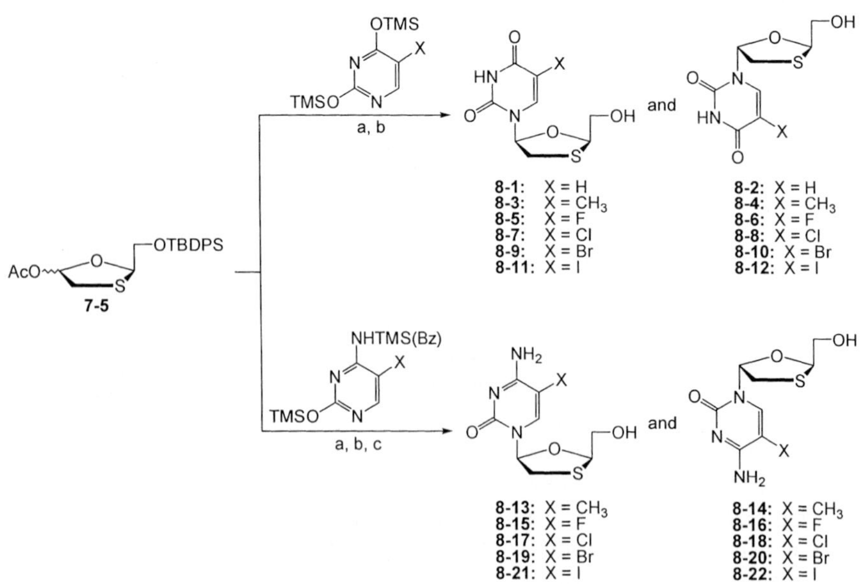

Reagents: a) TMSOTf, DCE, rt; b) TBAF, THF, rt; c) NH_3, MeOH, rt.

Scheme 8

In order to study the structure-activity relationships of various nucleobases derivatives, acetate **7-5** was condensed with pyrimidines (Scheme 8) and purines (Scheme 9). Upon evaluation of the anti-HIV aactivity of these nucleosides, 5-fluorocytosine derivative **8-15** was found to be the most potent (Table 2).[15]

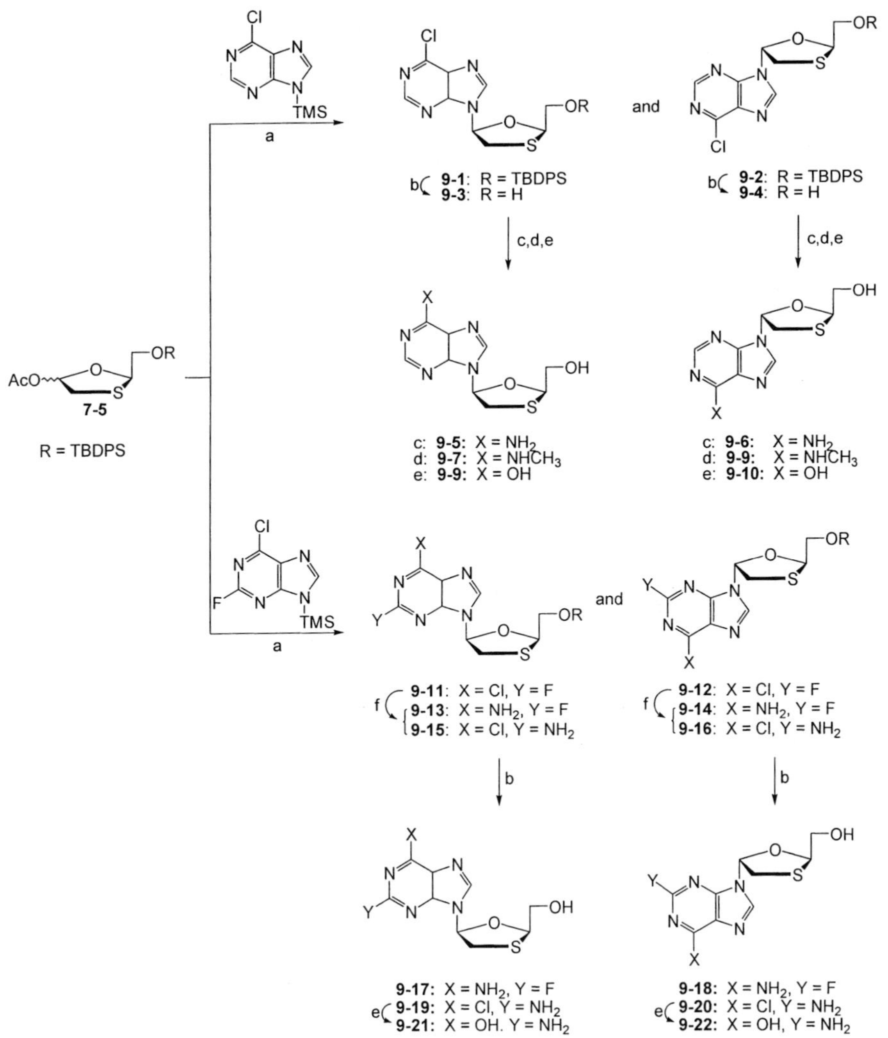

Reagents: a) TMSOTf, CH_2Cl_2, rt to reflux; b) TBAF, THF, rt; c) NH_3, MeOH, 80-90 °C; d) CH_3NH_2, MeOH, 80-90 °C; e) NaOMe, $HSCH_2CH_2OH$, MeOH, reflux. f) NH_3, DME, rt.

Scheme 9

Table 2. AntiL-HIVL-1 activity (EC_{50}) and inhibitory (IC_{50}) concentration of LL-oxathiolanylL-pyrimidine and purine nucleosides in PBM and Vero cells

Compd	Anomer	EC_{50} (μM) anti-HIV-1 (PBM)	IC_{50} (μM) cytotoxicity (PBM)	IC_{50} (μM) cytotoxicity (Vero)
7-8	(–)-β	0.0018	> 100	> 100
7-9	(+)-α	10.1	> 100	> 100
8-1	(–)-β	11.7	> 100	> 100
8-2	(+)-α	32.8	> 100	> 100
8-3	(–)-β	34.3	> 100	> 100
8-4	(+)-α	4.4	> 100	> 100
8-5	(–)-β	> 100	> 100	> 100
8-6	(+)-α	> 100	> 100	> 100
8-7	(–)-β	> 100	> 100	> 100
8-8	(+)-α	> 100	> 100	> 100
8-9	(–)-β	> 100	> 100	> 100
8-10	(+)-α	121.0	> 100	> 100
8-11	(–)-β	92.9	> 100	> 100
8-12	(+)-α	157	> 100	> 100
8-13	(–)-β	1.90	> 100	> 100
8-14	(+)-α	0.45	> 100	> 100
8-15	(–)-β	0.0013	> 100	> 100
8-16	(+)-α	0.43	> 100	> 100
8-17	(–)-β	31.8	> 100	> 100
8-18	(+)-α	> 100	> 100	31.1
8-19	(–)-β	2.51	> 100	> 100
8-20	(+)-α	> 100	> 100	> 100
8-21	(–)-β	0.14	> 100	> 100
8-22	(+)-α	> 100	> 100	> 100
9-3	(+)-β	1.44	> 100	20.2

Table 2. Continued

Compd	Anomer	EC_{50} (μM) anti-HIV-1 (PBM)	IC_{50} (μM) cytotoxicity (PBM)	IC_{50} (μM) cytotoxicity (Vero)
9-4	(+)-α	2.75	> 100	32.7
9-5	(+)-β	1.01	> 100	> 100
9-7	(+)-β	15.0	> 100	> 100
9-9	(+)-β	> 100	> 100	> 100
9-6	(+)-α	78.9	> 100	> 100
9-9	(+)-α	13.0	> 100	> 100
9-11	(+)-α	> 100	> 100	> 100
9-17	(+)-β	16.3	> 100	> 100
9-18	(+)-α	15.8	> 100	< 1.0
9-19	(-)-β	9.8	> 100	48.5
9-20	(+)-α	42.1	> 100	> 100
9-21	(+)-β	10.2	> 100	> 100
9-22	(–)-α	> 100	> 100	> 100
AZT		0.004	> 100	> 100

Humber *et al.*[16] reported a method for the synthesis of 3TC in four steps. Starting from commercially available (+)-thiolactic acid and 2-benzoylacetaldehyde in the presence of boron trifluoride/etherate, a 1:2 mixture of diastereomeric oxathiolane acids (**10-1** and **10-2**) was obtained in high yield (Scheme 10). Subsequent separation of diastereomers by silica gel column chromatography and treatment with lead tetraacetate furnished the key intermediate **10-3**. Condensation of **10-3** with silylated cytosine in the presence of iodotrimethylsilane gave a 1.3:1 mixture (β:α) of **10-4** and **10-5**. Separation of the anomeric mixture and deprotection using Amberlite IRA400 (OH) afforded 3TC (**7-8**) and its α-anomer **7-9**.

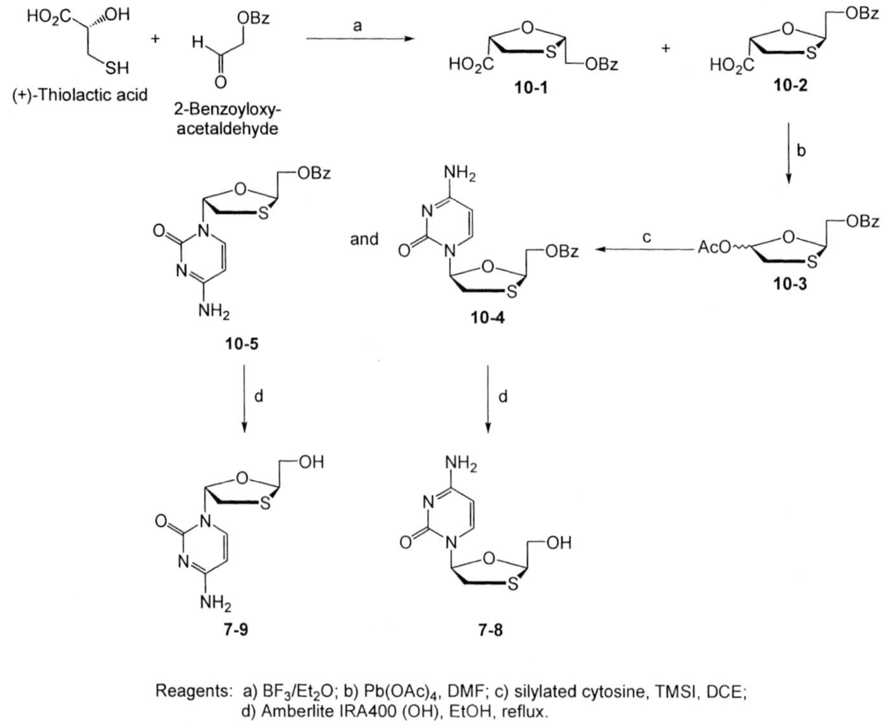

Scheme 10

Jin *et al.*[17] synthesized 3TC **7-8** and its enantiomer **3-16** from dithiane-1,4-diol in five steps (Scheme 11). Reaction of glyoxylic acid monohydrate and dithiane-1,4-diol **11-1** gave the hydroxy acid derivative **11-2**. Acetylation of the hydroxy acid derivative afforded a 1:2 mixture of the cis-acetoxy **11-3** and trans-acetoxy **11-4**. The racemic trans-isomer was isolated and esterified using (–)-L-menthol as a chiral auxiliary. The resulting diastereomeric mixture was successfully recrystallized to obtain enantiomerically pure menthyl ester **11-5**, which was condensed with persilylated cytosine in dichloromethane in the presence of TMSI affording the *cis* (**11-6**) and *trans* (**11-7**) nucleosides.

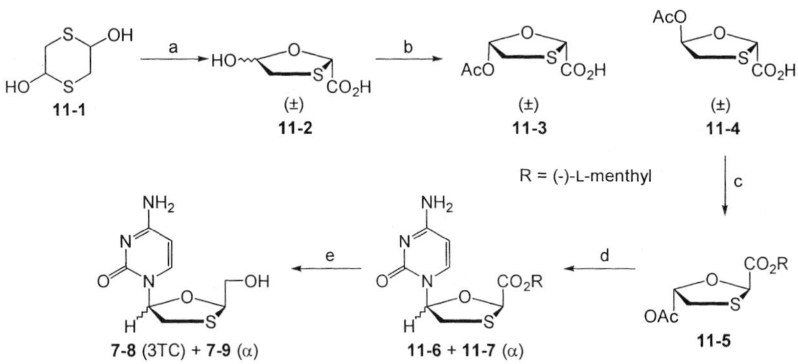

Reagents: a) glyoxylic acid hydrate, *t*-BuOMe, reflux; b) Ac_2O, CH_3SO_3H, then recrystallization; c) (-)-L-menthol, DCC, DMAP, CH_2Cl_2, then recrystallization; d) silylated cytosine, TMSI, CH_2Cl_2; e) LAH, THF.

Scheme 11

Another method to obtain the enantiomerically pure oxathiolane nucleosides was described by Liotta and co-workers, who accomplished an enantioselective enzyme-catalyzed hydrolysis of protected racemic nucleosides (Scheme 12).[18] 5'-*O*-Acyl derivatives of 3'-thiacytidine nucleosides (**12-7a-e** or **12-8a-e**) were prepared either by 5'-*O*-acylation of the 5'-hydroxyl group or by tin-mediated coupling of the corresponding acetate precursor **12-2~12-6** with silylated cytosine. Several enzymes were used to evaluate the reactivity and enantioselectivity of the 5'-*O*-acetyl group hydrolysis. Using porcine liver esterase (PLE), the butyrate ester of 2',3'-dideoxy-5-fluoro-3'-thiacytidine (**12-8d**, FTC) was hydrolyzed much faster than the 5'-*O*-acetate **12-8a**, but at a comparable rate to that of 5'-*O*-valerate and 5'-*O*-propionate esters, showing a high enantioselectivity. *O*-*t*-Butyl derivatives were poor substrates for PLE.

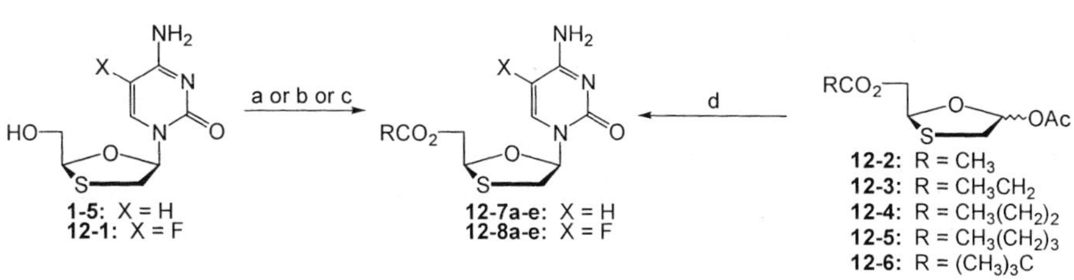

Reagents: a) HCl, then AcOH, AcCl, rt;
b) $(RCO)_2O$, reflux, then HCl, MeOH, rt;
c) RCOCl, Py, rt, then HCl, MeOH, rt;
d) silylated cytosine, $SnCl_4$.

a: R = CH_3
b: R = CH_3CH_2
c: R = $CH_3(CH_2)_2$
d: R = $CH_3(CH_2)_3$
e: R = $(CH_3)_3C$

Scheme 12

Enantioselective enzymatic synthesis of 3TC was also reported by Milton *et al.*,[19] who used an enzymatic resolution of an acetoxysulfide by a *Pseudomonas* fluorescens lipase (Scheme 13). The key intermediate **13-4** was prepared by two routes through the Pummerer reaction from methyl 2-mercaptoacetate or through condensation from bromoacetaldehyde diethylacetate.[20,21] α-Acetoxysulfide **13-4** was resolved using a lipase[22] in *t*-BuOMe resulting in high enantiomeric excess. Treatment of chiral acetoxysulfide **13-5** with HCl in dry ethanol induced acetate removal by transesterification to give the hemithioacetal **13-6**, which cyclized *in situ* to the oxathiolane **13-7** with a minor racemization. Completion of the synthesis of 3TC was accomplished by the conventional nucleoside condensation procedure.

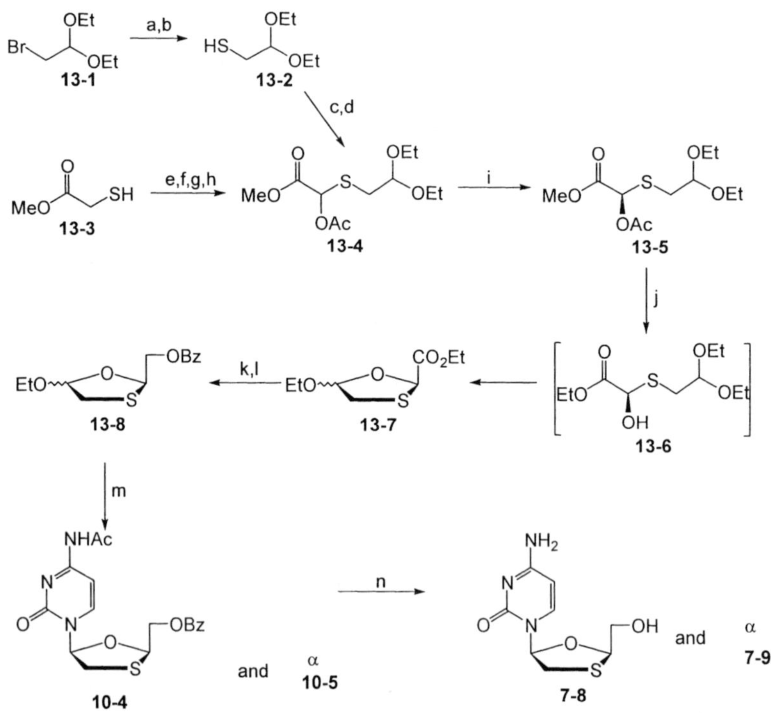

Reagents: a) KSC(S)OEt, acetone; b) $H_2NCH_2NH_2$; c) MeOC(O)CHO, CH_2Cl_2, M.S.; d) Ac_2O, Py, DMAP; e) Na, DMF; f) $BrCH_2CH(OEt)_2$; g) mCPBA, CH_2Cl_2; h) Ac_2O, NaOAc, 90 °C; i) *Pseudomonas* fluorescens lipase, pH 7 phosphate buffer, 30 °C; j) EtOH, HCl; k) $LiBH_4$, *i*-PrOH, THF; l) BzCl, Py, CH_2Cl_2; m) silylated *N*^{4}-acetylcytosine, TMSOTf, CH_3CN; n) NH_3, MeOH.

Scheme 13

Cousins *et al.*[23] carried out the enantiomeric enrichment of a (-)-*trans*-propionate by enantioselective hydrolysis using *Mucor* miehei lipase. Racemic *trans*-propionate

(±)-**14-2** was prepared in good yield from hydroxyoxathiolane (±)-**14-1** (Scheme 14). The enantiomerically enriched oxathiolane propionate, prepared by enzymatic removal of the opposite isomer, was coupled with silylated cytosine using TMSI as a Lewis acid with a *cis:trans* ratio of 1.3:1. After deprotection of nucleosides **10-4** and **10-5** by a basic resin in methanol, *trans*-diastereomer **7-9** could not be purified even by chiral HPLC, but the *cis*-diastereomer **7-8** (3TC) was purified displaying an *ee* of 70%.

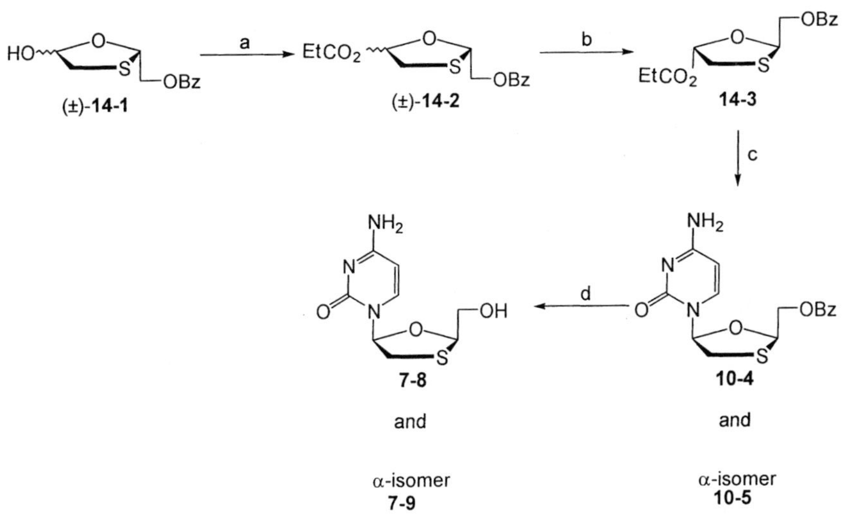

Reagents: a) EtCOCl, Py, CH_2Cl_2, 0 °C; b) *Mucor* miehei, buffer, 28 °C; c) silylated cytosine, TMSI, DCE; d) Amberite IRA400 (OH), MeOH, reflux.

Scheme 14

The synthesis of optically pure 3TC by enzymatic resolution was also carried out by Mahmoudian *et al.*[24] Cytidine deaminase from *Escherichia* coli only deaminated the D-form of 2'-deoxy-3'-thiacytidine, leaving the optically pure L-form (3TC). This enzymatic procedure has been used for the synthesis of multikilogram quantities of 3TC.

In view of the interesting antiviral activity of ribavirin,[25] Faury *et al.*[26] synthesized the tetrazole analogs of 1,3-oxathiolane nucleosides (Scheme 15). Condensation of benzoyloxyacetaldehyde **1-1** and mercaptoacetaldehyde di-[2-methoxyethyl]-acetal **15-1** using *p*-toluenesulfonic acid as a catalyst gave 2-benzoyloxymethyl-5-[2-methoxyethyloxy]-1,3-oxathiolane **15-2**. Condensation of silylated tetrazoles with **15-2** in the presence of TMSOTf, followed by silica gel column chromatography and deprotection in methanolic ammonia gave the final nucleosides **15-7~15-10**. Unfortunately, the introduction of a tetrazole ring on the oxathiolane moiety abolished the anti-HIV-1 activity and significantly increased the cytotoxicity.

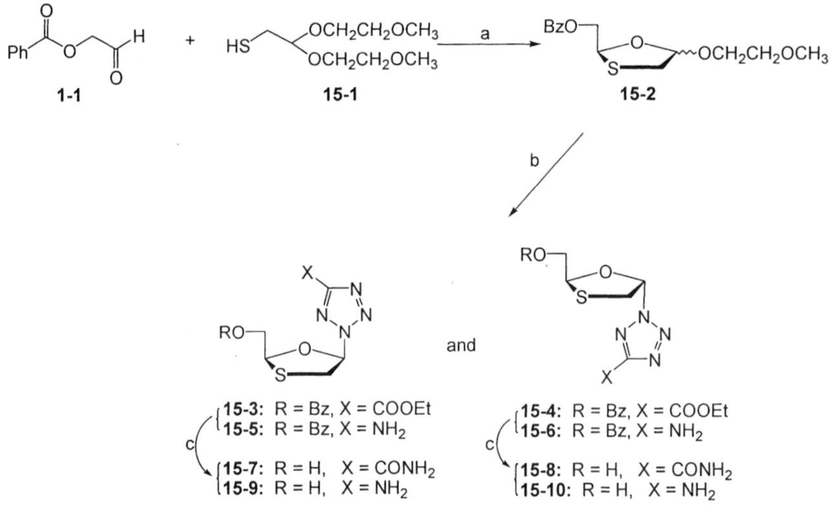

Reagents: a) *p*-TsOH, Tol; b) silylated tetrazole, TMSOTf or $TiCl_4$, CH_3CN; c) NH_3/MeOH.

Scheme 15

Belleau *et al.*[27] developed another class of (racemic) 1,3-oxathiolane nucleosides, where the position of the heteroatoms in the ring of BCH-189 is inverted (Scheme 16). The 1,3-oxathiolane moiety **16-4**, prepared from benzoyloxyacetaldehyde in five steps, was coupled with various bases. These compounds were evaluated for anti-HIV activity in the MT-4 cell lines. Among them, *cis*-5-fluorocytosine **16-7** (BCH-1089) and **16-11** (BCH-371) exhibited an EC_{50} value of 2.5 and 2.1 μg/mL, respectively.

2,4-Disubstituted 1,3-dioxolanes and -1,3-oxathiolane nucleosides containing (*E*)-5-(2-bromovinyl)-uracil were synthesized by Bednarski *et al*,[28] who used acetate derivatives **17-1** and **17-4** (Scheme 17) as starting materials for the coupling reaction with persilylated 5-bromovinyluracil.[29] The usual condensation of the sugar moiety and silylated base followed by deprotection afforded the diastereomeric nucleosides **17-2/17-3** and **17-5/17-6**. These compounds were assayed for antiviral activity against HSV-1, HSV-2 and HCMV replication *in vitro*. Among them, the β-D-oxathiolane **17-2** showed potent activity against HSV-2.

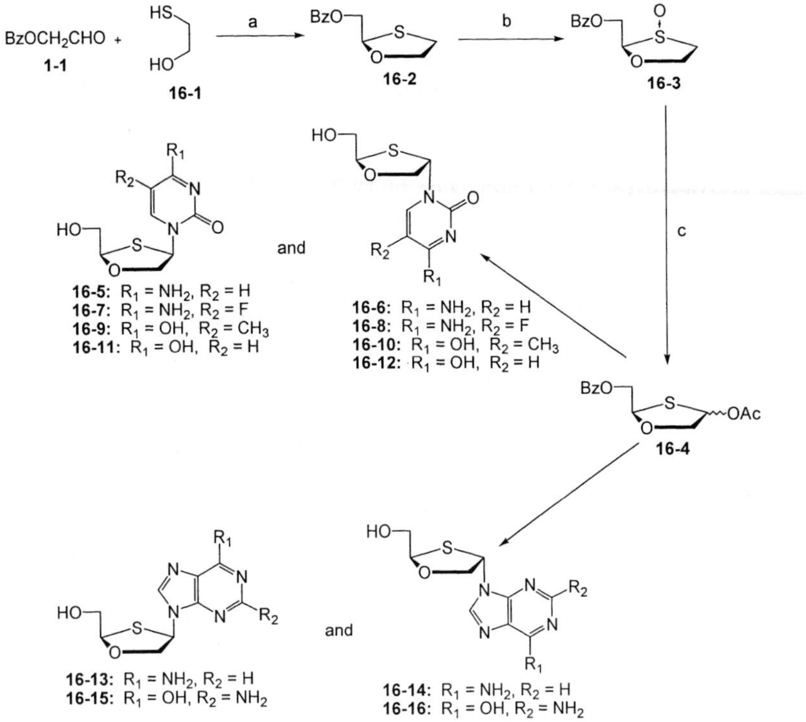

Reagents: a) *p*-TsOH, Tol, reflux; b) MMPP, CH_2Cl_2/H_2O; c) Ac_2O, $(n\text{-}Bu)_4NOAc$.

Scheme 16

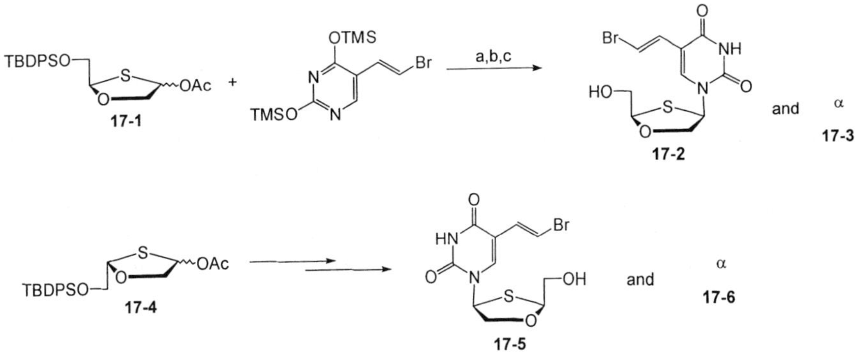

Reagents: a) TMSOTf, lutidine, DCE; b) chromatography; c) TBAF, THF.

Scheme 17

The synthesis of optically active 2'-deoxy-3'-oxa-4'-thiocytidine (dOTC) and 2'-deoxy-3'-oxa-4'-thio-5-fluorocytidine (dOTFC) derivatives were reported by Mansour and co-workers.[30] The key chiral syntons were optically active oxathiolanes having an acetoxy group at C-3 (**18-5** and **18-19**, Scheme 18).[31,32] Reductive replacement of the acetoxy group by hydrogen was achieved by treating **11-5** with TMSOTf in triethylsilane at rt. The resulting compound **18-1** was further reduced with sodium borohydride to the alcohol derivative **18-2**. Silyl protection of **18-2** gave compound **18-3**, which was then subjected to oxidation with mCPBA to yield a diastereomeric mixture of sulfoxides **18-4**. Pummerer reaction of **18-4** in acetic anhydride in the presence of *tert-n*-butylammonium acetate readily afforded the desired oxathiolane **18-5** as a mixture of *cis* and *trans* isomers. Condensation of **18-5** with nucleobases afforded β-D and α-D-2',3'-dideoxy-3'-oxa-4'-thio-ribonucleosides. The same reaction sequence was utilized to transform the chiral oxathiolane **18-18** to the key intermediate **18-19**, which was condensed with various bases to afford β-L and α-L series of nucleoside analogs **18-20~18-27**.

Anti-HBV and anti-HIV activities of **18-8** [(-)-dOTC], **18-12**, **18-22** and **18-26** were assessed in 2.2.14 and MT4 cells, respectively. In these assays, all nucleosides exhibited similar antiviral potency (Tables 3 and 4).[31,33]

Table 3. AntiL-HBV activities (IC_{50}) and cytotoxicity (CC_{50}) in 2.2.15 cells

Compound	IC_{50} (μM)	CC_{50} (μM)
(–) dOTC (**18-8**)	> 45	> 45
(–) dOTFC (**18-12**)	> 45	> 45
(+) dOTC (**18-22**)	17.5	> 50
(+) dOTFC (**18-26**)	> 45	> 45
3TC (Lamivudine)	0.02	> 45

Table 4. AntiL-HIV activities (IC_{50}) and cytotoxicity (CC_{50}) in MTL-4 Cells

Compound	IC_{50} (μM)	CC_{50} (μM)	Selective Index
(–) dOTC (**18-8**)	2.8	> 500	> 178
(–) dOTFC (**18-12**)	3.2	> 500	> 156
(+) dOTC (**18-22**)	0.9	> 500	> 555
(+) dOTFC (**18-26**)	3.0	> 500	> 166
AZT	0.005	110	22,000

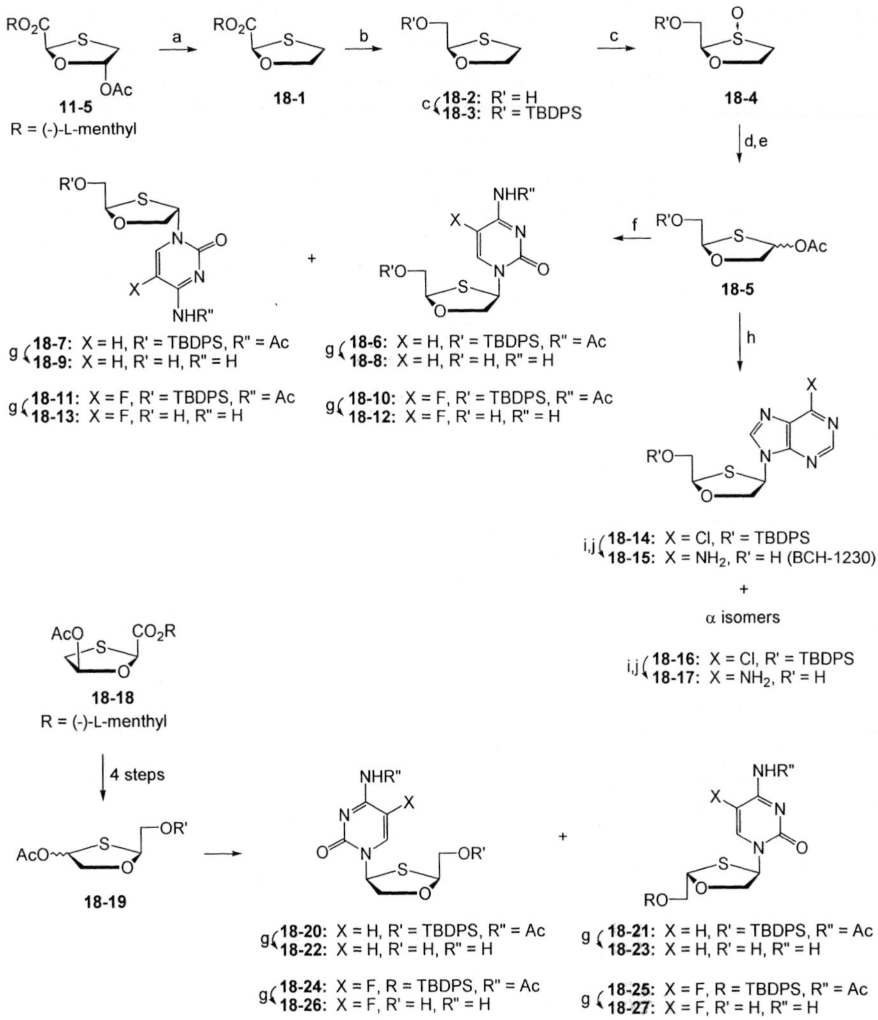

Reagents: a) Et_3SiH, TMSOTf; b) $NaBH_4$, EtOH, MeOH; c) TBDPSCl, imidazole, THF; d) mCPBA, CH_2Cl_2; e) $(n\text{-}Bu)_4NOAc$, Ac_2O, 120 °C; f) silylated N^4-acetylcytosine or -5-fluorocytosine, TMSOTf, DCE, reflux; g) TBAF, HOAc, THF, K_2CO_3; h) silylated 6-chloropurine, TMSOTf, DCE; i) TBAF; j) NH_3, MeOH.

Scheme 18

The synthesis of a N^4-substituted analog of 2',3'-dideoxy-3'-thiacytosine **19-4** was described by Camplo *et al.* (Scheme 19).[34] The prodrug was tested as activator of human polymorphonuclear leukocytes (EC_{50} value of 10^{-5} M was determined from the dose-response curve for superoxide production) as well as an inhibitor of the syncytia formation caused by HIV-1 in MT-4 cells ($IC_{50} = 8.0 \pm 0.8$ μM). The prodrug was designed for targeting specific receptors located on leukocytes membranes.

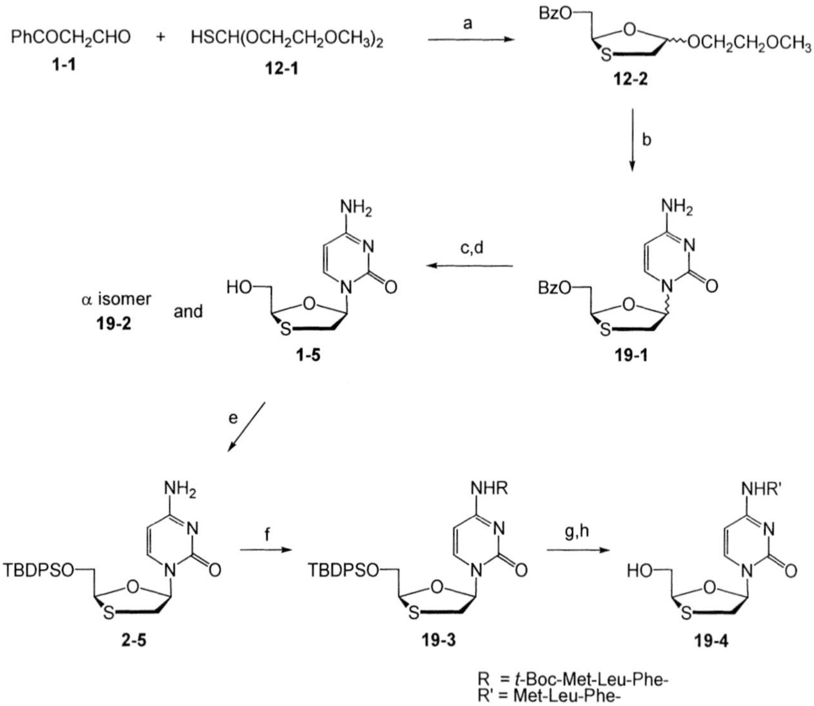

Reagents: a) p-TsOH, toluene; b) silylated cytosine, DCE, $TiCl_4$; c) Ac_2O, Py d) NH_3/MeOH; e) TBDPSCl, Py; f) Boc-Met-Leu-PheOH, DCC/HOBT/CH_2Cl_2, Py; g) EEDQ, $CHCl_3$; h) TBAF, THF.

Scheme 19

Phosphonate analogs of 3'-thia-2',3'-dideoxycytidine were synthesized by Kraus[35] in five steps *via* cyclocondensation of 2-mercaptoacetaldehyde di-(2-methoxyethyl)acetal with 3-diethylphosphonoaldehyde, followed by a Lewis acid-coupling with the appropriate nucleobase (Scheme 20). To obtain both α and β isomers for biological testing, $TiCl_4$ was used as a Lewis acid in the cytosine condensation instead of $SnCl_4$.[10]

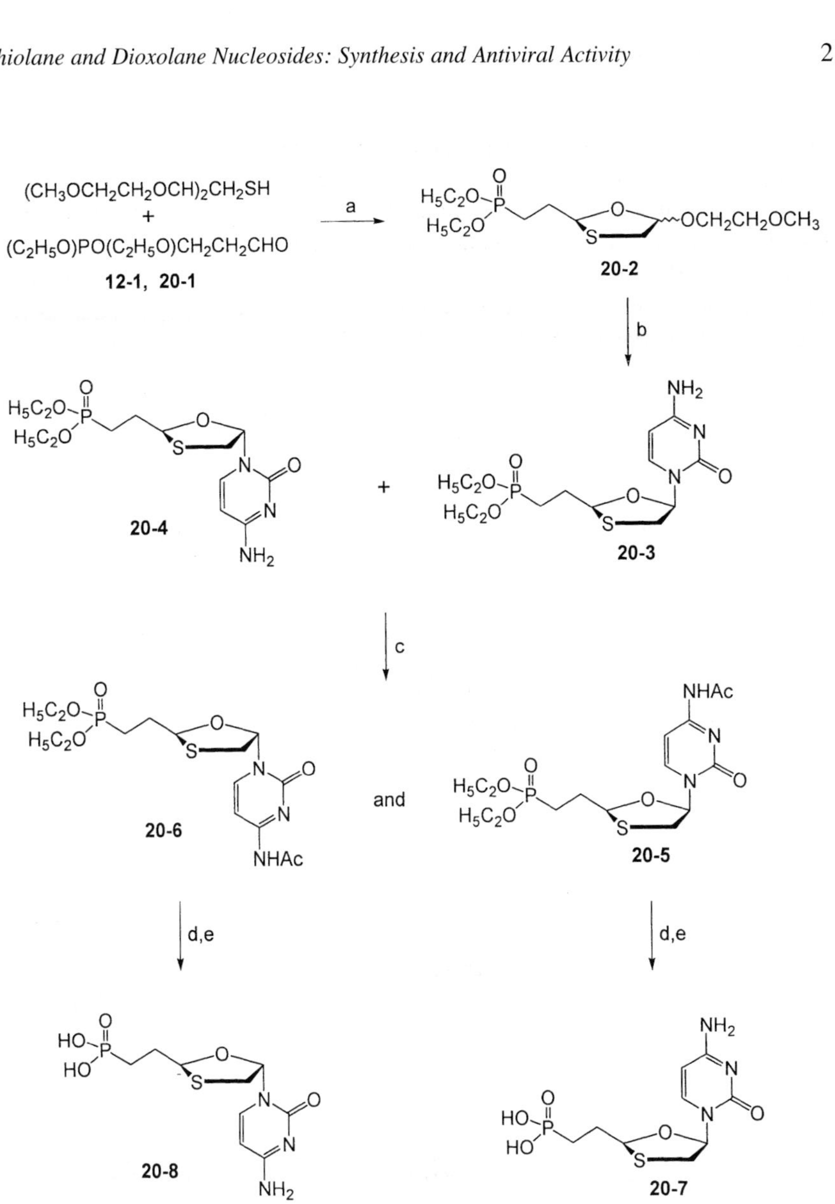

Reagents: a) *p*-TsOH, *Tol*, reflux; b) silylated cytosine, $TiCl_4$, DCE; c) Ac_2O, DMF; d) TBSBr, DMF; e) NH_3, MeOH.

Scheme 20

Separation of the mixture of α and β was carried out after N^4-acetylation, accomplished by using acetic anhydride in DMF. Pure anomers **20-5** and **20-6** were isolated in 80% yield in a 1/1 ratio. Hydrolysis of phosphonic ethyl esters **20-5** and **20-6**, followed by treatment with methanolic ammonia afforded the phosphonate nucleosides **20-7** and **20-8**, respectively. Anti-HIV evaluation of these analogs showed that the α-form **20-8** was inactive while the β-form **20-7** was found to be less potent than the parent compound (BCH-189), probably because the phosphorylated modified analog **20-7** was not a good substrate for nucleotide kinases.

Storer *et al.*[36] reported that enzymatic resolution of the monophosphate derivative of (±)-*cis*-1-[2-(hydroxymethyl)-1,3-oxathiolan-5-yl]cytosine using the 5'-nucleotidase from *Crotalus atrox* venom allowed facile access to the individual enantiomers. This method was applied to the resolution of other carbocyclic nucleoside analogs (Scheme 21).

Treatment of racemic **1-5** with phosphorous oxychloride in trimethyl phosphate at 0 °C followed by appropriate work-up and sequential chromatography on charcoal and DEAE Sephadex (HCO_3^--form) gave the racemic monophosphate of **21-1** as the ammonium salt. A solution of the racemic monophosphate of **21-1** at 37 °C in aqueous buffer prepared from glycine and magnesium chloride was treated with the 5'-nucleotidase [EC 3.1.3.5] and the resulting two-component mixture was separated by chromatography to give enantiomerically pure (+)-BCH-189 and (–)-BCH-189 monophosphate **21-2**. The latter was dephosphorylated to BCH-189 by means of an alkaline phosphatase.

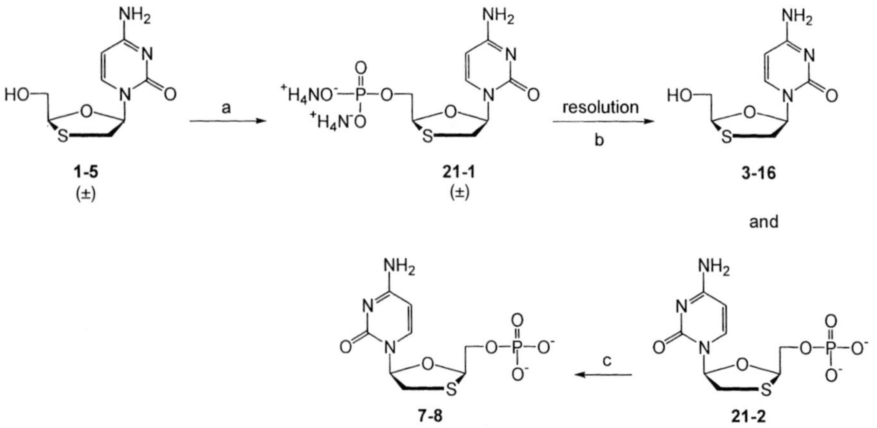

Reagents: a) $POCl_3$, $(CH_3)_3PO_4$; b) 5'-nucleotidase [EC 3.1.3.5]; c) alkaline phosphatase.

Scheme 21

Nokami *et al.*[37] reported the preparation of 1,3-oxathiolane *via* electrochemical chemoselective α-acetoxylation of β'-hydroxy sulfides (Scheme 22). The 1,3-oxathiolane **22-6** was converted to 4-acetoxy-1,3-oxathiolanes **22-7** by Pummerer reaction *via* sulfoxide. Ethyl (2,3-dihydroxypropylthio)acetate **22-1** was prepared by treatment of thioglycerol with ethyl bromoacetate in refluxing acetone in the presence of sodium carbonate. Further chemical manipulation afforded sulfide **22-4**, which underwent electrolysis to give α-acetoxy-β'-hydroxy sulfide **22-5**. These sulfides were converted to 1,3-oxathiolanes **22-6** by treatment with boron trifluoride.

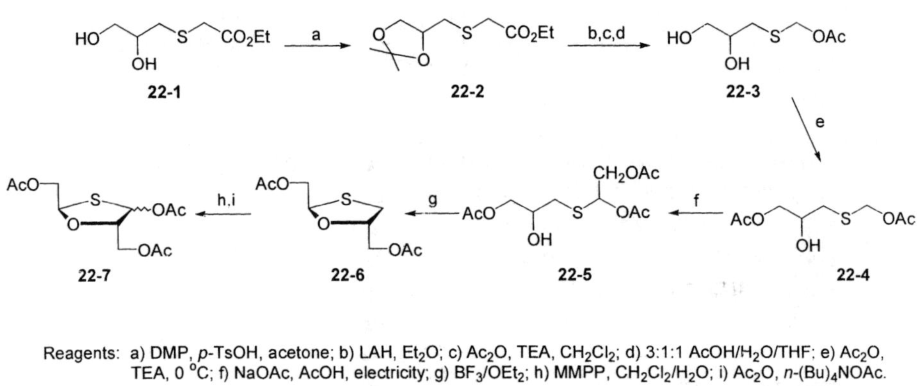

Scheme 22

Chao and Nair reported the synthesis and antiviral evaluation of 4'-hydroxymethyl oxathiolane nucleosides.[38] The synthetic approach started with the 1,3-dihydroxyacetone dimer **23-1** (Scheme 23), which was acetylated to 1,3-diacetoxyacetone **23-2**. This was cyclized to oxathiolane **23-3** by treatment with 2-mercaptoacetaldehyde diethyl acetal in acid catalysis. Compound **23-3** was condensed directly with persilylated thymine or acetylated cytosine to give, after deprotection, nucleosides **23-5**. These were also converted to the spyrocyclic monophosphates **23-6**. None of the synthesized compounds **23-5** and **23-6** showed anti-HIV activity.

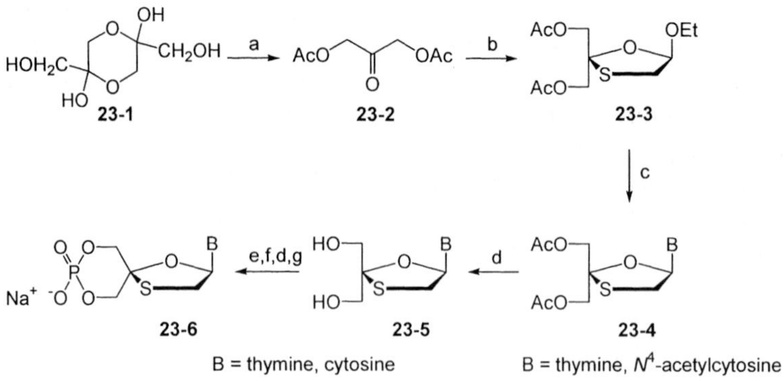

Reagents: a) Ac_2O, Py; b) $HSCH_2CH(OEt)_2$, PTSA, PhH; c) silylated base, TMS triflate, MeCN; d) NH_3/MeOH; e) 1*H*-tetrazole, 2-cyanoethyl tetraisopropylphosphorodiamidite, MeCN; f) I_2, THF, H_2O, 2,6-lutidine; g) Dowex 50Wx4-400 (Na^+).

Scheme 23

3.2.2. Antiviral activity

(±)-BCH-189 was found to be effective against HIV-1 in MT-4 cells with an EC_{50} range of 0.37 to 1.31 μM, whereas AZT had an EC_{50} range of 0.0048 to 0.0217 μM. Although it only exhibited moderate activity against HIV-1, the cytotoxicity in MT-4 cells of (±)-BCH-189 was considerably less than that of AZT with an EC_{50} of 405 μM, while the EC_{50} of AZT was 37.6 μM. (±)-BCH-189 is also a potent *in vitro* anti-HIV agent in both T-lymphoid and monocytoid cell lines as well as in human peripheral lymphocytes (EC_{50} 0.06 μM in PBM cells). The anti-HIV and anti-HBV activities and cytotoxicities of the four isomers of (±)-BCH-189 (**3-16**, **3-17**, **7-8**, and **7-9**) have been determined and are summarized in Table 5.[5,39,40]

The L-enantiomer 3TC and its D-enantiomer (+)-BCH-189 have comparable antiviral activities against HIV-2. 3TC shows no or little cytotoxicity, while (+)-BCH-189 has cytotoxicity (IC_{50} 2.7 μM) in human PBM cells, indicating that the D-isomer is more toxic. 3TC also shows less cytoxicity than (+)-BCH-189 in bone marrow progenitor cells, granulocyte macrophage colony-forming units (CFU-GM) and erythroid burst-forming units (BFU-E).[41] 3TC has no effect on mitochondrial DNA synthesis while (+)-BCH-189 exhibits a dose-dependent effect on mitochondrial DNA synthesis and lactic acid production.[42,43]

The half-life of 3TC-5'-triphosphate in mock-infected cells is 12 to 15.5 h and in HIV-1-infected cells 10.5 to 13 h. On the other hand, the 5'-triphosphate of (+)-β-D-1,3-oxathiolanylcytosine [(+)-BCH-189] shows shorter half-lives in mock-infected cells (3.5 h) and HIV-1-infected cells (5-7 h). 3TC is resistant to deamination or phosphorolysis, whereas (+)-BCH-189 is deaminated to 2'-deoxy-3'-thiauridine (Scheme 24).[44]

Table 5. AntiL-HIV and AntiL-HBV activity and cytotoxicity of the racemic and enantiomers of 3TC

	Anti-HIV activity EC_{50} (μM)			Anti-HBV activity EC_{50} (μM)	Cytotoxicity IC_{50} (μM)		
Compound	PBM cells	CEM cells	CEM-TK^- cells	2.2.15 cells	PBM cells	CEM cells	Vero cells
(+)-β-D (**3-16**)	0.2	0.1	0.08	0.5	> 100	2.7	> 100
(–) α-D (**3-17**)	> 100	71.3		> 5	> 100	> 100	> 100
(–)-β-L (**7-8**) (**3TC**)	0.0018	0.07	0.02	0.01	> 100	> 50	> 100
(+)-α-L (**7-9**)	10.1	79.8		> 5	> 100	> 100	> 100
(±) (**1-5**)	0.02-0.06	0.07	0.09	0.05	> 100	20-40	> 100

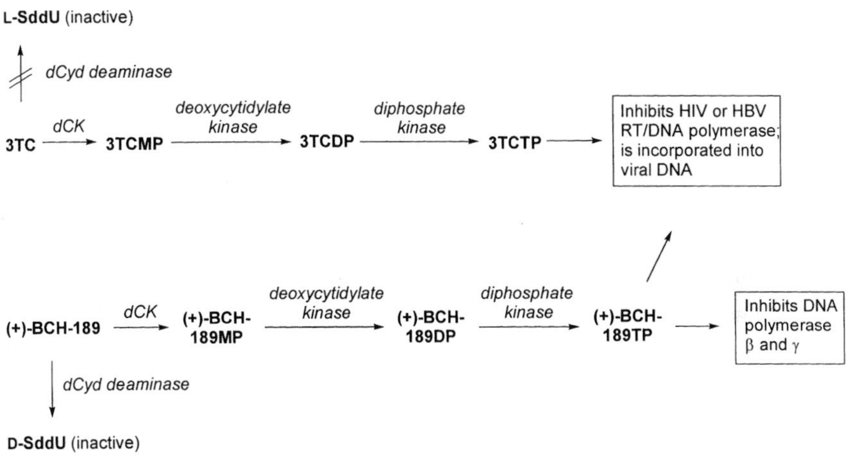

Scheme 24

3TC is an effective inhibitor of HIV-1 and HIV-2 replication *in vitro* acting at an early stage in the virus life cycle. In common with other nucleoside analogs, the 5'-triphosphate derivative of 3TC inhibits HIV-1 RT *in vitro* and also acts as a chain terminator.[45] The triphosphate of 3TC has a quite high IC_{50} value of 43.8 ± 16.4 mM against human DNA polymerase γ at concentrations of dCTP equal to its K_M value. In comparison with the triphosphates of (±)-BCH-189 and D-BCH-189, which have IC_{50} of 0.067 ± 0.030 and 0.049 ± 0.005 mM respectively, 3TC weakly inhibits DNA polymerase γ.[46] This may explain why 3TC is significantly less toxic than either the racemic or the D-enantiomer.

2',3'-Dideoxy-2',3'-didehydro-β-L-(–)-5-fluoro-cytidine [L-(–)-d4FC] has been evaluated and compared with 3TC, with particular regard to its behavior toward deoxycytidine kinase and the interaction of L-(–)-d4FC 5'-triphosphate with virion-associated HBV DNA polymerase.[47] L-(–)-d4FC was found to be at least 10 times more potent than 3TC. However, its cytotoxicity against HepG2 growth in culture is also greater than for 3TC. Like 3TC, L-(–)-d4FC can be phosphorylated to mono, di and triphosphates. The level of formation of phosphorylated metabolites in cells is higher for L-(–)-d4FC than that of 3TC. It was found that for both β-L-(–) nucleoside analogs, the step of conversion of the diphosphate metabolite to the triphosphate metabolite is not efficient.

Lisignoli *et al.*[48] compared the *in vitro* effect of (±)-BCH-189 and AZT on the immune function of lymphocytes from 10 normal and 12 HIV-1-positive (HIV-1$^+$) patients. The effect of various doses of (±)-BCH-189 and AZT were determined *in vitro* by: a) evaluation of T-cell proliferation after stimulation with concanavalin A or anti-CD3 MoAβ; b) B-cell proliferation and immunoglobulin production after stimulation with pokeweed mitogen; c) cytokine production (IL-2, IL-6, GM-CSF, tumor necrosis factor-alpha (TNF-α), interferon-γ) from lymphocytes stimulated with anti-CD3 MoAβ or phytohaemagglutinin. (±)-BCH-189 inhibited the proliferation of B and T-lymphocytes in HIV-1$^+$ subjects less than AZT, and HIV-1$^+$ subjects produced higher levels of IL-6 and TNF-α demonstrating that neither (±)-BCH-189 nor AZT interfered with cytokine release. Immunoglobulin production from B-lymphocytes was inhibited only by a high concentration (50 μM) of (±)-BCH-189 or AZT. These results show that (±)-BCH-189 affects lymphocyte proliferation *in vitro* less than AZT.

3TC is well absorbed after oral administration, with a mean bioavailability of 82% up to a dose of 8 mg/kg. 3TC has poor penetration into the cerebrospinal fluid (CSF), with CSF-to-serum ratios of 0.06 and 0.11 two hours after oral doses of 8 and 20 mg/kg per day, respectively.[49] Adverse reactions observed with 3TC include neutropenia (7%), anemia (3%), and thrombocytopenia (2%); CNS effects such as depression (9%), insomnia (11%), dizziness (10%), neuropathy (12%), headache (35%), and malacia/fatigue (27%); GI disturbances characterized by nausea (33%), vomiting (12%), and diarrhea (18%); other side effects (< 5%) including cough, rash, and pruritus.[50] Because of some side effects associated with 3TC, dosing adjustment is needed in the treatment of HIV patients with renal dysfunction because renal clearance accounts for approximately 70% of the excretion of unchanged drug during a 24 hour period.[51] All of these effects were reported in patients receiving combination therapy with AZT, and the incidence of these effects in general was greater than in patients receiving AZT alone. In pediatric trials, pancreatitis (15%) and peripheral neuropathy (13%) have been reported.[52]

The woodchuck (*Marmota monax*) has proven to be a suitable animal model for studying HBV infection because there are similarities in the course of infection between woodchuck hepatitis virus (WHV) in woodchucks and HBV in humans. The percent of 3TC excreted unchanged in the urine in woodchucks (26%) was lower than that reported for rats (75%), monkeys (32 to 59%), and humans (60 to 80%). Interspecies scaling of the pharmacokinetics parameters of 3TC showed good correlation between clearance, apparent volume of distribution, steady-state volume of distribution, and species body weight. The allometric relationships for clearance and volume of distribution at steady state predicted the observed pharmacokinetics parameters in humans quite well; however the apparent volume of distribution was underestimated in humans.[53]

In pregnant women, 3TC concentrations in maternal serum, amniotic fluid, umbilical cord and neonatal serum are comparable, indicating that the drug diffuses freely across the placenta. In postpartum women lamivudine is secreted into breast milk. The concentration of 3TC in cerebrospinal fluid is low to modest, being 4 to 8% of serum concentrations in adults and 9 to 17% in children; about 5% is metabolized to the *trans*-sulfoxide, which is pharmacologically inactive.[54]

As with HIV inhibition, 3TC-triphosphate acts as a chain terminator in the inhibition of duck HBV DNA polymerase, which was shown by [^{32}P] 3TC-triphosphate incorporation into duck HBV DNA.[55] Through the use of computer-aided DNA and protein sequence analyses, it has been shown that HBV DNA polymerase and reverse transcriptase from retroviruses share homology.[56] This suggests that if a compound has anti-HIV activity, it may also have anti-HBV activity. For example, 2',3'-dideoxycytidine (ddC), is a potent inhibitor of HIV replication in cell culture[57] and also exhibits potent antiviral activity against human HBV *in vitro*[58] and duck HBV *in vitro* and *in vivo*.[59] Unfortunately, ddC also causes latent peripheral neuropathy in patients.

In search for effective agents against hepadnaviruses with an improved therapeutic index, 2',3'-dideoxy-3-thiacytidine analogs were tested against HBV. FTC and 3TC were found to be the most effective in blocking the production of HBV in 2.2.15 cells *in vitro* (Table 6).[39,40] In comparison with other known potent inhibitors of HBV replication, their mitochondrial toxicity was lower. This indicates that peripheral neuropathy may not develop after long-term usage.

Table 6. AntiL-HIV and antiL-HBV activity and cytotoxicity of DL- and LL-oxathiolaneL-cytosine analogs

	(+)-D-BCH-189	3TC	(+)-D-FTC	(-)-L-FTC
Anti-HIV-1 (EC_{50}, μM)				
PBM	0.2	0.002	0.84	0.008
CEM	0.1	0.07	1.4	0.009
Anti-HBV-1 (EC_{50}, μM)				
2.2.15	0.5	0.01	0.96	0.01
Cytotoxicity (ID_{50}, μM)				
PBM	2.7	> 100	> 100	> 100
CEM	> 100	> 100	> 100	> 100

In vivo studies confirm that 3TC is effective in lowering the HBV DNA baseline.[60] In a one-year trial of 3TC for chronic hepatitis B, groups receiving 25 mg and 100 mg of 3TC ($P = 0.04$ and 0.08, respectively) had reduced necroinflammatory effect ($P < 0.001$ and $P = 0.001$ for the 100 mg and 25 mg doses, respectively, in comparison with the placebo group). After one year of treatment, 13 percent and 16 percent of the population was HbeAg negative for 25-mg and 100-mg groups respectively, *vs.* only 4 percent for the placebo group. The incidence of genotypic mutations in the YMDD motif (leading to resistance, *vide infra*) was 14 percent for the 25-mg and 100-mg groups. There was no significant difference in adverse events between 3TC groups and the placebo. The 100 mg group had sustained levels of alanine aminotransferase (68 percent of patients) and limited progression of fibrosis. Because of its effectiveness, 3TC has been recommended for approval by U.S. FDA Advisory Committee for the treatment of chronic hepatitis B in 1998, and is now available throughout the world for the treatment of chronically infected HBV patients.

As mentioned above, monotherapy with 3TC leads to the appearance of a drug-resistant mutant of HIV-1 with a substitution of valine for methionine at position 184 within the highly conserved YMDD motif of HIV-1 RT, which results in the development of resistance up to more than 1000-fold.[61] Despite this resistance, the treatment for more than 48 weeks is associated with a lower plasma viral level than that of the baseline, and mutant strains containing only the M184V substitution do not display heightened sensitivity neither cross-resistance to AZT, d4T, nevirapine, delavirdine, and saquinavir. Furthermore, this mutation increases the fidelity of nucleotide insertion by the mutant compared to the wild-type enzyme.[62]

The incidence of resistance to 3TC in HBV-infected patients is caused by a point mutation involving the substitution of valine or isoleucine for methionine at the 204 position of the HBV DNA polymerase rt domain (which shares high homology with the HIV-1 RT), as well as point mutations outside the motif.[63] It is worth mentioning that the 204 position (rtM204V/I), corresponding to the M184V/I mutation of HIV RT, has also been named 552, 550, 539 or 549, depending on different HBV genotypes considered. In this chapter, we will use the recently proposed genotype-independent nomenclature system.[64] It was reported that the efficiency of viral DNA replication was 100 times less than that of the wild-type with a point mutation rtM204I within the YMDD motif, but other mutations outside the YMDD motif, such as rtL180M (previously described as L526M), seem to enhance viral replication of the rtM204V-mutated HBV DNA polymerase. This finding indicates that mutations outside the YMDD motif are an essential part in causing clinical viral resistance.[63]

Gao and co-workers reported *in vitro* HIV-1 variants resistant to cytidine analogs such as ddC and (±)-BCH-189. The median effective concentrations of ddC and (±)-BCH-189 obtained for the resistant virus ranged between 10 and 50 times above those for parental wild type strains, and extensive cross-resistance was observed against ddI but not against AZT.[65] Other *in vitro* studies indicate that 3TC may lead to double resistance to AZT and 3TC, but the two-drug combination limit the emergence of resistance to 3TC.[66]

Another 3TC-resistant variant, which generates by T-C substitution (184Thr, 28%), has also been observed. The RT enzyme of the 184Thr variant is less than 10% active compared with the wild-type enzyme, and the replication capacity of this variant is severely reduced.[67]

As in the case of HBV DNA polymerase, 3TC-resistance has also been associated with the substitution of isoleucine for methionine at position 184 of the YMDD motif of HIV-1 RT.[68] As a result of this substitution, cross-resistance to (-)-FTC was noted to occur, however, 3TC resistant viruses retain susceptibility to both ddI and ddC. 3TC-resistant viruses were also susceptible to (+)-enantiomers of 3TC and FTC.[67] *In vitro* assays indicate that there is a possible cross-resistance between AZT and (–)-FTC as well as with other cytidine analogs such as ddC and 3TC in PBMC cultures.[69]

Gu *et al.* have attempted to relate genetic recombination involving human immunodeficiency virus type 1(HIV-1) to multiple drug resistance by using PEG to fuse subclones of U937 cells that carried HIV-1 recombinant strains resistant to either AZT or 3TC. These studies indicated that viral recombination had occurred, and established a theoretical basis on which to conclude that the acquisition of multiple drug resistance on the part of HIV-1 may be related to its ability to promote cell fusion.[70]

A non-culture-based assay for rapid analysis of phenotype resistance to 3TC of HIV-1 in plasma showed that 5 μM 3TC triphosphate (3TC-TP) inhibited activity of wild-type, zidovudine-resistant, or nevirapine-resistant HIV-1 but not of HIV-1 carrying either the M184V mutation or multi-drug (MD) resistance mutations (77L/116Y/151M or 62V/75I/77L/116Y/151M). Mixing experiments showed a detection threshold of 10% for 3TC-resistant virus (M184V) in a background of WT HIV-1.[71]

Several liver transplant patients who were undergoing treatment with 3TC for HBV infection experienced a breakthrough of virus while on 3TC. The predominant virus found in sera contained either the rtM204V or the rtM204I mutation. These studies confirm that the "Met-total" substitution in the YMDD nucleotide binding site of HBV polymerase is responsible for sufficient decrease in the sensitivity of HBV to the antiviral effects of 3TC. It was also found that transiently transfected cells were approximately 330-fold less sensitive to the antiviral effects of 3TC and produced 7-fold less viral DNA than the wild type.[72]

It has been reported of two patients with chronic hepatitis B treated with 48- and 52-weeks administration of 3TC whose serum HBV DNA became positive during therapy.[73] The amino acid sequence of the polymerase region of HBV from these patients revealed a YI/VDD mutation of the YMDD motif and a (223) Leu to Met mutation, which had been reported previously in immune-suppressed patients during long-term 3TC therapy for HBV. After the cessation of therapy, serum HBV-DNA returned to wild-type level immediately. This is an indication that emergence of mutant virus resistance to 3TC should be monitored during long-term 3TC therapy.[73]

In a phase II study of adefovir dipivoxil treatment in HIV-infected patients for 6-12 months, 8 of 29 patients developed RT mutation potentially attributable to adefovir dipivoxil therapy.[74] Recombinant viral DNA from pre- and post-treatment plasma samples from these 8 patients showed no change or minor decrease in adefovir susceptibility, consistently with the durable antiviral effect observed. Additionally, these patients developed the M184V RT mutation because of concomitant 3TC use. Recombinant viral DNA pairs from 4 patients with zidovudine-resistant HIV showed statistically significant increase in adefovir susceptibility of 3- to 4-fold in comparison with the wild type, and viral DNA pairs from 2 of 4 patients with zidovudine-sensitive HIV showed a 2- to 3-fold increase in susceptibility. In growth kinetics studies, expression of the M184V RT mutation resulted in attenuated viral growth in peripheral blood mononuclear cell cultures.

3.2.3. Combination therapy

As previously mentioned, 3TC is active against AZT-resistant strains. Furthermore, the combination of 3TC with AZT provide more thorough viral suppression and may limit the emergence of drug resistance.[75,76] In the 3TC/AZT combination therapy (Combivir®)[77], a two log decrease in viral burden was observed which was sustained for nearly one year. Also, this combination treatment was well tolerated and provided greater and more sustained increase in $CD4^+$ cell counts of the HIV-1 infected patients. Adverse events were no more frequent with the combination therapy than with AZT alone.[78,79]

Triple combination therapy against HIV is more effective. The combination of 3TC, AZT and indinavir[80,81] reduced plasma levels of HIV-RNA to less than 500 copies per cubic millimeter for HIV-infected patients for as long as one year while their CD4 counts increased. The combination of loviride (a NNRTI), 3TC and AZT[82] had an excellent safety/tolerability profile and demonstrated a higher level of *in vivo* efficacy, as measured by changes in surrogate markers, when compared to the loviride plus AZT combination arm.[83] The *in vivo* efficacy was evident despite the relatively high level of antiretroviral pretreatment history in the patient population.[82]

Genotypic and phenotypic resistance to 3TC develops within 12 weeks in most patients treated with this drug. Nevertheless, there is a significantly greater and more sustained decrease in plasma HIV-1 RNA level among patients who receive AZT in combination with 3TC as compared to those who received AZT mono-therapy.[84] Despite the rapid emergence of 3TC resistance, 3TC and AZT had similar antiviral activity as single agents over a reported 240 weeks study period.[84] This is a surprising result, since median susceptibility to 3TC decreased approximately 1000-fold in isolates from the patients who received 3TC monotherapy, whereas the AZT susceptibility of isolates from patients treated with AZT monotherapy appeared unchanged. A possible explanation for the limited activity of AZT is the emergence of the resistance mutation at codon 70 (K70R). This substitution is the first resistance mutation to emerge in isolates of HIV-1 from AZT-treated individuals. In the study, the K70R codon mutation was detected at week 12 in 44% of HIV-1 isolates taken from patients who received AZT monotherapy. In contrast to the findings in HIV-1 isolates from patients treated with AZT monotherapy, the K70R mutation was detected in only 9% of isolates obtained at week 12 from patients who received the 3TC/AZT combination. The sustained activity of the 3TC/AZT combination may be explained by prevention of this mutation.[85] The mechanism by which the K70R mutation is prevented by co-administration of AZT plus 3TC is unknown. Possible explanations include unfavorable interaction between the M184V and K70R substitutions, causing greater fidelity of the M184V mutant RT, or more effective suppression of virus replication by the combination regimen. An alternative explanation for the sustained activity of the 3TC/AZT combination is that 3TC-resistant variants are attenuated and thus more susceptible to inhibition by other RT inhibitors.

In a metabolic study of combination of 3TC and other RTIs, 3TC had no effect on the intracellular phosphorylation of AZT but interacted with ddC phosphorylation.[86] 3TC significantly inhibited ddC phosphorylation at concentration of 0.6 μM (38% inhibition of total ddC phosphates), 6 μM (72% inhibition) and 60 μM (97% inhibition). Intracellular ddC concentration remained unaltered with increasing concentrations of

3TC (0.47 ± 0.05 *vs.* 0.52 ± 0.08 pmoles/10^6 cells at 0 and 60 μM 3TC, respectively), thereby indicating that 3TC did not affect ddC transport into the cells. Similar results were seen in MOLT-4 and U937 cells with 60 and 84% inhibition by ddC phosphates and, at 3TC concentrations of 6 and 60 μM in MOLT-4 and U937 cells, 82 and 95% inhibition, respectively, was noted.[86] This interaction of 3TC with ddC is not surprising when we consider the structural similarity of the two drugs and the enzyme involved in their intracellular phosphorylation. The initial steps in the phosphorylation of both ddC and 3TC are carried out by deoxycytidine kinase. Therefore, when administered together, the two drugs are in direct competition for the catalytic site of this enzyme in HIV-infected lymphocytes. As AZT is phosphorylated by thymidine kinase, an enzyme for which 3TC is not a substrate, 3TC does not exert any inhibitory effects on AZT intracellular metabolism.

The result of randomized controlled studies was that after 24 weeks of treatment, d4T/3TC was at least as effective as AZT/3TC in reducing virus load and changing the $CD4^+$ cell count.[87,88] Both combinations were well-tolerated and the frequency of side-effects was comparable in both treatment groups. Although d4T/3TC resulted in a significantly higher $CD4^+$ cell count than AZT/3TC, the absolute difference at week 24 was small (median, 30 x 10^6/L). From all nucleoside analogs used in antiretroviral drug regimens, AZT has often been preferred because of its penetration into the CSF. In a neurological study, it was found that d4T and 3TC also have good CSF penetration, and that both combination regimens result in significant lower CSF virus load.[89] Even patients with relatively low baseline plasma HIV RNA levels (10000-45000 copies/mL) are at risk of developing 3TC resistance. The chance that viral resistance will develop can be minimized if plasma HIV RNA levels are reduced to undetectable levels during the use of double nucleoside therapy in combination with a protease inhibitor or a non-nucleoside RT inhibitor.[90]

Passage of wild-type virus with a combination of AZT/FTC appreciably delayed emergence of FTC-resistance virus. Also in the case of FTC, DNA sequence analysis of the RT encoding region from FTC-resistant virus revealed changes at codon 184 in the YMDD region. When the mutation M184V was introduced into the infectious clone HXB2, this change alone accounted for the resistance (> 1000-fold) seen with both 3TC and FTC, and for a 5- to 15-fold reduction in sensitivity to their (+) enantiomer, while it had no effect on susceptibility to AZT or nevirapine and minimal effect on susceptibility to 2',3'-dideoxyinosine (ddI) and 2',3'-dideoxycytidine (ddC). In order to determine the influence of this mutation and of others conferring resistance to AZT and NNRTIs, a series of HIV-1 variants were created by site-directed mutagenesis. All mutants with M184V were cross-resistant to 3TC and FTC. The M184V mutation did not influence nevirapine resistance, but resistance to AZT was suppressed. Similar suppression of AZT resistance was seen with Y181C. Interestingly, when both M184V and Y181C substitutions were present, highly resistant virus reverted to complete AZT sensitivity.[61]

Nelfinavir, a protease inhibitor, when used in two-drug combination with 3TC and three-drug combination with 3TC and AZT, resulted in statistically significant synergistic interaction with lower drug concentrations and reduced cytotoxic effects against HIV-1 strain RF infection of CEM-SS cells.[91] The three-drug combination with zidovudine, lamivudine and indinavir seems to be very effective as HIV post-exposure prophylaxis.

The proposed regimen consists of zidovudine 200 mg every 8 hours, lamivudine 150 mg every 12 hours and indinavir 800 mg every 8 hours for 4 to 6 weeks.[92] Ritonavir, another protease inhibitor, in a 3TC/AZT triple combination elicited high antiviral activity in both plasma and lymphoid tissue of HIV patients. The steep decline in RNA levels (2.87 log decline) coupled with an increase in CD4 count (152 x 10^6/L) indicated the clinical benefits of this particular triple combination.[93]

HIV RNA and DNA levels in blood and lymph node biopsies obtained from ten HIV-infected subjects who received 36-52 weeks of the combination indinavir (IDV)/AZT/3TC, IDV alone or the combination AZT/3TC confirmed the presence of mutations leading to resistance. After one year of therapy, viral RNA levels in lymph node of individuals remained detectable but were four log lower than in subjects on the triple drug regimen with interruption of therapy. When plasma virus suppression was incomplete, lymph node and PBMC cultures were positive and drug resistance detected. The persistence of even modest level plasma virus after one-year treatment reflects ongoing viral replication, the emergence of drug resistance and the maintenance of high burdens of virus in the lymph nodes.[94]

Mutants of feline immunodeficiency virus (FIV) resistant to 3TC were selected by culturing the virus in the presence of stepwise increasing concentrations of 3TC. Two variants were isolated from the original mutant population, and both of these mutants were resistant to 3TC. Surprisingly, these mutants were also phenotypically resistant to AZT and to the combination of 3TC and AZT. Purified RT from one of these plaque-purified mutants was resistant to the 5'-triphosphates of 3TC and AZT. DNA sequence analysis of RT-encoding region of the pol gene amplified from the plaque-purified mutants revealed a Pro-to-Ser mutation at position 156 of RT. A site-directed mutant of FIV engineered to contain this Pro-156-Ser mutation was resistant to 3TC, AZT and the combination of 3TC and AZT confirming the role of the Pro-156-Ser mutation in the resistance of FIV to these two nucleoside analogs. This represents the first report of a retroviral mutant resistant to the combination of AZT and 3TC due to a single unique point mutation.[95] Eventually, AZT-3TC resistant viruses were also found to have a mutation at position 333 of HIV-1 RT. This mutation is responsible for the reduction in potency of AZT and 3TC in combination therapy.[96]

Schinazi *et al.* reported the anti-HIV activity of racemates (Racevir) and enantiomers (Coviracil) of FTC.[97] (±)-FTC exhibited potent anti-HIV-1 activity (EC_{50} 0.03 μM in PBM cells), and it was also active against HIV-2, simian immunodeficiency virus (SIV) and FIV in various cell cultures. The relative *in vitro* potency of nine HIV-1 RT inhibitors was evaluated in a co-culture assay which measured the frequencies of infectious primary cells from HIV-positive patients by the limiting dilution technique and determined their apparent reduction under increasing concentrations of drugs. Potency ranking placed (-)-FTC (EC_{90} 55nM) before ddC (74 nM), 3TC (300 nM), AZT (530 nM), TIBO R82913 (670 nM) and ddI (6,400 nM). (–)-FTC showed a 20-fold greater potency than (+)-FTC against HIV-1 in human PBM cells and a 156-fold in CEM cells infected with HIV-1. (+)-FTC is significantly more toxic than (-)-FTC toward myeloid progenitor cells. Racemic FTC and each enantiomer were also evaluated against AZT-resistant and susceptible HIV-1.[98] The 5'-triphosphates of the (-)- and (+)- enantiomers equally inhibited the production of full-length minus-strand DNA in an endogenous RT reaction, each competitively inhibited DNA synthesis, and each was used as a chain-terminating substrate.[99]

The two enantiomers of BCH-189 and their 5-fluoro analogs (FTC) were found to be good substrates for human deoxycytidine kinase with K_M values in the 5.7 to 42.1 μM range.[100] The affinity of the (-)-enantiomers was greater than that of the (+)-compounds. These results may explain the greater *in vitro* antiviral potency against HIV and HBV of the (-)-enantiomers when compared to their (+)-counterparts. The (+)- and (-)-enantiomers of FTC and BCH-189 are the first nucleoside analogs found to have lower apparent kinetic constants for this enzyme in the presence of ATP compared to UTP.

Neither D- nor L-isomer of FTC showed significant cytotoxicity in human bone marrow progenitor cell assays, or any detectable hepatotoxic effects at concentrations above their antiviral activities.[101] (-)-FTC is currently undergoing phase III clinical evaluation for the treatment of HIV and HBV infection and (±)-FTC is also undergoing phase I clinical trials on HIV patients.[102]

(-)-FTC has a bioavailability of 73% after oral administration, and penetrates the blood-brain barrier. Pharmacokinetic studies indicate that the renal excretion of the racemic FTC is the major route of elimination.[103] (+)-FTC is slowly deaminated to the 5-fluorouracil analog (+)-FTU *via* cellular cytidine-deoxycytidine deaminase, while (-)-FTC is essentially resistant to this enzyme. Both enantiomers are not metabolized by thymidine phosphorylase to 5-FU.

As mentioned above, (-)-FTC is also a potent inhibitor of HBV with an apparent 50% inhibitory concentration of 10 nM, but (+)-FTC is significantly less potent.[104] In a murine model developed to investigate the *in vivo* activity of anti-HBV agents,[105] orally administered (-)-FTC showed reductions in levels of HBV in serum and in intracellular levels of HBV. (-)-FTC also shows strong inhibition to the replication of DHBV *in vivo* in chronically infected ducks,[106] and both the (+) and (-) enantiomers inhibit HBV replication *in vitro* in HepG2 cells.[101] (-)-FTC and (+)-FTC were anabolized to 5'-monophosphate, 5'-diphosphate and 5'-triphosphate in this cell line. (-)-FTC was more efficiently phosphorylated to the 5'-triphosphate than (+)-FTC. The intracellular half-life of (-)-FTC 5'-triphosphate was 2.4 h. The metabolic results are consistent with the conclusion that (-)-FTC 5'-triphosphate mediates the anti-HBV activity of (-)-FTC (Scheme 25).

Scheme 25

(-)-FTC has been assessed for its efficacy in woodchucks with naturally acquired WHV infection. Pharmacokinetics and *in vitro* anabolism were also determined, showing that (–)-FTC is anabolized to the 5'-triphosphate in a dose-related fashion, reaching a maximum concentration at about 24 hours in cultured woodchuck hepatocytes. Following administration of a dose of 10 mg/kg of body weight intraperitoneally (i.p.), the clearance of (–)-FTC from plasma was monoexponential, the terminal half-life was 3.76 ± 1.4 h, and the systemic clearance was 0.12 ± 0.06 liters/h/kg. The antiviral efficacy of (-)-FTC in the woodchuck model was assessed by quantitation of serum WHV DNA levels and by WHV particle-associated DNA polymerase activity at two dosages, 30 and 20 mg/kg given i.p. two times a day (b.i.d.), respectively. The level of WHV DNA in serum was reduced 20- to 150-fold (average, 56-fold) in the 30-mg/kg-b.i.d. treatment group and 6- to 49-fold (average, 27-fold) in the 20-mg/kg-b.i.d. treatment group. Viral DNA polymerase levels diminished accordingly. One week after treatment was discontinued, WHV levels returned to pretreatment levels in both studies. Liver biopsies before and following treatment with 30 mg of (–)-FTC per kg showed a mild increase in cytoplasmic lipid levels, but this change was not associated with altered liver enzyme levels. Serum chemistry and hematology results were within the normal ranges for all treated animals.[107]

A recent phase I/II clinical study assessed the safety, antiviral activity and pharmacokinetics of FTC therapy administered once daily for 8 weeks to HBV-infected patients.[108] The drug was well tolerated and all the 49 patients subject to therapy experienced significant, dose-dependent decrease in serum HBV DNA. At least 30% of patients receiving doses of 50 mg/day or more achieved a reduction in HBV DNA below the limit of detection. This study also suggested that the therapeutic dose does not need to be higher than 200 mg/day, at which dose over 90% of the maximal activity was achieved.[108]

A group of enantiomeric nucleosides with β-D or β-L configuration, which represent potential candidates for the treatment of HBV infection, were incubated in human hepatoblastoma HepG2 cells at concentrations between 0.1 and 10 μM for 4-14 days. Then the effect on mitochondrial DNA (mtDNA) content, lactic acid production, lipid droplet formation, and mitochondrial morphology were evaluated.[101] An effect on lactic acid production was not detected in cells treated with 3TC, FTC, D-FTC, (±)-FTC, and 2,4-diamino-7-(2,3-dideoxy-2-fluoro-β-D-arabinofuranosyl) pyrrolo[2',3'-d]pyrimidine (T70178), whereas a slight increase was associated with 1-β-D-2,6-diaminopurine dioxolane (DAPD) and 4-amino-7-(2-deoxy-2-fluoro-β-D-arabinofuranosyl) pyrrolo[2,3-d]pyrimidine (T70182) at 10 μM. A concentration-dependent increase in lactic acid was observed in the case of (+)-BCH-189, (±)-BCH-189, D-FddC, L-FddC, β-D-OddFC, 2,4-diamino-7-(2-deoxy-2-fluoro-β-D-arabinofuranosyl) pyrrolo[2,3-d]pyrimidine (T70080), and 4-amino-7-(2-deoxy-2-fluoro-β-D-arabinofuranosyl)pyrrolo [2,3-d]pyrimidine (T70179). Inhibition of mtDNA content was demonstrated to be concentration-dependent with (+)-BCH-189, D-FddC, and T70080, whereas 3TC, (±)-BCH-189, L- FTC, D-FTC, (±)-FTC, L-FddC, D-DAPD, T70178, T70179, and T70182 had no effect. β-D-OddFC resulted in a marked inhibition of mtDNA synthesis at 10 μM but not at lower concentrations. Cells treated with 3TC, (±)-BCH-189, L-FTC, D-FTC, (±)-FTC, L-FddC, D-DAPD, T70178, T70179, and T70182 did not show morphological changes compared with the control. In contrast, increased cytoplasmic lipid droplets associated with a loss of cristae in mitochondria were detected in cells treated with either D-FDOC, D-FddC, or T70080. (+)-BCH-189 treatment also resulted in loss of cristae in mitochondria.

Variants of FIV possessing a methionine to threonine mutation within the YMDD motif of RT were selected by culturing virus in the presence of inhibitory concentrations of (–)-FTC.[109] These mutants were resistant to (–)-FTC and 3TC and additionally exhibited low-level resistance to ddC. Purified RT from the mutants was also resistant to the 5'-triphosphate forms of 3TC, (–)-FTC, and ddC. Site-directed mutants of FIV have been engineered which contain either this methionine to threonine mutation or the methionine to valine mutation seen in oxathiolane nucleoside-resistant HIV-1. Both mutants displayed resistance to 3TC, thus confirming the role of these mutations in the resistance of FIV to β-L-3'-thianucleosides.[109]

Reversed oxathiolane nucleosides, such as 2'-deoxy-3'-oxa-4'-thiocytidine (dOTC), have also been found active against HSV-1, HSV-2, HBV and HIV-1. The BVU analog has demonstrated potent activity against HSV-2 and the cytosine and 5-F-cytosine derivatives have exhibited appreciable antiviral activity in cord blood mononuclear cells (CBMCs) and human monocyte U937 cell lines.[33,110] (+)-dOTC is moderately active against HBV in 2.2.15 cells. In cell culture experiments, (±)-dOTC is a potent inhibitor of primary isolates of HIV-1 with an IC_{50} for viruses resistant to 3TC and viruses resistant to 3TC and AZT of 2.53 and 2.5 μM, respectively.[111] The favorable cross-resistance profile of (-)-dOTC (BCH-10618) makes it a promising candidate as an anti-HIV agent. It is currently undergoing phase I clinical trials.

(±)-dOTC (**16-5**) is phosphorylated within cells *via* the dCyd kinase pathway and approximately 2 to 5% is converted to the racemic triphosphate derivatives. Both 5'-triphosphate derivatives (TP) of (±)-dOTC are more potent than 3TCTP in inhibiting HIV-1 reverse transcriptase (RT) *in vitro* (Scheme 26).[111]

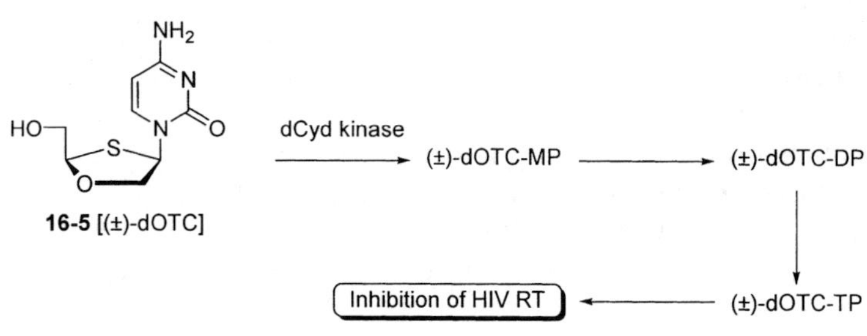

Scheme 26

3.3. Dioxolane nucleosides

3.3.1. Synthesis

Dioxolane nucleosides are 3'-heteroatom-substituted analogs in which the 3'-carbon of the furanose sugar moiety is replaced by an oxygen. The first synthesis of (±)-dioxolane-cytosine was reported by Belleau and co-workers.[9] Norbeck and co-workers[112] subsequently reported the synthesis of (±)-dioxolane-thymine *via* condensation of benzyloxyacetaldehyde dimethyl acetal **27-1** with (±)-methyl glycerate **27-2** affording the ester **27-3** as a mixture of diastereoisomers (Scheme 27). Saponification with LiOH in aqueous THF followed by acidification gave the diastereoisomeric mixture of carboxylic acids **27-4**. This underwent oxidative decarboxylation to the diastereoisomeric mixture of acetates **27-5** as a 1.7:1 mixture of *cis* and *trans* isomers, together with anhydride **27-6** as a side product. Acetates **27-5** were condensed with silylated thymine in the presence of TMSOTf, which afforded the dioxolane nucleosides **27-7** and **27-8** as a 1:1 mixture of diastereomers. The (±)-cis-isomer **27-7** was purified by silica gel column chromatography followed by deprotection and crystallization to obtain (±)-dioxolane-T **27-9**. The racemic mixture of (±)-dioxolane-T was found to exhibit moderate anti-HIV type I activity *in vitro* (EC_{50} 20 μM in ATH8 cells).

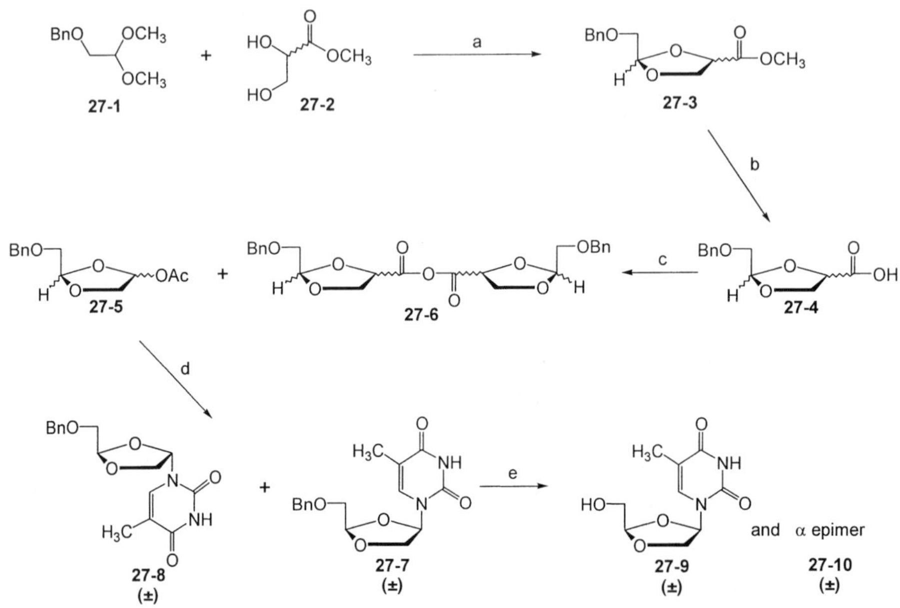

Reagents: a) *p*-TsOH, CH_3CN; b) LiOH, 4:1 THF/H_2O; c) $Pb(OAc)_4$, Py, H^+; d) silylated thymine, TMSOTf; e) H_2, $Pd(OH)_2$/C, EtOH.

Scheme 27

In order to synthesize 1,3-dioxolane nucleosides, Liotta and co-workers used lactone derivatives **28-3** or **28-9** as key intermediates (Scheme 28).[113] The starting materials in their synthesis were alkene **2-1** or **28-6**, which were converted to aldehyde **3-2** or **28-7** by ozonolysis. The latter were condensed with glycolic acid to afford lactones **28-2** or **28-8** as racemic mixtures. These lactones were then reduced to lactols, the acetylation of which afforded **28-3** or **28-9**. The acetate intermediates were condensed with various heterocyclic bases in the presence of a Lewis acid to afford 5-substituted uridine (**28-5**) and cytidine (**28-11**) analogs. The Lewis acid rendered the α-face inaccessible, which led to exclusive β-attack by the base.[10] Dioxolanyl cytosine and its 5-fluoro analog were effective against HIV-1 but display a high degree of cytotoxicity in Vero, CEM and MT-4 cells.[113]

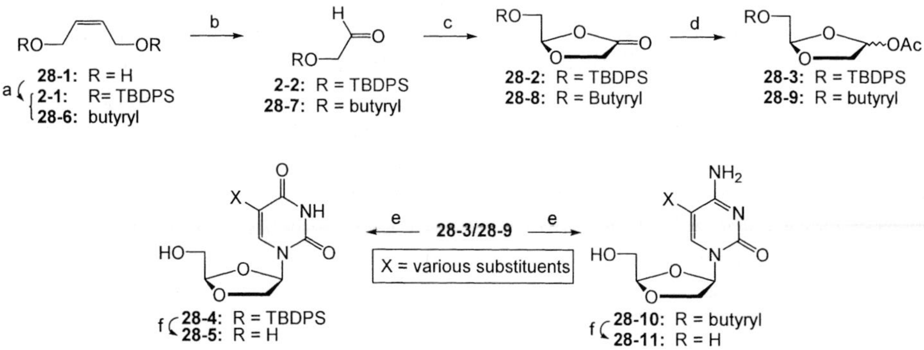

Reagents: **(R = TBDPS)** a) RCl, TEA, DMAP; b) O_3, CH_2Cl_2, DMS; c) $HOCH_2CO_2H$, DCE, *p*-TsOH; d) Dibal-H, Ac_2O; e) silylated pyrimidine, $TiCl_3$(O-iPr); f) TBAF, THF.
(R = Butyryl) a) RCl, Py, DMAP; b) O_3, CH_2Cl_2, DMS; c) $TMSOCH_2CO_2TMS$, DCE, TMSOTf; d) Dibal-H, Ac_2O; e) silylated pyrimidine, $TiCl_3$(O-iPr); f) NaOMe, MeOH.

Scheme 28

Jin *et al.*[114] reported that complexation of strong Lewis acids with the oxygen atom of the dioxolane ring during condensation may influence ring opening and closing, which leads to partial racemization of the *cis*-isomer. Partial racemization was observed when $TiCl_4$ and $SnCl_4$ were employed in the condensation step of dioxolanyl nucleosides synthesis. Lewis acids such as $TiCl_2(Oi\text{-}Pr)_2$, TMSOTf and TMSI did not to cause racemization, but exhibited lower selectivity, giving 1:1 mixtures of α and β anomers (Table 7).

Table 7. Effect of Lewis acids on the diastereoselectivity of formation of 2'L-deoxy 3'L-oxacytidines

Lewis acid	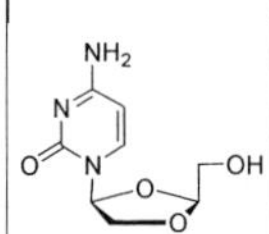 (−)-BCH-204	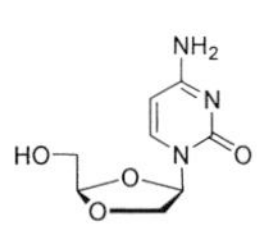 (+)-BCH-204	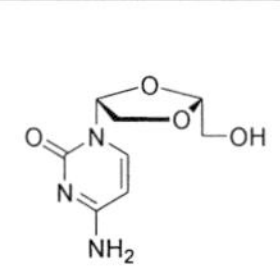 (+)-BCH-203	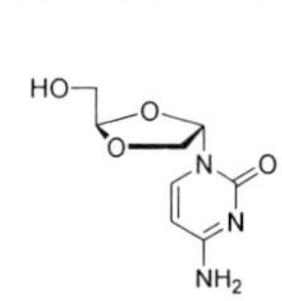 (−)-BCH-203
$TiCl_2(O\text{-}iPr)_2$	50	-	50	-
$TiCl_4$	36.5	36.5	21	6
$SnCl_4$	41.7	11.7	38.5	8.1
TMSOTf	50	-	50	-
TMSI	50	-	50	-

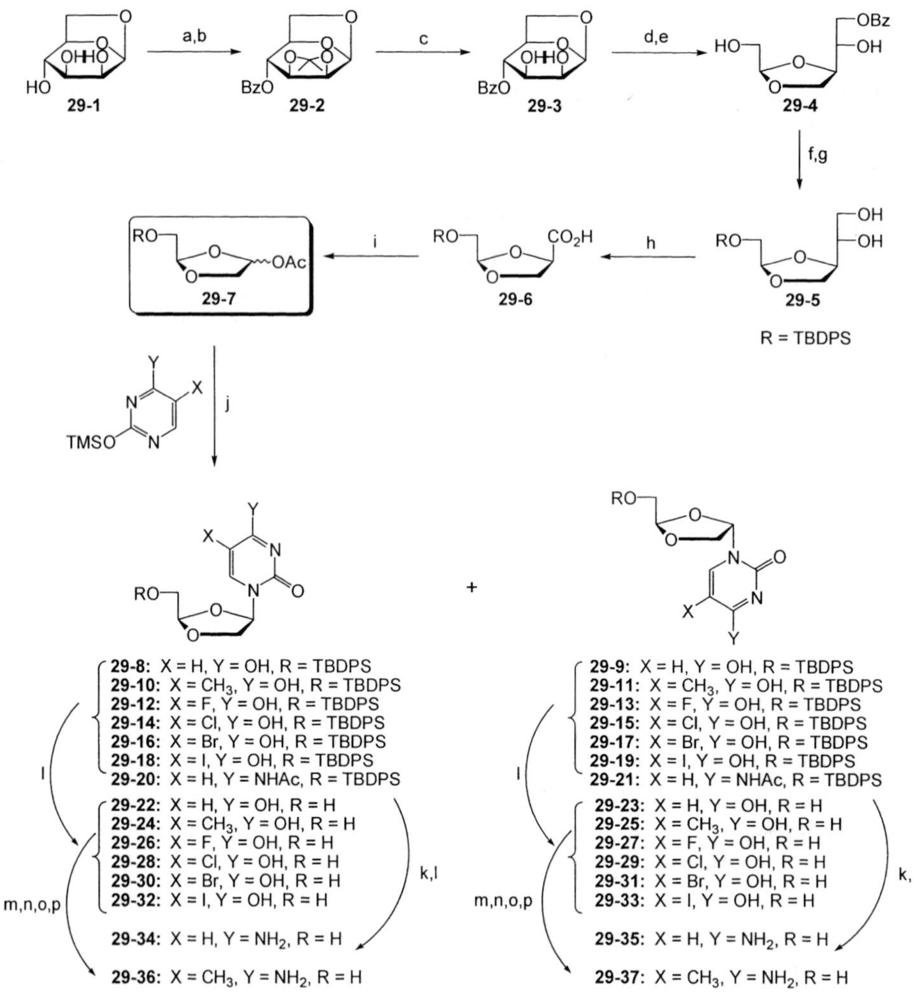

Reagents: a) DMP, H^+; b) BzCl; c) H^+; d) $NaIO_4$; e) $NaBH_4$; f) TBDPSCl; g) NaOMe; h) $NaIO_4$, RuO_2; i) $Pb(OAc)_4$; j) TMSOTf, DCE, rt; k) TBAF, THF, rt; l) NH_3, MeOH, rt; m) Ac_2O, Py; n) $(4\text{-Cl-Ph})Cl_2PO_4$, 3-nitrotriazole, Py; o) NH_4OH, dioxane (1:3); p) NH_3, MeOH.

Scheme 29

Chu and co-workers[115] published the asymmetric synthesis of D-dioxolane-pyrimidine nucleosides using D-mannose as a chiral starting material (Scheme 29). 1,6-Anhydro-D-mannose **29-1** was protected as the isopropylidene acetal, whose treatment with benzoyl chloride afforded the fully protected 1,6-anhydro-D-mannose **29-2**. The isopropylidene group of **29-2** was then selectively removed by dilute sulfuric acid in 60% aqueous dioxane to give 1,6-anhydro-4-*O*-benzoyl-D-mannose **29-3**. Consecutive treatment of **29-3** with $NaIO_4$ followed by reduction with sodium borohydride and migration of benzoyl group gave dioxolane derivative **29-4** without racemization. The primary hydroxyl group of **29-4** was then selectively protected with *tert*-butyldiphenylsilyl (TBDPS) group followed by removal of the benzoyl group using NaOMe to give **29-5**. Oxidation of the diol **29-5** to the acid **29-6** followed by oxidative decarboxylation afforded the key intermediate **29-7** for the synthesis of dioxolane nucleosides.

Pyrimidine nucleosides **29-22~29-37** were synthesized by condensation of **29-7** with the corresponding silylated pyrimidine bases in DCE catalyzed by TMSOTf, followed by deprotection. Cytosine (α and β isomers), thymine, 5-chlorouracil, 5-bromouracil, and 5-fluorouracil derivatives exhibited anti-HIV potency (Table 8). The (+)-β-cytosine analog **29-34** was found to be the most potent compound. However, its cytotoxicity profile was not as favorable as the other active compounds.

Table 8. AntiL-HIVL-1 activity (EC_{50}) and inhibitory (IC_{50}) concentration of DL-dioxolaneL- pyrimidine nucleosides in PBM cells and cytotoxicity in Vero cells

Compd	Anomer	EC_{50} (μM) anti-HIV-1 (PBM)	IC_{50} (μM) cytotoxicity (PBM)	IC_{50} (μM) cytotoxicity (Vero)
29-22	(–)-β	> 100	> 100	> 100
29-23	(–)-α	> 100	> 100	> 100
29-24	(–)-β	0.39	> 100	> 100
29-25	(+)-α	> 100	> 100	> 100
29-26	(–)-β	69.6	> 100	> 100
29-27	(–)-α	> 100	> 100	> 100
29-28	(–)-β	6.8	> 100	> 100
29-29	(–)-α	> 100	> 100	> 100
29-30	(–)-β	9.3	> 100	> 100
29-31	(+)-α	> 100	> 100	> 100
29-32	(–)-β	> 100	> 100	> 100
29-33	(+)-α	> 100	> 100	> 100
29-34	(+)-β	0.016	62.0	8.3
29-35	(–)-α	2.4	> 100	> 100
29-36	(+)-β	> 100	> 100	> 100
AZT		0.009	> 100	28.0
dioxolane-T	(±)	0.09	> 100	> 100

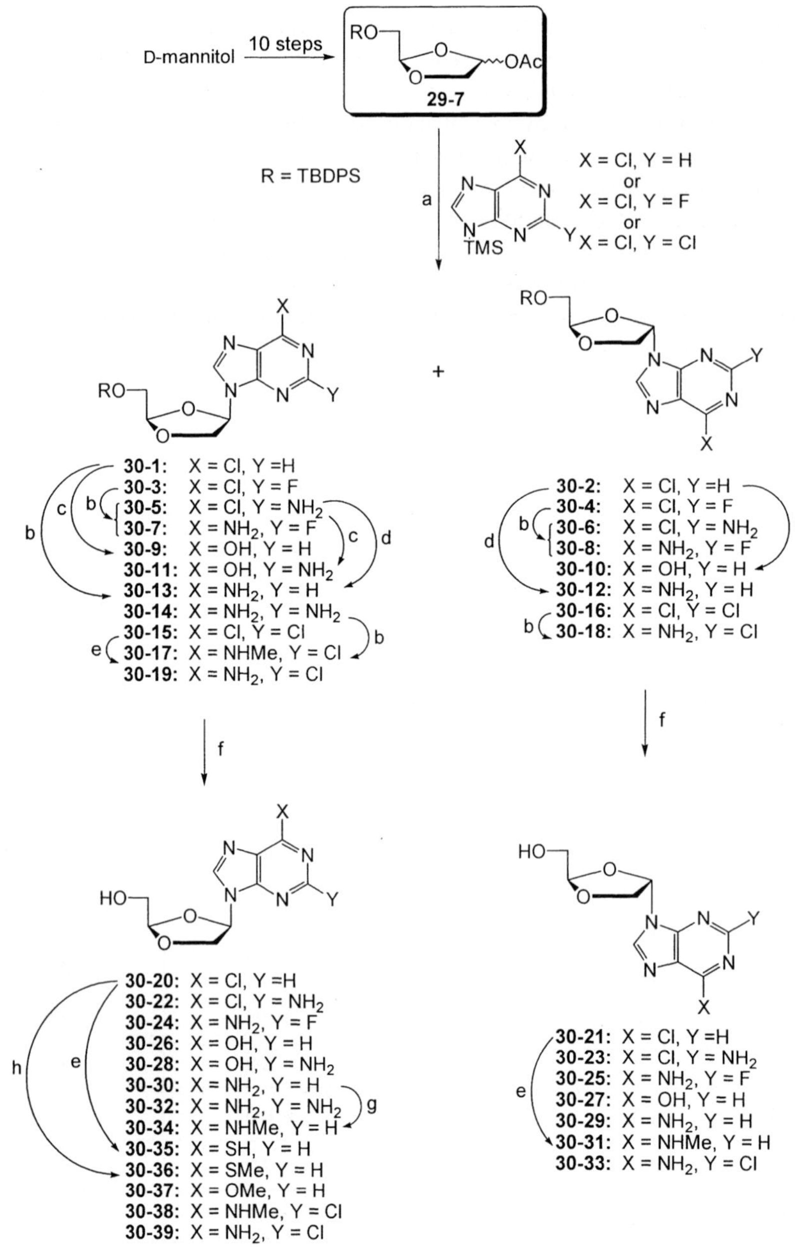

Reagents: a) TMSOTf, CH_2Cl_2; b) NH_3, DME; c) $HSCH_2CH_2OH$, NaOMe, MeOH; d) NH_3, EtOH e) NH_2CH_3, MeOH; f) TBAF, THF; g) MeI, NaOMe, MeOH; h) NaSH, MeOH.

Scheme 30

The key intermediate **29-7** was also used for the synthesis of various D-1,3-dioxolanylpurine nucleosides (Scheme 30).[116] Condensation of **29-7** with 6-chloropurine, 6-chloro-2-fluoropurine and 2,6-dichloropurine in the presence of TMSOTf afforded the initial dioxolanylpurine nucleosides. Transformation of the chloro or fluoro substituents to amino, *N*-methlyamino, hydroxy, methoxy, thiol, and methylthio was carried out under the appropriate reaction conditions.

Of the compounds synthesized, the guanine derivative exhibited the most potent anti-HIV-1 activity with low degree of cytotoxicity (Table 9). The order of activity for the β isomer of dioxolanylpurine nucleosides was as follows: guanine > 6-chloro-2-aminopurine > 2-fluoroadenine □adenine □2,6-diaminopurine > 2-chloroadenine > hypoxanthine > N^6-methyladenine □6-chloropurine □6-mercaptopurine □6-(methylthio) purine (Figure 29 and Table 9).

Table 9. AntiL-HIVL-1 activity (EC_{50}) and inhibitory (IC_{50}) concentration of DL-dioxolaneL-purine nucleosides in PBM cells and cytotoxicity in Vero and CEM cells

Compd	Anomer	EC_{50} (μM) anti-HIV-1 (PBM)	IC_{50} (μM) cytotoxicity (PBM)	IC_{50} (μM) cytotoxicity (Vero)
30-20	(–)-β	22.8	> 100	> 100
30-21	(+)-α	11.1	> 100	> 100
30-22	(–)-β	0.09	> 100	> 100
30-23	(+)-α	23.1	> 100	> 100
30-24	(–)-β	0.3	> 100	93.3
30-25	(+)-α	2.5	> 100	> 100
30-26	(–)-β	5	> 100	> 100
30-27	(+)-α	> 100	> 100	> 100
30-28	(–)-β	0.03	> 100	> 100
30-29	(+)-α	6.2	> 100	> 100
30-30	(–)-β	0.5	> 100	> 100
30-31	(+)-α	30.3	> 100	> 100
30-32	(–)-β	0.7	> 100	> 100
30-33	(+)-α	39.9	> 100	> 100
30-34	(–)-β	14.3	> 100	> 100
30-35	(–)-β	26.5	> 100	> 100

Table 9. Continued

Compd	Anomer	EC_{50} (μM) anti-HIV-1 (PBM)	IC_{50} (μM) cytotoxicity (PBM)	IC_{50} (μM) cytotoxicity (Vero)
30-36	(–)-β	25.1	> 100	> 100
30-37	(–)-β	58.1	> 100	> 100
30-38	(–)-β	40	> 100	> 100
30-39	(–)-β	1.7	> 100	> 100
AZT		0.004	> 100	28.0

In the same fashion as the D-isomers, the L-isomers of 1,3-dioxolane-pyrimidine nucleosides were synthesized using a chiral synthon (Scheme 31), L-gulose, which was dehydrated to 1,6-anhydro-L-gulose **31-1**.[117,118] The anhydro sugar was converted to dioxolane **31-2** by selective oxidation of the vicinal dihydroxyl group with *cis* -relationship with $NaIO_4$ followed by reduction with sodium borohydride and isopropylidene protection of the remaining 1,2-diol. Benzoylation of the primary hydroxyl group and deprotection of the isopropylidene group gave the diol **31-3**, which was oxidized with $NaIO_4$ to the acid **31-4**. Decarboxylative acetylation of **31-4** afforded the key intermediate **31-5**, which was coupled to 5-substituted pyrimidines, 6-chloropurine, and 2,6-disubstituted purines to obtain various dioxolanylpyrimidine (Scheme 31) and -purine (Scheme 32) nucleosides.[117,118]

The anti-HIV activities of the synthesized nucleosides were evaluated in human peripheral blood mononuclear (PBM) cells (Table 10).[118] Among the synthesized compounds, the (–)-β-5-fluorocytosine derivative **31-26** was found to have the most potent anti-HIV activity (EC_{50} 0.0012 μM) although it showed relevant toxicity ($IC_{50} = 10.0\ \mu$M). Also (-)-β-L-(2*S*,4*S*)-dioxolanylcytosine **31-22** exhibited potent anti-HIV activity, but also cellular toxicity.

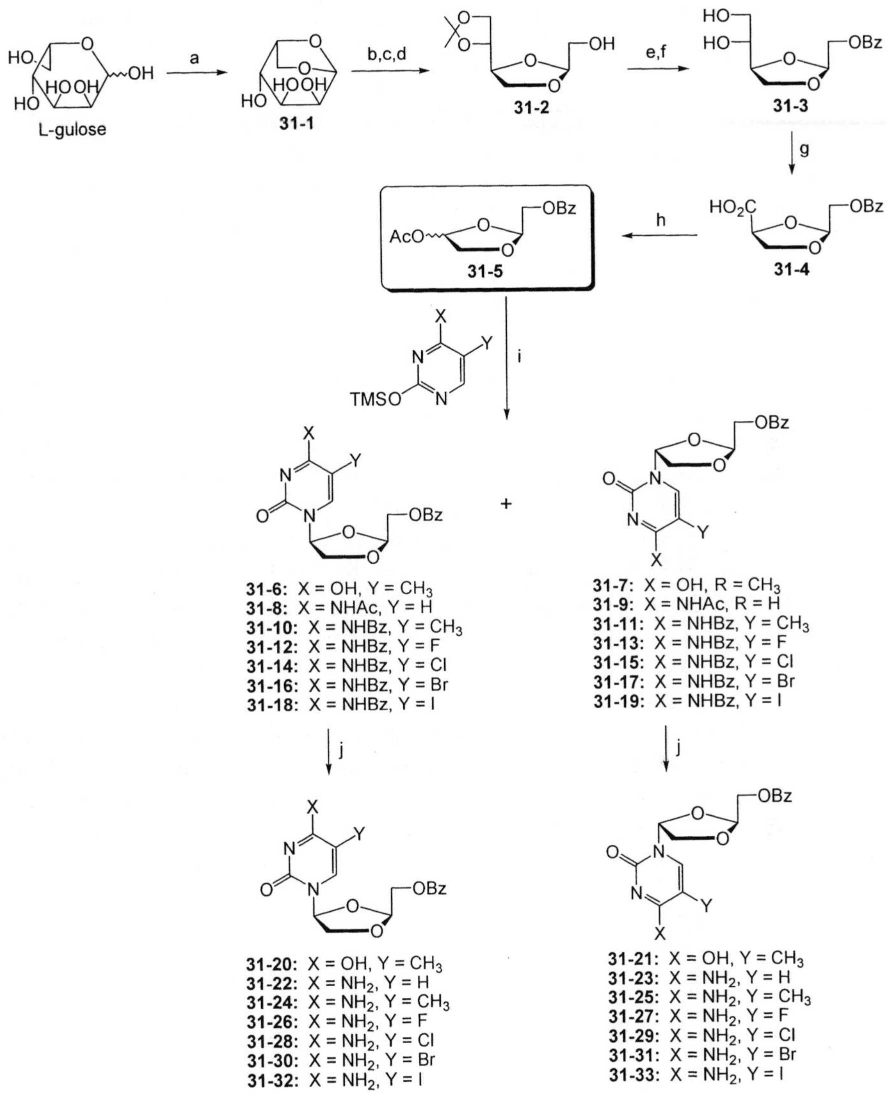

Reagents: a) 0.5 N HCl; b) $NaIO_4$, MeOH; c) $NaBH_4$; d) *p*-TsOH, acetone; e) BzCl, Py, CH_2Cl_2; f) *p*-TsOH, MeOH; g) $NaIO_4$, RuO_2, 2:2:3 $CH_3CN/CCl_4/H_2O$; h) $Pb(OAc)_4$, THF; i) silylated cytosine or thymine, TMSOTf, DCE; j) NH_3, MeOH.

Scheme 31

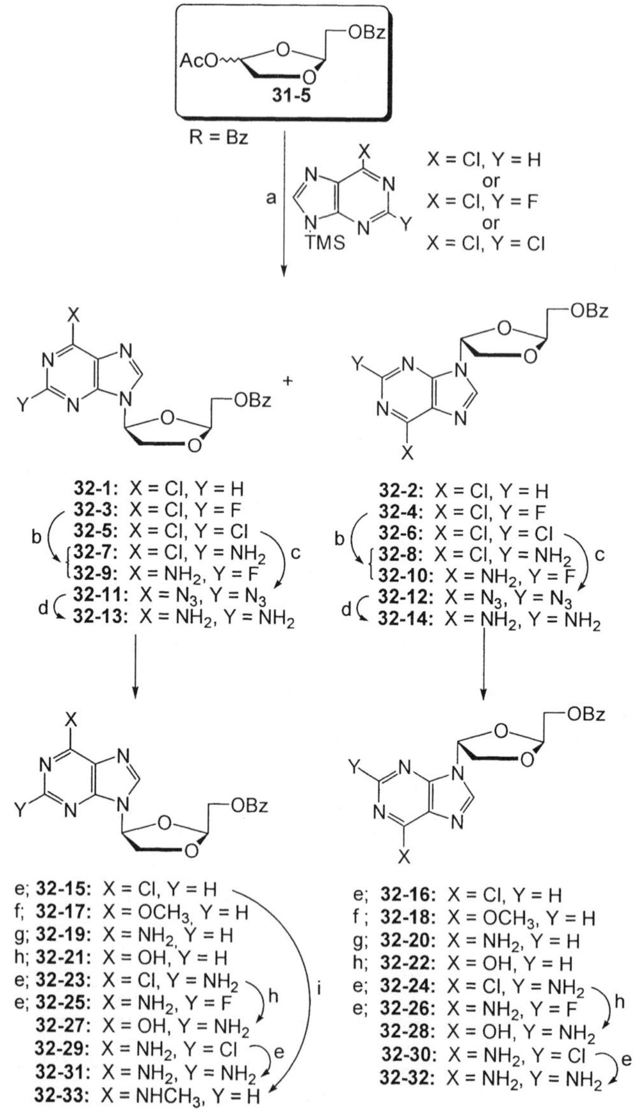

Reagents: a) TMSOTf, DCE, reflux; b) NH_3, DME, rt; c) NaN_3, EtOH, reflux; d) H_2, 10% Pd/C, EtOH, rt; e) NH_3, MeOH, rt; f) NaOMe, MeOH, rt; g) NH_3, MeOH, steel bomb, 90 °C; h) NaOMe, $HSCH_2CH_2OH$, MeOH, reflux. i) $NHCH_3$, MeOH, 85 °C.

Scheme 32

Table 10. AntiL-HIVL-1 activity (EC_{50}) and inhibitory (IC_{50}) concentration of LL-dioxolaneL-pyrimidine and purine nucleosides in PBM and Vero cells

Compd	Anomer	EC_{50} (μM) anti-HIV-1 (PBM)	IC_{50} (μM) cytotoxicity (PBM)	IC_{50} (μM) cytotoxicity (Vero)
31-20	(+)-β	4.8	> 100	> 100
31-21	(–)-α	> 100	> 100	> 100
31-22	(–)-β	0.002	> 10	0.1
31-23	(+)-α	1.3	> 10	16.8
31-24	(–)-β	45.9	> 100	> 100
31-25	(+)-α	18.9	> 100	> 100
31-26	(–)-β	0.0012	10.0	< 1.0
31-27	(+)-α	0.063	> 100	49.1
31-28	(–)-β	34.3	> 100	> 100
31-29	(+)-α	> 100	> 100	> 100
31-30	(–)-β	1.8	> 100	> 100
31-31	(+)-α	> 100	> 100	> 100
31-32	(–)-β	0.41	56.4	22.9
31-33	(+)-α	28.8	> 100	> 100
32-17	(+)-β	> 100	> 100	> 100
32-19	(+)-β	3.8	> 100	> 100
32-20	(–)-α	29.0	> 100	> 100
32-21	(+)-β	> 100	> 100	> 100
32-33	(+)-β	62.6	> 100	> 100
32-29	(+)-β	13.4	> 100	> 100
32-30	(–)-α	8.1	> 100	□100
32-25	(+)-β	1.6	> 100	> 100
32-26	(–)-α	23.7	> 100	□100
32-27	(+)-β	17.5	> 100	> 100
32-28	(–)-α	101.9	> 100	> 100
32-23	(+)-β	34.7	n.d.	n.d.
32-24	(–)-α	1.27	n.d.	n.d.

Table 10. Continued

Compd	Anomer	EC_{50} (μM) anti-HIV-1 (PBM)	IC_{50} (μM) cytotoxicity (PBM)	IC_{50} (μM) cytotoxicity (Vero)
32-31	(+)-β	0.014	n.d.	n.d.
32-32	(−)-α	42.12	n.d.	n.d.
AZT		0.004	> 100	28.0

Based on the previous work of Abushanab,[119,120] Belleau *et al.*[121,122] reported the divergent synthesis of both enantiomerically pure D- and L-dioxolanes, starting from L-ascorbic acid (Scheme 33). Condensation of L-ascorbic acid with benzyloxyacetaldehyde dimethylacetal catalyzed by *p*-TsOH gave a mixture of diasteromers **33-1** and **33-2**, which were separated by fractional crystallization. Oxidative degradation of **33-1** gave acid **33-3**, the oxidative decarboxylation of which afforded the key intermediate **33-5** that was used in the condensation step for the synthesis of D-dioxolanyl nucleosides. The remaining 1:4 mixture of mother liquor was subjected to oxidative degradation to afford acids **33-4** and **33-3**, from which **33-4** was isolated in pure form by silica gel chromatography. Oxidative decarboxylation of **33-4** gave the key intermediate **33-6**, which was used in the synthesis of the L-series of dioxolanyl nucleosides.

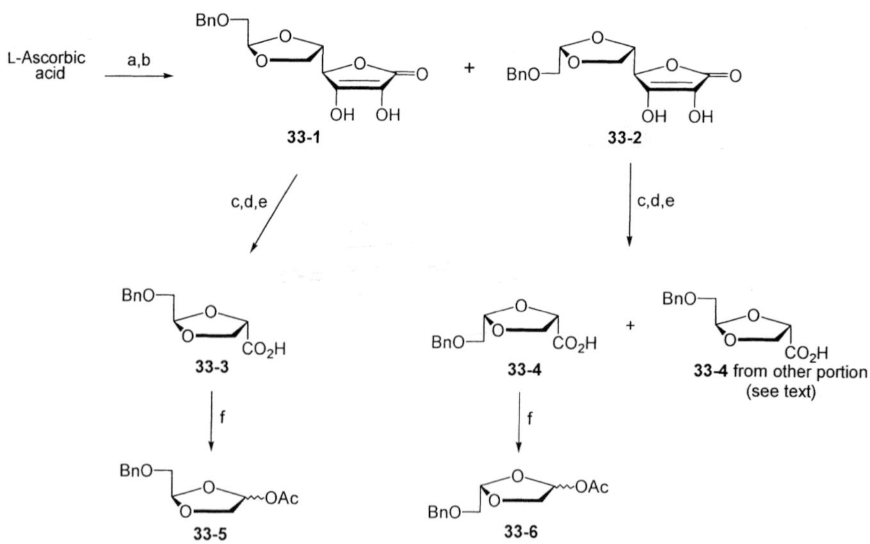

reagents: a) $PhCH_2OCH_2CH(OCH_3)_2$, TsOH, CH_3CN; b) fractional recrystallization; c) 30% H_2O_2, K_2CO_3, EtOH; d) $RuCl_3$ hydrate, NaOCl, DCE, CH_3CN, H_2O, $BnEt_3Cl$, pH 8; e) H^+; f) $Pb(OAc)_4$, CH_3CN, DCM, Py.

Scheme 33

Another divergent synthesis of dioxolanyl nucleosides was reported by Mansour and co-workers[123] using D-mannitol as the starting material (Scheme 34). D-mannitol and benzyloxyacetaldehyde dimethyl acetal **27-1** were condensed in the presence of 1.0 equivalent of $SnCl_2$ under reflux in DME to yield the bis-acetal mixture **34-1** which was subjected to oxidation to afford 1:1 mixture of acids **34-2** and **34-3**. Acids **34-2** and **34-3** were separated in the pure form by silica gel chromatography. Oxidative decarboxylation of **34-2** and **34-3** afforded the key intermediates **33-5** and **33-6**, which were used in the synthesis of D- and L-dioxolanyl nucleosides, respectively.

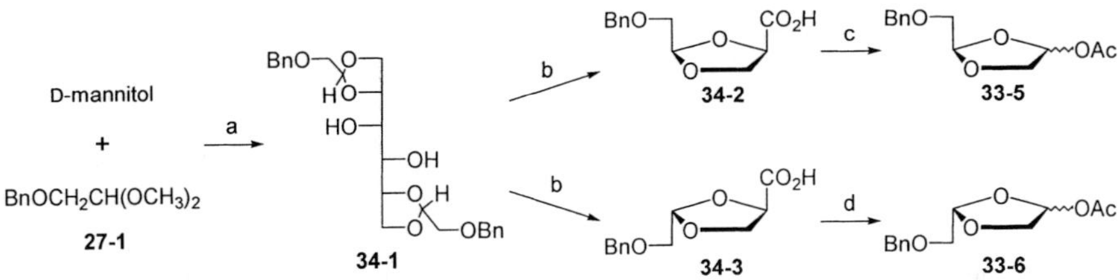

Reagents: a) 1.0 eq. $SnCl_2$; b) $RuCl_3$ hydrate, $NaOCl_3$; c) $Pb(OAc)_4$, CH_3CN, Py; d) $Pb(OAc)_4$, CH_3CN/CH_2Cl_2, Py.

Scheme 34

Enzymatic resolution of racemic dioxolane purine nucleosides by adenosine deaminase was first reported by Siddiqui *et al.* (Scheme 35).[124] Selective deamination of (±)-2,6-diaminopurine dioxolane nucleoside **35-1** produced the (-)-guanine analog **30-28** having the 2*R*,4*R* stereochemistry. The (±)-adenine analog **35-2** gave the 2*R*,4*R*-hypoxanthine derivative **30-26**.

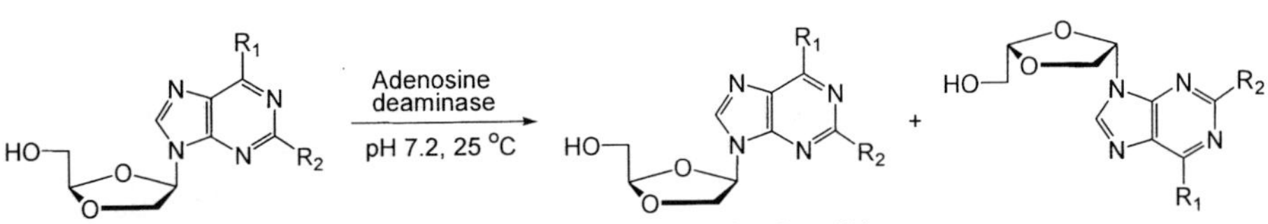

35-1: R_1, R_2 = NH_2
35-2: R_1 = NH_2, R_2 =H
(racemic)

30-28: R_1 = OH, R_2 = NH_2
30-26: R_1 = OH, R_2 = H

32-31: R1, R2 = NH_2
32-19: R1 = NH_2, R_2 = H

Scheme 35

A different class of dioxolane nucleosides in which the base and sugar moiety are connected by a methylene group was reported by Efimtseva and co-workers (Scheme 36).[125] Bromoacetaldehyde underwent acid-catalyzed transacetylation with 1-*O*-benzoylglycerol to afford a mixture of diastereoisomers **36-1** and **36-2**. Alkylation of the of the sodium salt of uracil with the mixture of **36-1** and **36-2** gave N^1-alkylated and N^1,N^3-bis-alkylated uracil. In similar fashion, alkylation of the sodium salt of adenine gave N^3- and N^9-alkylated adenine. Separation of the diastereomeric pairs **36-3/36-4** and **36-5/36-6** was accomplished by reverse phase chromatograpy. Debenzoylation in NH_3/MeOH gave the nucleosides analogs **36-7~36-10**.

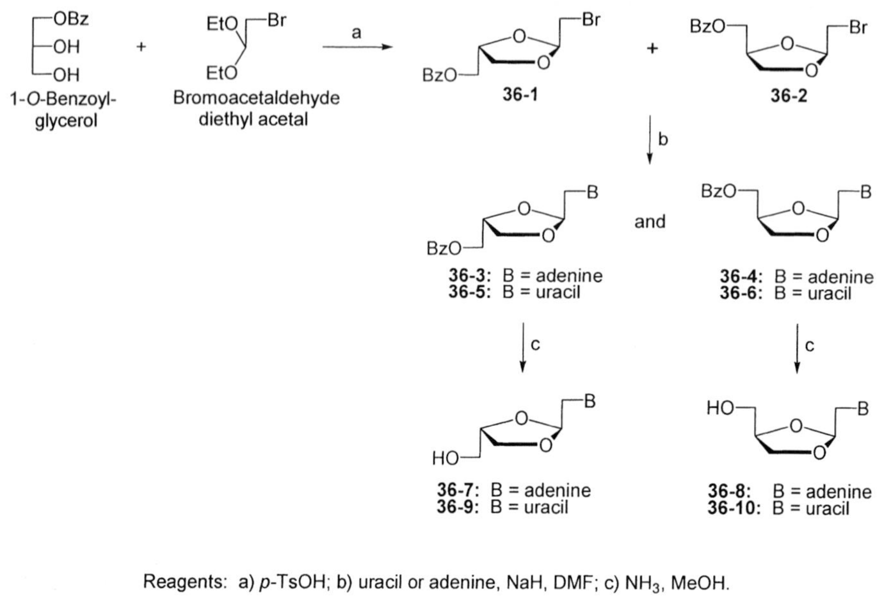

Scheme 36

Efimtseva and co-workers[126] also used D-tartaric acid as the starting material to obtain the key intermediate 2-bromomethyl-(4*R*,5*R*)-dibenzyloxymethyl-1,3-dioxolane **37-4**, which was condensed with the sodium salts of thymine, adenine, and N^2-palmitoylguanine to give, after deprotection, the desired chiral nucleoside analogs (Scheme 37).

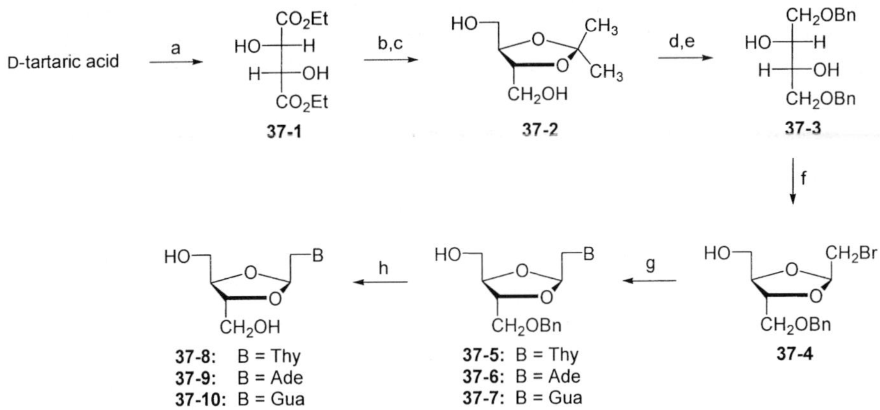

Reagents: a) EtOH, HCl; b) acetone, DMP, HCl; c) LAH; d) NaH, BnCl, THF; e) 0.5 HCl, MeOH; f) $BrCH_2CH(OEt)_2$, *p*-TsOH; g) sodium salts of thymine, adenine or N^2-palmitoylguanine, K_2CO_3, DMF; h) 10% Pd/C, HCO_2NH_4, MeOH.

Scheme 37

Later, Efimtseva and co-workers[127] also synthesized phosphonates of homologous of dioxolane nucleosides (Scheme 38). These compounds were inactive against HIV-1, HSV, HCMV, VZV, and HCMV. Although no activity was observed in these compounds, it was found that incorporation of their triphosphate into DNA by both DNA polymerase and RT occurred.

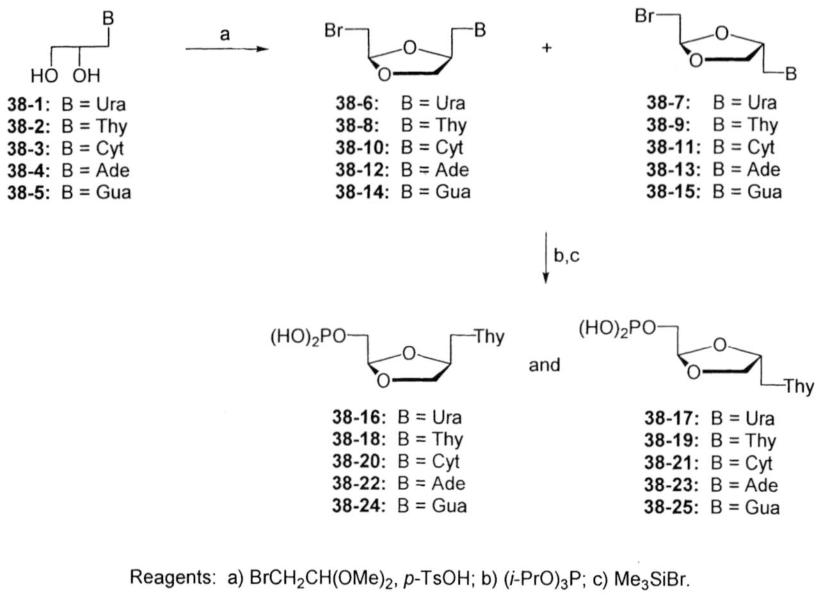

Scheme 38

2,4-Disubstituted 1,3-dioxolanes containing (*E*)-5-(2-bromovinyl)-uracil were synthesized by Bednarski *et al.* [28,110] with substantial increase in yield (75%) by debenzylation and acetylation of key intermediates **33-6** and **33-5** (Scheme 39). The usual condensation of the resulting acetates **39-1** and **39-4** with the silylated base followed by deprotection afforded the diastereomeric nucleosides **39-2/39-3** and **39-5/39-6**. These compounds were assayed for antiviral activity against HSV-1, HSV-2 and HCMV replication *in vitro*. The β-L-dioxolane nucleoside **39-2** displayed significant activity against HSV-1, whereas β-D-dioxolane **39-5** demonstrated potent activity against HSV-2. The α-L-dioxolane **39-3** was moderately active against HSV-1 and HSV-2.

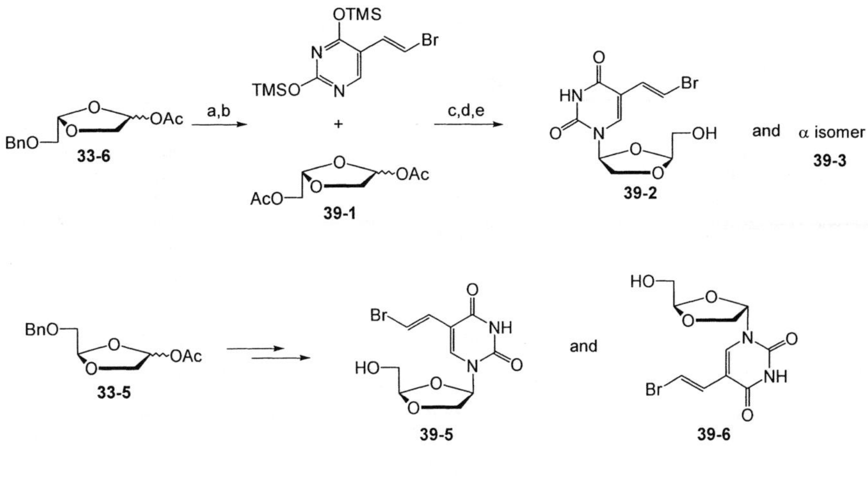

Reagents: a) H_2, Pd/C, EtOH, 48h; b) Ac_2O, Py, DMAP; c) TMSOTf, CH_2Cl_2; d) chromatography; e) K_2CO_3, MeOH.

Scheme 39

Chu and co-workers[128] synthesized 1,3-dioxolanyl-*C*-nucleosides (Scheme 40). 1,2:5,6-*O* -Diisopropylidene-D-mannitol was used as the starting material to synthesize the D-isomers **40-4a** and **40-5a** *via* the thioamide **40-2**, and L-gulonic γ-lactone was used as the starting material to synthesize the L-isomers **40-4b** and **40-5b** using the same procedure as the D-counterpart. Using similar reactions, they also synthesized selenazole (**42-9a**, **42-9b**, **42-10a** and **42-10b**) and triazole *C*-nucleosides (**42-11a**, **42-11b**, **42-12a** and **42-12b**) (Scheme 41, 42).[129]

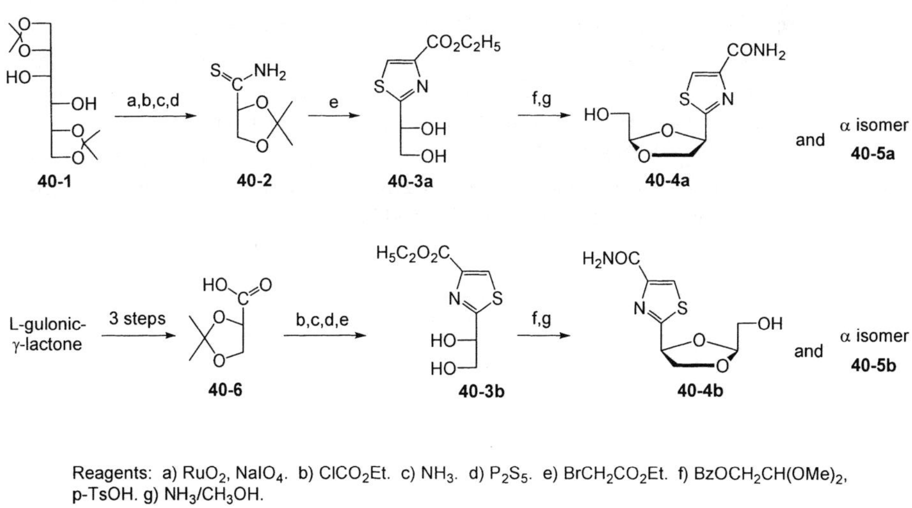

Reagents: a) RuO_2, $NaIO_4$. b) $ClCO_2Et$. c) NH_3. d) P_2S_5. e) $BrCH_2CO_2Et$. f) $BzOCH_2CH(OMe)_2$, p-TsOH. g) NH_3/CH_3OH.

Scheme 40

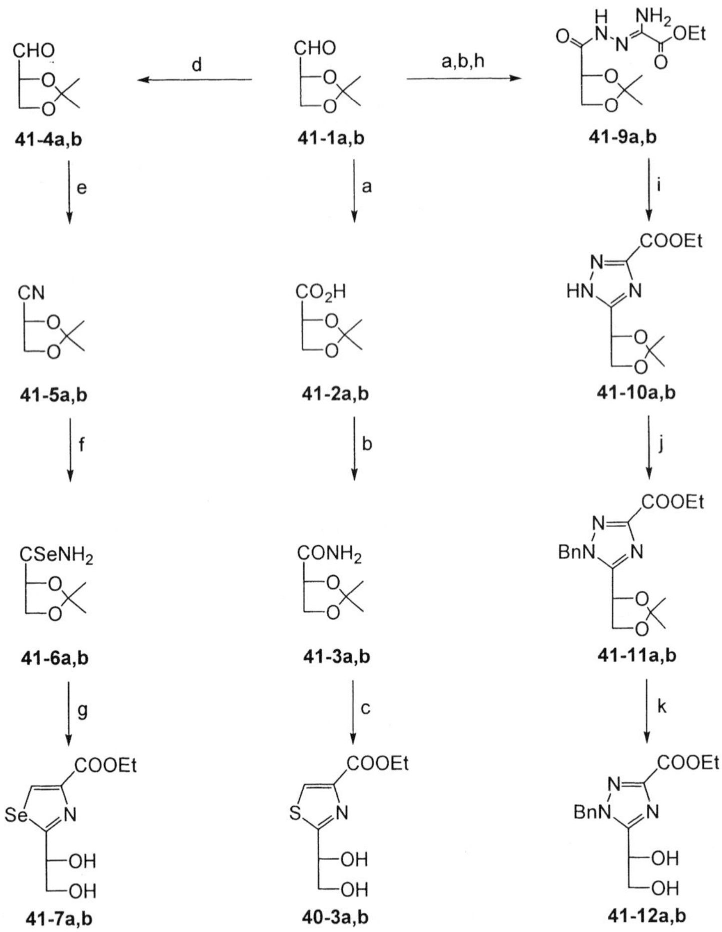

Reagents: a) (1) $KMnO_4$, KOH, then 0.5 N H_2SO_4; b) ClCOOEt, Et_3N, then NH_4OH; c) P_2S_5, then $BrCH_2COCO_2Et$, EtOH; d) NH_2OH, $NaCO_3$, EtOH; e) $(CF_3O)_2O$, -78 °C; f) Al_2Se_3, reflux, 2 h; g) $BrCH_2COCO_2Et$; h) amidrazonate, [H_2N-N=C(NH_2)-CO_2Et]; i) xylene, reflux, 4 h; j) BnBr, NaH, DMF. k) CF_3CO_2H, THF/H_2O (2:1), 50 °C, 8 h.

* "**a**" numbers are D-nucleoside related compounds.
(yield is given only for "**a**" compounds).
"**b**" numbers are L-nucleoside related compounds.

Scheme 41

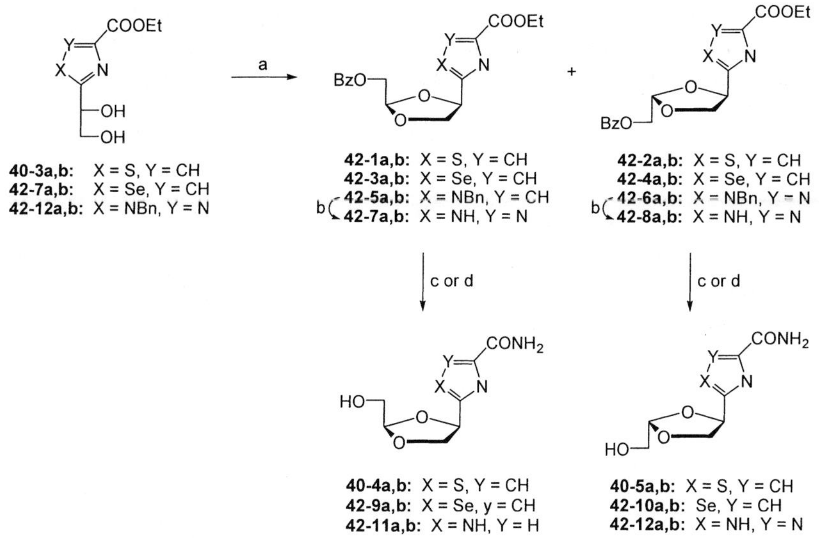

Reagents: a) $BzOCH_2CH(OMe)_2$, *p*-TsOH, benzene, reflux; b) H_2, $PdCl_2$, EtOH, 50 psi, 6 h; c) NH_3/CH_3OH, rt, 24 h;d) NH_3/CH_3OH, steel bomb, 110 °C, 24 h (for **42-11** and **42-12**).

Scheme 42

3.3.2. Antiviral activity

Extensive studies of the structure-activity relationships of dioxolane nucleosides led to the synthesis of β-L-dioxolanylcytosine **31-22** (L-OddC) and its 5-fluoro derivative **31-26** (L-OddFC), which exhibited potent activity against HIV-1 and HBV activities *in vitro*. Although L-OddC has an extremely potent anti-HIV (EC_{50} 2 nM in PBM cells) and anti-HBV activity (EC_{50} 0.5 nM in 2.2.15 cells), it is also rather toxic (IC_{50} 0.1 μM and IC_{50} 0.26 μM in Vero and CEM cells, respectively).[118] For this reason, it is not considered as a potential antiviral agent. However, *in vitro* and *in vivo* data demonstrate that L-OddC has selective toxicity towards a number of tumor cell lines, which makes it a potential anticancer candidate.[130,131]

The metabolism of L-OddC in tumor cells has been studied extensively.[132] L-OddC is rapidly transported into cells by both equilibrative-sensitive and -insensitive nucleoside transport systems. Unlike its D-enantiomer and the anticancer agent citarabine, it is not susceptible to degradation by deoxycytidine deaminase, which strengthens its biological activity. After entering the cells, L-OddC is metabolized by deoxycytidine kinase to its monophosphate, and subsequently to the di- and triphosphate. The latter inhibits DNA polymerase α, β, and γ and can also be incorporated into DNA, which directly correlates with its cytotoxicity (Scheme 43).

L-OddC accumulates within the cells as the diphosphate as occurs with 3TC. The accumulation of L-OddC diphosphate suggests that the nucleotide diphosphate kinases involved in phosphorylation are more chiral specific than other nucleoside kinases.

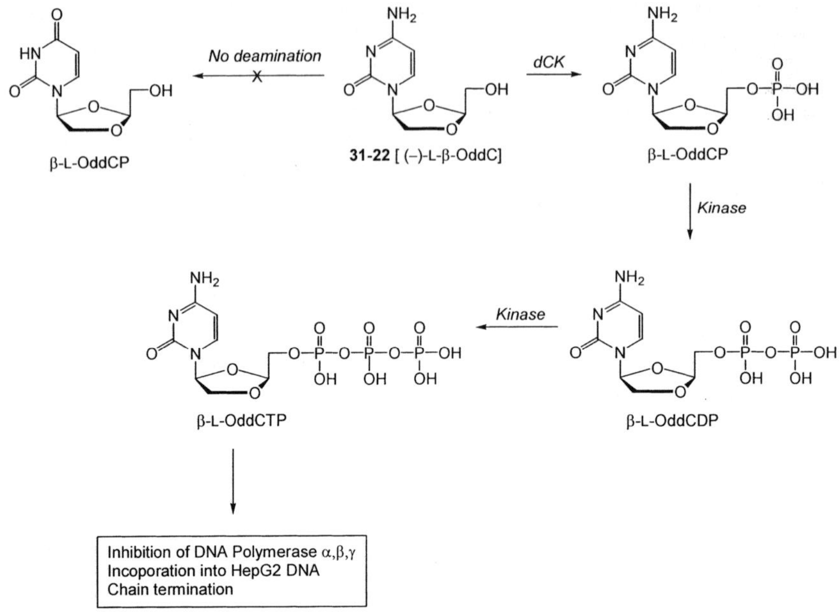

Scheme 43

The first L-nucleoside analog ever found to have antineoplastic activity, L-OddC (troxacitabine) is currently undergoing Phase II/III clinical trials as an anticancer agent. Important features of troxacitabine include: 1) it is active in solid tumors that are usually unresponsive to nucleoside analogs, such as prostatic, renal, hepatic and colon tumors; 2) its stability against deoxycytidine deaminase can prolong its effect as a drug. In fact, phase I and pharmacokinetic studies have defined protocols in which L-OddC has been administered as a single 30-minute infusion every 3 weeks.[133]

In vitro, L-OddC was also found to be effective in limiting the rate of proliferation of keratinocytes by 50% at a 50 nM concentration, which indicates that it may have potential as a therapeutic agent in the treatment of hyperproliferative skin diseases.[134] D-OddC, L-OddC, and L-OTC were tested against EBV DNA replication *in vitro* in P3HR-1 cells. The ED_{50} for EBV DNA were 0.83, 1.5, 8.3 and 14 μM respectively. L-OddC is the first nucleoside analog with L-configuration exhibiting significant activities against EBV, HBV and HIV.[135]

The structure-activity relationships of various enantiomerically pure dioxolanyl purine nucleosides have also been reported.[136] The enantiomers of 2,6-diaminopurine-dioxolane analogs, L- (**32-31**) and D-DAPD (**30-32**) show potent anti-HIV and anti-HBV activities. Interestingly, L-DAPD exhibits more potent anti-HIV activity (EC_{50} 0.014 μM) than D-DAPD (EC_{50} 0.7 μM) in PBM cell, while D-DAPD has greater anti-HBV activity (EC_{50} 0.009 μM in 2.2.15 cells) than its L-isomer (EC_{50} 8.3 μM), with favorable toxicity profiles. In *in vitro* studies involving human hepatoblastoma HepG2 cells at concentration between 0.1 and 10 μM, D-DAPD demonstrated no quantitative alteration in mtDNA synthesis.[101]

Pharmacokinetic studies suggest that D-DAPD (**30-32**) is the prodrug of the corresponding guanine derivative, dioxolane-guanine (DXG, **30-28**).[137,138,139] D-DAPD, as well as β-D-2-amino-6-chloropurine dioxolane (ACPD, **30-22**), are converted to DXG by adenosine deaminase (ADA) and β-D-2-aminopurine dioxolane (APD, **44-1**) is converted to DXG by xanthine oxidase (Scheme 44).[137,138] By using calf ADA, DAPD showed a comparable K_M to the natural substrate adenosine, although the k_{cat} for DAPD was 540-fold slower than the one for adenosine, giving a 360-fold overall substrate efficiency (k_{cat}/K_M) for the natural substrate.[139] The anti-HIV activity of DXG is higher than DAPD in CBM cells (EC_{50} 0.032 μM *vs.* 0.05 μM),[140] PBM cells (EC_{50} 4.0 μM *vs.* 0.25 μM)[139] and MT2 cells (EC_{50} 12.5 μM *vs.* 3.4 μM).[139] A phase I/II 14-day DAPD monotherapy clinical trial showed no change in HIV RT genotypes and good tolerability in all the 18 treated subjects.[141]

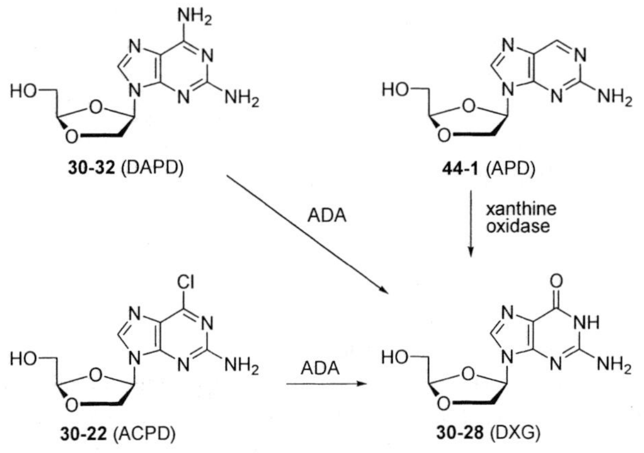

Scheme 44

The most important feature of DAPD is its activity against mutants which are resistant to other anti-HIV[142] and anti-HBV[143] agents. DXG, the active form of DAPD, is active against M184V mutants in HIV RT[142] and rtM204V/I in HBV DNA polymerase.[144] Although three RT mutations have been reported which confer resistance to DXG (namely, L74V, K65R and the multidrug resistance Q151M),[142,145,146] they cause only a moderate loss of activity (EC_{50} 3.5-, 5.6- and 9.6-fold, respectively).[142] More importantly, the most common of these mutations, K65R, reverses AZT resistance.[142,146] Furthermore, DAPD is active against K103N mutants, fully resistant to the NNRTI efavirenz, and the combination of K103N and K65R mutations only causes a slight, non-significant increase in the EC_{50} for DXG.[142] Currently, D-DAPD (amdoxovir) is undergoing phase II clinical evaluation as a potential anti-HIV for those patients who have developed drug resistance against AZT and 3TC.

According to recent reports, dioxolane derivatives L-OddT **31-20** and its 5-iodo analog L-I-OddU show activity against EBV. Particularly, L-I-OddU shows an EC_{50} value of 0.03 μM and an EC_{90} value of 0.16 μM against EBV and no cytotoxicity up to 100 μM.[147] The selectivity of L-I-OddU seems to be due to activation by EBV TK.[148] Subsequent phosphorylations to the active triphosphate may be carried out by human dTMP kinase and NDP kinase. The triphosphate of L-I-OddU (L-I-OddUTP) may inhibit EBV DNA polymerase and/or act as a chain terminator. Human DNA polymerases α, β, and γ do not seem to be inhibited by L-I-OddUTP.[147] Structure-activity studies revealed that the antiviral activity of 5-halo-ddU analogs can be related to the size of the halogens (EC_{50} Cl 0.15; Br 0.07; I 0.033 μM).[147]

Recently, the anti-VZV and anti-EBV activities of 5-halovinyl dioxolane uridine analogs have been described.[149] Also in this series, the antiviral potency can be related to the size of the halogen atoms (chlorovinyl < bromovinyl < iodovinyl against VZV and iodovinyl < bromovinyl < chlorovinyl against EBV). β-L-(*E*)-5-(2-Iodovinyl)uracil dioxolane is 60-fold more potent against VZV than ACV. No inhibition of CEM cell growth or mt-DNA synthesis is observed for any compounds at concentrations up to 200 μM. This has been explained, in the case of β-L-(*E*)-5-(2-bromovinyl)uracil dioxolane (L-BV-OddU), in terms of selective phosphorylation by viral TK, but not human TK.[150] Unlike other D-configuration BVU analogs, such as BVdU and BVaraU, L-BV-OddU is metabolized only to its corresponding monophosphate instead of the di- or triphosphate, which suggests a unique inhibitory mechanism other than DNA chain termination.[150b]

3.4. References

1. Mitsuya. H.; Weinhold, K. J.; Furman, P. A.; St Clair, M. H.; Lehrman, S. N.; Gallo, R. C.; Bolognesi, D.; Barry, D. W.; Broder, S. 3'-Azido-3'-deoxythymidine (BW A509U) – An antiviral agent that inhibits the infectivity and cytopathic effect of human lymphotropic-T virus type III lymphadenopathy-associated virus *in vitro*. *Proc. Natl. Acad. Sci. USA*. **1985**, *82*, 7096-7100.
2. Mitsuya, H.; Broder, S. Inhibition of the in vitro infectivity and cytopathic effect of human T-lymphotropic virus type III/lymphadenopathy-associated virus (HTLV-III/LAV) by 2',3'-dideoxynucleosides. *Proc. Natl. Acad. Sci. USA* **1986**, *83*, 1911-1915.
3. Lin, T.-S.; Schinazi, R. F.; Prusoff, W. H. Potent and selective in vitro activity of 3'-deoxythymidine-2'-ene (3'-deoxy-2',3'-didehydrothymine) against human immunodeficiency virus. *Biochem. Pharmacol.* **1987**, *36*, 2713-2718.
4. Riddler, S.A., Anderson, R.E., Mellors, J.W. Antiretroviral activity of stavudine (2',3'-didehydro-3'-deoxythimidine, D4T). *Antiviral Res.* **1995**, *27*, 189-203.
5. Schinazi, R. F.; Chu, C. K.; Peck, A.; McMillan, A.; Mathis, S.; Cannon, D.; Jeong, L. S.; Beach, J. W.;

Choi, W. B.; Yeola, S.; Liotta, D. C. Activities of the four optical isomers of 2',3'-dideoxy-3'-thiacytidine (BCH-189) against human immunodeficiency virus type I in human lymphocytes. *Antimicrob. Agents Chemother.* **1992**, *36*, 672-676.

6. Dobkin J. F. Abacavir enters the clinic. *Infect. Med.* **1999**, *16*, 7-7.
7. Daluge S. M.; Good S. S.; Faletto M. B.; Miller W. H.; St Clair M. H.; Boone L. R.; Tisdale M.; Parry N. R.; Reardon J. E.; Dornsife R. E.; Averett D. R.; Krenitsky T. A. 1592U89, A novel carbocyclic nucleoside analog with potent, selective anti-human immunodeficiency virus activity. *Antimicrob. Agents Chemother.* **1997**, *41*, 1082-1093.
8. Balzarini, J.; Aquaro, S.; Perno, C. -F.; Witvrouw, M.; Holy, A.; De Clercq, E. Activity of the (*R*)-enantiomers of 9-(2-phosphonylmethoxypropyl)adenine and 9-(2-phosphonylmethoxypropyl)-2,6-diaminopurine against human immunodeficiency virus in different human cell systems. *Biochem. Biophys. Res. Comm.* **1996**, *219*, 337-341.
9. Belleau, B.; Dixit, D.; Nguyen-Ba, N.; Kraus, J. L. Design and activity of a novel class of nucleoside analogs effective against HIV-1. *5th International Conference on AIDS* **1989**, T.C.O.1.
10. Choi, W. B.; Wilson, L. J.; Yeola, S.; Liotta, D. C. *In situ* complexation directs the stereochemistry of *N*-glycosylation in the synthesis of oxathiolanyl and dioxolanyl nucleoside analogs. *J. Am. Chem. Soc.* **1991**, *113*, 9377-9379.
11. Chu, C. K.; Beach, J. W.; Jeong, L. S.; Choi, B. G.; Comer, F. I.; Alves, A. J.; Schinazi, R. F. Enantiomeric synthesis of (+)-BCH-189 [(+)-(2S,5R)-1-[2-(hydroxymethyl)-1,3-oxathiolan-5-yl]cytosine] from D-mannose and its anti-HIV activity. *J. Org. Chem.* **1991**, *56*, 6503-6505.
12. Jeong L. S.; Alves A. J.; Carrigan S. W.; Kim H. O.; Beach J. W.; Chu C. K. An efficient synthesis of enantiomerically pure (+)-(2S,5R)-1-[2-(hydroxymethyl)-1,3-oxathiolan-5-yl]-cytosine [(+)-BCH-189] from D-galactose. *Tetrahedron Lett.* **1992**, *33*, 595-598.
13. Jeong, L. S.; Schinazi, R. F.; Beach, J. W.; Kim, H. O.; Shanmuganathan, K.; Nampalli, S.; Chun, M. W.; Chung, W. K.; Choi, B. G.; Chu, C. K. Structure-activity relationships of β-D-(2*S*,5*R*)- and α-D-(2*S*,5*S*)-1,3-oxathiolanyl nucleosides as potential anti-HIV agents. *J. Med. Chem.* **1993**, *36*, 2627-2638.
14. Beach, J. W.; Jeong, L. S.; Alves, A. J.; Pohl, D.; Kim, H. O.; Chang, C. -N.; Doong, S. -L.; Schinazi, R. F.; Cheng, Y. -C.; Chu, C. K. Synthesis of enantiomerically pure 2'*R*,5'*S*-(-)-[2-(hydroxymethyl)-oxathiolan-5-yl]cytosine as a potent antiviral agent against hepatitis B virus (HBV) and human immunodeficiency virus (HIV). *J. Org. Chem.* **1992**, *57*, 2217-2219.
15. Jeong, L. S.; Schinazi, R. F.; Beach, J. W.; Kim, H. O.; Nampalli, S.; Shanmuganathan, K.; Alves, A. J.; McMillan, A.; Chu, C. K.; Mathis, R. Asymmetric synthesis and biological evaluation of β-L-(2*R*,5*S*)- and α-L-(2*R*,5*R*)-1,3 oxathiolane-pyrimidine and -purine nucleosides as potential anti-HIV agents. *J. Med. Chem.* **1993**, *36*, 181-195.
16. Humber, D. C.; Jones, M. F.; Payne, J. J.; Ramsay, M. V. J.; Zacharie, B.; Jin, H.; Siddiqui, M. A.; Evans, C. A.; Tse, H. L. A.; Mansour, T. S. Expeditious preparation of (-)-2'-deoxy-3'-thiacytidine (3TC). *Tetrahedron Lett.* **1992**, *32*, 4625-4628.
17. Jin, H.; Siddiqui, M. A.; Evans, C. A.; Tse, H. L. A.; Mansour, T. S. Diastereoselective synthesis of the potent antiviral agent (-)-2'-deoxy-3'-thiacytidine and its enantiomer. *J. Org. Chem.* **1995** *60*, 2621-2623.
18. Hoong, L. K.; Strange, L. E.; Liotta, D. C. Enzyme-mediated enantioselective preparation of pure enantiomers of the antiviral agent 2',3'-dideoxy-5-fluoro-3'-thiacytidine (FTC) and related compounds. *J. Org. Chem.* **1992**, *57*, 5563-5565.
19. Milton, J.; Brand, S.; Jones, M. F.; Rayner, C. M.; Walcott, S. M. Enantioselective enzymatic synthesis of the antiviral agent lamivudine (3TC™). *Tetrahedron Lett.* **1995** *36*, 6961-6964.
20. Djerassi, C.; Gorman, M.; Markley, F. X.; Oldenberg, E. B. Studies in organic sulfur compounds. 7.

Lithium aluminum hydride reduction of xanthates to mercaptans - synthesis of substituted beta-mercaptoethanols. *J. Am. Chem. Soc.* **1955**, *77*, 568-571.

21. Mori, K.; Nakamura, Y. Thiols and sulfides from xanthogenic acid ester. *J. Org. Chem.* **1969**, *34*, 4170-4172.
22. Drescher, M.; Li, Y. F.; Hammerschmidt, F. Enzymes in organic chemistry. 2. Lipase-catalyzed hydrolysis of 1-acyloxy-2-arylethylphosphonates and synthesis of phosphonic acid analogs of L-phenylalanine and L-tyrosine. *Tetrahedron* **1995**, *51*, 4933-4946.
23. Cousins, R. P. C.; Mahmoudian, M.; Youds, P. M. Enzymatic resolution of oxathiolane intermediates – An alternative approach to the antiviral agent lamivudine (3TC™). *Tetrahedron Asymmetry* **1995**, *6*, 393-396.
24. Mahmoudian, M.; Baines, B. S.; Drake, C. S.; Hale, R. S.; Jones, P.; Piercery, J. E.; Montgomery, D. S.; Purvis, I. J.; Storer, R.; Dawson, M. J.; Lawrence, G. C. Enzymatic production of optically pure (2'*R-cis*)-2'-deoxy-3'-thiacytidine (3TC, lamivudine): a potent anti-HIV agent. *Enzym. Microb. Technol.* **1993**, *15*, 749-755.
25. Gilbert, B. E.; Knight, V. Biochemistry and clinical applications of ribavirin. *Antimicrob. Agents Chemother.* **1986**, *30*, 201-205.
26. Faury, P.; Camplo, M.; Charvet, A. S.; Chermann, J. C.; Kraus, J. L. Synthesis of tetrazole oxathiolane nucleoside analogs and their evaluation as HIV-1 antiviral agents. *Nucleosides Nucleotides* **1992**, *11*, 1481-1488.
27. Belleau, B.; Brasili, L.; Chan, L.; Di Marco, M. P.; Zacharie, B.; Nguyen-Ba, N.; Jenkinson, H. J.; Coates, J. A. V.; Cameron, J. M. A novel class of 1,3-oxathiolane nucleoside analogs having potent anti-HIV activity. *Bioorg. Med. Chem. Lett.* **1993**, *3*, 1723-1728.
28. Bednarski, K.; Dixit, D. M.; Wang, W.; Evans, C. A.; Jin, H. L.; Yuen, L.; Mansour, T. S.; Knight, D. J.; Walcott, S.; Coates, J. A. V. 5'-(*E*)-(2"-bromovinyl)uracil 1,3-dioxolanes and 1,3-oxathiolanes as potential antiherpetic agents. *Abstr. Papers Am. Chem. Soc.* **1994**, MEDI 024.
29. Jones, A.S., Rahim, S.G., Verhelst, G., Walker, R.T. In *Nucleic Acid Chemistry, Improved and New Synthetic Procedures, Methods and Techniques, Part 3;* Townsend, L. B., Tipson, R. S, Eds.; John Wiley & Sons, Inc.: New york, **1986**, 61-64.
30. Mansour, T. S.; Jin, H. L.; Wang, W.; Dixit, D. M.; Evans, C. A.; Tse, H. L. A.; Belleau, B., Gillard, J. W.; Hooker, E. U.; Ashman, C., Cammack, N., Salomon, H., Belmonte, A. R.; Wainberg, M. A. Structure-activity relationships among a new class of antiviral heterosubstituted 2',3'-dideoxynucleoside analogs. *Nucleosides Nucleotides* **1995**, *14*, 627-635.
31. Siddiqui, M. A.; Jin, H. L.; Evans, C. A.; Di Marco, M. P.; Tse, H. L. A.; Mansour, T. S. (1'*R*,2'*S*,5'*R*)-Menthyl-(5*R*)-acetox-1,3-oxathiolan-(2*R*)-carboxylate – Synthesis and isomeric purity determination by HPLC. *Chirality* **1994**, *6*, 156-160.
32. Wang, W.; Jin, H. L.; Mansour, T. S. Synthesis of optically active 2',3'-dideoxy-3'-oxa-4'-thio-ribonucleoside analogs by transposition of a leaving group on chiral oxathiolanes via a reductive-oxidative process. *Tetrahedron Lett.* **1994**, *35*, 4739-4742.
33. Mansour, T. S.; Jin, H.; Wang, W.; Hooker, E. U.; Ashman, C.; Cammack, N.; Salomon, H.; Belmonte, A. R.; Wainberg, M. A. Anti-human immunodeficiency virus and anti-hepatitis B virus activities and toxicities of the enantiomers of 2'-deoxy-3'-oxa-4'-thiocytidine and their 5-fluoro analogs *in vitro*. *J. Med. Chem.* **1995**, *38*, 1-4.
34. Camplo, M.; Faury, P.; Charvet, A. S.; Lederer, F.; Chermann, J. C.; Kraus, J. L. Synthesis and biological evaluation of a new N^4-(*N*-formylpeptide)-2',3'-dideoxy-3'-thiacytidine as anti-HIV prodrug. *Nucleosides Nucleotides* **1993**, *12*, 631-641.
35. Kraus, J. L. New phosphonate analogs of 3'-thia-2',3'-dideoxycytidine (BCH-189) – Synthesis and

anti-HIV evaluation. *Nucleosides Nucleotides* **1993**, *12*, 157-162.

36. Storer, R.; Clemens, I. R.; Lamont, B.; Noble, S. A.; Williamson, C.; Belleau, B. The resolution and absolute stereochemistry of the enantiomers of *cis*-1,2-hydroxymethyl-1,3-oxathiolan-5-yl-cytosine BCH-189: equipotent anti-HIV agents. *Nucleosides Nucleotides* **1993**, *12*, 225-236.
37. Nokami, J.; Ryokume, K.; Inada, J. Synthesis of 1,3-oxathiolane derivatives as novel precursors of 2',3'-dideoxy-3'-oxa-4'-thioribonucleosides. *Tetrahedron Lett.* **1995**, *36*, 6099-6100.
38. Chao, Q.; Nair, V. Synthesis and antiviral evaluation of 4'-hydroxymethyl-2',3'-dideoxy-3'-thianucleosides and their cyclic monophosphates. *Bioorg. Med. Chem. Lett.* **1997**, *7*, 1199-2002.
39. Chang, C. -N.; Doong, S. -L.; Zhou, J. H.; Beach, J. W.; Jeong, L. S.; Chu, C. K.; Tsai, C. H.; Cheng, Y. -C. Deoxycytidine deamination-resistant stereoisomer is the active form of (-)-2',3'-dideoxy-3'-thiacytidine in the inhibition of hepatitis B virus replication. *J. Biol. Chem.* **1992**, *267*, 13938-13942.
40. Choi, W. B.; Yeola, S.; Liotta, D. C.; Schinazi, R. F.; Painter, G. R.; Davis, M.; St. Clair, M.; Furman, P. A. Synthesis, anti-human immunodeficiency virus, and anti-hepatitis B virus activity of pyrimidine oxathiolane nucleosides. *Bioorg. Med. Chem. Lett.* **1993**, *3*, 693-696.
41. Sommadossi, J. P.; Schinazi, R. F.; Chu, C. K.; Xie, M. Y. Comparison of cytotoxicity of the (-)-enantiomer and (+)-enantiomer of 2',3'-dideoxy-3'-thiacytidine in normal human bone marrow progenitor cells. *Biochem. Pharmacol.* **1992**, *44*, 1921-1925.
42. Parker, W. B.; Cheng, Y. -C. Mitochondrial toxicity of antiviral nucleoside analogs. *J. NIH Res.* **1994**, *6*, 57-61.
43. Cui, L.; Locatelli, L.; Xie, M. Y.; Sommadossi, J. P. Effect of nucleoside analogs on neurite regeneration and mitochondrial DNA synthesis in PC-12 cells. *J. Pharmacol. Exp. Ther.* **1997**, *280*, 1228-34.
44. Cammack, N.; Rouse, P.; Marr, C. L. P.; Reid, P. J.; Boehme, R. E.; Coates, J. A. V.; Penn, C. R.; Cameron, J. M. Cellular metabolism of (-) enantiomeric 2'-deoxy-3'-thiacytidine. *Biochem. Pharmacol.* **1992**, *43*, 2059-2064.
45. Coates, J. A. V.; Cammack, N.; Jenkinson, H. J.; Jowett, A. J.; Jowett, M. I.; Pearson, B. A.; Penn, C. R.; Rouse, P. L.; Viner, K. C.; Cameron, J. M. (-)-2'-Deoxy-3'-thiacytidine is a potent, highly selective inhibitor of human immunodeficiency virus type 1 and type 2 replication *in vitro*. *Antimicrob. Agents Chemother.* **1992**, *36*, 733-739.
46. Hart, G. J.; Orr, D. C.; Penn, C. R.; Figueiredo, H. T.; Gray, N. M.; Boehme, R. E.; Cameron, J. M. Effects of (-)-2'-deoxy-3'-thiacytidine (3TC) 5'-triphosphate on human immunodeficiency virus reverse transcriptase and mammalian DNA polymerase α, β, and γ. *Antimicrob. Agents Chemother.* **1992**, *36*, 1688-1694.
47. Zhu, Y. L.; Dutschman, G. E.; Liu, S. H.; Bridges, E. G.; Cheng, Y. -C. Anti-hepatitis B virus activity and metabolism of 2',3'-dideoxy-2',3'-didehydro-β-L(-)-5-fluorocytidine. *Antimicrob. Agents Chemother.* **1998**, *42*, 1805-1810.
48. Lisignoli, G.; Monaco, M. C. G.; De Grassi, A.; Toneguzzi, S.; Ricchi, E.; Costigliola, P.; Facchini, A. *In vitro* immunotoxicity of 2'-deoxy-3'-thiacytidine, a new anti-HIV agent. *Clin. Exp. Immunol.* **1993**, *92*, 455-459.
49. Van Leeuwen, R.; Lange, J. M. A.; Hussey, E. K.; Donn, K. H.; Hall, S. T.; Harker, A. J.; Jonker, P.; Danner, S. A. The safety and pharmacokinetics of a reverse transcriptase inhibitor, 3TC, in patients with HIV infection – A phase I study. *AIDS* **1992**, *6*, 1471-1475.
50. Beach, J. W. Chemotherapeutic agents for human immunodeficiency virus infection: Mechanism of action, pharmacokinetics, metabolism and adverse reactions. *Clin. Ther.* **1998**, *20*, 2-25.
51. Heald, A. E.; Hsyu, P. H.; Yuen, G. J.; Robinson, P.; Mydlow, P.; Bartlett, J. A. Pharmacokinetics of lamivudine in human immunodeficiency virus-infected patients with renal dysfunction. *Antimicrob.*

Agents Chemother. **1996**, *40*, 1514-1519.

52. Epivir product information. Glaxo Wellcome Inc., Research Triangle Park, NC, **1996**.
53. Rajagopalan, P.; Boudinot, F. D.; Chu, C. K.; Tennant, B. C.; Baldwin, B. H.; Schinazi, R. F. Pharmacokinetics of (-)-2'-3'-dideoxy-3'-thiacytidine in woodchucks. *Antimicrob. Agents Chemother.* **1996**, *40*, 642-645.
54. Johnson, M. A.; Moore, K. H. P.; Yuen, G. J.; Bye, A.; Pakes, G. E. Clinical pharmacokinetics of lamivudine. *Clin. Pharmacok.* **1999**, *36*, 41-66.
55. Severini, A.; Liu, X.-Y.; Wilson, J. S.; Tyrrell, D. L. J. Mechanism of inhibition of duck hepatitis B virus polymerase by (-)-β-L-2',3'-dideoxy-3'-thiacytidine. *Antimicrob. Agents Chemother.* **1995**, *39*, 1430-1435.
56. Miller, R. H.; Robinson, W. S. Common evolutionary origin of hepatitis B virus and retroviruses. *Proc. Natl. Acad. Sci. USA.* **1986**, *83*, 2531-2535.
57. Gosselin, G.; Schinazi, R. F.; Sommadossi, J. P.; Mathé, C.; Bergogne, M. C.; Aubertin, A. M.; Kirn, A.; Imbach, J. -L. Anti-human immunodeficiency virus activities of the β-L enantiomer of 2',3'-dideoxycytidine and its 5-fluoro derivative *in vitro*. *Antimicrob. Agents Chemother.* **1994**, *38*, 1292-1297.
58. Schinazi, R. F.; Gosselin, G.; Faraj, A.; Korba, B. E.; Liotta, D. C.; Chu, C. K.; Mathé, C.; Imbach, J. -L.; Sommadossi, J. -P. Pure nucleoside enantiomers of β-2',3'-dideoxycytidine analogs are selective inhibitors of hepatitis B virus *in vitro*. *Antimicrob. Agents Chemother.* **1994**, *38*, 2172-2174.
59. Lee, B.; Luo, W. X.; Suzuki, S.; Robins, M. J.; Tyrrell, D. L. J. *In vitro* and *in vivo* comparison of the abilities of purine and pyrimidine 2',3'-dideoxynucleosides to inhibit duck hepadnavirus. *Antimicrob. Agents Chemother.* **1989**, *33*, 336-339.
60. Lai, C. L.; Chien, R. N.; Leung, N. W. Y.; Chang, T. T.; Guan, R.; Tai, D. I.; Ng, K. Y.; Wu, P. C.; Dent, J. C.; Barber, J.; Stephenson, S. L.; Fraser, G. A one-year trial of lamivudine for chronic hepatitis B. *N. Engl. J. Med.* **1998**, *339*, 61-68.
61. Tisdale, M.; Kemp, S. D.; Parry, N. R.; Larder, B. A. Rapid *in vitro* selection of human immunodeficiency virus type 1 resistant to 3'-thiacytidine inhibitors due to a mutation in the YMDD region of reverse transcriptase. *Proc. Natl. Acad. Sci. USA.* **1993**, *90*, 5653-5656.
62. Wainberg, M. A.; Drosopoulos, W. C.; Salomon, H.; Hsu, M.; Borkow, G.; Parniak, M. A.; Gu, Z. X.; Song, Q. B.; Manne, J.; Islam, S.; Castriota, G.; Prasad, V. R. Enhanced fidelity of 3TC-selected mutant HIV-1 reverse transcriptase. *Science* **1996**, *271*, 1282-1285.
63. Fu, L.; Cheng, Y. -C. Role of additional mutations outside the YMDD motif of hepatitis B virus polymerase in L(-)SddC (3TC) resistance. *Biochem. Pharmacol.* **1998**, *55*, 1567-1572.
64. Stuyver, L. J.; Locarnini, S. A.; Lok, A.; Richman, D. D.; Carman, W. F.; Diesntag, J. L.; Schinazi, R. F. Nomenclature for antiviral-resistant human hepatitis B virus mutations in the polymerase region. *Hepatology* **2001**, *33*, 751-757.
65. Gao, Q.; Gu, Z. X.; Hiscott, J.; Dionne, G.; Wainberg, M. A. Generation of drug-resistant variants of human immunodeficiency virus type 1 by *in vitro* passage in increasing concentrations of 2',3'-dideoxycytidine and 2',3'-dideoxy-3'-thiacytidine. *Antimicrob. Agents Chemother.* **1993**, *37*, 130-133.
66. Rusconi, S.; De Pasquale, M. P.; Milazzo, L.; Bulgheroni, E.; Cittero, P.; Kurtagic, S.; D'Arminio-Monforte, A.; Galli, M. *In vitro* effects of continuous pressure with zidovudine (ZDV) and lamivudine on a ZDV-resistant HIV-1 isolate. *AIDS* **1997**, *11*, 1406-1410.
67. Keulen, W.; Back, N. K.; Van Wijk, A.; Boucher, C. A. B.; Berkhout, B. Initial appearance of the 184Ile variant in lamivudine-treated patients is caused by the mutational bias of human immunodeficiency virus type 1 reverse transcriptase. *J. Virol.* **1997**, *71*, 3346-3350.
68. Wainberg, M. A.; Salomon, H.; Gu, Z.; Montaner, J. S. G.; Cooley, T. P.; McCaffrey, R.; Ruedy, J.; Hirst, H. M.; Cammack, N.; Cameron, J.; Nicholson, W. Development of HIV-1 resistance to (-)2'-deoxy-

3'-thiacytidine in patients with AIDS or advanced AIDS-related complex. *AIDS,* **1995**, *9*, 351-357.

69. Mathez, D.; Schinazi, R. F.; Liotta, D. C.; Leibowitch, J. Infectious amplification of wild-type human immunodeficiency virus from patients lymphocytes and modulation by reverse transcriptase inhibitors *in vitro*. *Antimicrob. Agents Chemother.* **1993**, *37*, 2206-2211.
70. Gu, Z.; Gao, Q.; Faust, E. A.; Wainberg, M. A. Possible involvement of cell-fusion and viral recombination in generation of human immunodeficiency virus variants that display dual resistance to AZT and 3TC. *J. Gen. Virol.* **1995**, *76*, 2601-2605.
71. García Lerma, J. G.; Schinazi, R. F.; Juodawlkis, A. S.; Soriano, V.; Lin, Y. L.; Tatti, K.; Rimland, D.; Folks, T. M.; Heneine, W. A rapid non-culture-based assay for clinical monitoring of phenotypic resistance of human immunodeficiency virus type 1 to lamivudine (3TC). *Antimicrob. Agents Chemother.* **1999**, *43*, 264-270.
72. Ladner, S. K.; Miller, T. J.; King, R. W. The M539V polymerase variant of human hepatitis B virus demonstrates resistance to 2 '-deoxy-3 '-thiacytidine and a reduced ability to synthesize viral DNA. *Antimicrob. Agents Chemother.* **1998**, *42*, 2128-2131.
73. Fujioka, S.; Shimomura, H.; Fujio, K.; Ikeda, F.; Miyake, M.; Ishii, Y.; Itoh, M.; Sakaguchi, K.; Tsuji, T. Two cases of chronic hepatitis B with emergence of lamivudine-resistant virus during long-term therapy. *Hepatol. Res.* **1999**, *13*, 97-104.
74. Miller, M. D.; Anton, K. E.; Mulato, A. S.; Lamy, P. D.; Cherrington, J. M. Human immunodeficiency virus type 1 expressing the lamivudine-associated M184V mutation in reverse transcriptase shows increased susceptibility to adefovir and decreased replication capability *in vitro*. *J. Infect. Dis.***1999**, *179*, 92-100.
75. Bartlett, J. A.; Benoit, S. L.; Johnson, V. A.; Quinn, J. B.; Sepulveda, G. E.; Ehmann, W. C.; Tsoukas, C.; Fallon, M. A.; Self, P. L.; Rubin, M. Lamivudine plus zidovudine compared with zalcitabine plus zidovudine in patients with HIV infection. *Ann. Intern. Med.* **1996**, *125*, 161-172.
76. Staszewski, S.; Loveday, C.; Picazo, J. J.; Dellamonica, P.; Skinhoj, P.; Johnson, M. A.; Danner, S. A.; Harrigan, P. R.; Hill, A. M.; Verity, L.; McDade, H. Safety and efficacy of lamivudine-zidovudine combination therapy in zidovudine-experienced patients - A randomized controlled comparison with zidovudine monotherapy. *JAMA* **1996**, *276*, 111-117.
77. Murray, L.; Kelly, G. L.; Deutsch, M.; Wyble, C. Combivir® tablets (lamivudine/ zidovudine tablets), Physicians' Desk Reference, 55th Ed., **2001**, *55*, 1365-1368.
78. Eron, J. J.; Benoit, S. L.; Jemsek, J.; MacArthur, R. D.; Santana, J.; Quinn, J. B.; Kuritzkes, D. R.; Fallon, M. A.; Rubin, M. Treatment with lamivudine, zidovudine, or both in HIV-positive patients with 200 to 500 CD4+ cells per cubic millimiter. *N. Engl. J. Med.* **1995**, *333*, 1662-1669.
79. Katlama, C.; Ingrand, D.; Loveday, C.; Clumeck, N.; Mallolas, J.; Staszewski, S.; Johnson, M.; Hill, A. M.; Pearce, G.; MaDade, H. Safety and efficacy of lamivudine-zidovudine combination therapy in antiretroviral-naive patients - A randomized controlled comparison with zidovudine monotherapy. *JAMA* **1996**, *276*, 118-125.
80. Gulick, R. M.; Mellors, J. W.; Havlir, D.; Eron, J. J.; Gonzalez, C.; McMahon, D.; Richman, D. D.; Valentine, F. T.; Jonas, L.; Meibohm, A.; Emini, E. A.; Chodakewitz, J. A. Treatment with indinavir, zidovudine, and lamivudine in adults with human immunodeficiency virus infection and prior antiretroviral therapy. *N. Engl. J. Med,* **1997**, *337*, 734-739.
81. Hammer, S. M.; Squires, K. E.; Hughes, M. D.; Grimes, J. M.; Demeter, L. M.; Currier, J. S.; Eron, J. J.; Feinberg, J. E.; Balfour, H. H.; Dayton, L. R.; Chodakewitz, J. A.; Fischl, M. A. A controlled trial of two nucleoside analogs plus indinavir in persons with human immunodeficiency virus infection and CD4 cell counts of 200 per cubic millimeter or less. *N. Engl. J. Med.* **1997**, *337*, 725-733.

82. Staszewski, S.; Miller, V.; Rehmet, S.; Stark, T.; De Cree, J.; De Brabander, M.; Peeters, M.; Andries, K.; Moeremans, M.; De Raemaeker, M.; Pearce, G.; Van Den Broeck, R.; Verbiest, W.; Stoffels, P. Virological and immunological analysis of a triple combination pilot study with loviride, lamivudine and zidovudine in HIV-1-infected patients. *AIDS* **1996**, *10*, F1-F7.
83. Youle, M.; Davies, S.; Andries, K.; Moeremans, M.; Van den Broek, R.; Joyce, T.; Stoffels, P.; Gazzard, B. Randomized double-blind trial of loviride [R89439] (L), zidovudine (Z) and the combination in HIV-1 infected patients. *2nd National Conference on Human Retroviruses and Related Infections* **1995**, LB6A.
84. Kuritzkes, D. R.; Quinn, J. B.; Benoit, S. L.; Shugarts, D. L.; Griffin, A.; Bakhtiari, M.; Poticha, D.; Eron, J. J.; Fallon, M. A.; Rubin, M. Drug resistance and virologic response in NUCA 3001, a randomized trial of lamivudine (3TC) versus zidovudine (ZDV) versus ZDV plus 3TC in previously untreated patients. *AIDS* **1996**, *10*, 975-981.
85. Larder, B. A.; Kemp, S. D.; Harrigan, P. R. Potential mechanism for sustained antiretroviral efficacy of AZT-3TC combination therapy. *Science* **1995**, *269*, 696-699.
86. Veal, G. J.; Hoggard, P. G.; Barry, M. G.; Khoo, S.; Back, D. J. Interaction between lamivudine (3TC) and other nucleoside analogs for intracellular phosphorylation. *AIDS* **1996**, *10*, 546-548.
87. Foudraine, N. A.; De Jong, J. J.; Weverling, G. J.; Van Benthem, B. H. B.; Maas, J.; Keet, I. P. M.; Jurriaans, S.; Roos, M. T. L.; Vandermeulen, K.; De Wolf, F.; Lange, J. M. A. An open randomized controlled trial of zidovudine plus lamivudine versus stavudine plus lamivudine. *AIDS* **1998**, *12*, 1513-1519.
88. Havlir, D. V.; Lange, J. M. A. New antiretrovirals and new combinations. *AIDS* **1998**, *12*, S165-S174.
89. Foudraine, N. A.; Hoetelmans, R.; Lange, J. M. A.; De Wolf, F.; Van Benthem, B. H. B.; Mass, J. J.; Portegies, P. Cerebrospinal-fluid HIV-1 RNA and drug concentrations after treatment with lamivudine plus zidovudine or stavudine. *Lancet* **1998**, *351*, 1547-1551.
90. Massari, F.; Conant, M.; Mellors, J.; Steigbigel, R.; Mildvan, D.; Greenberg, R.; Carpenter, C.; Murphy, R.; Squires, K.; Rigsb, M.; Drusano, G.; McKinley, G.; Gilde, L.; Nessly, M. A phase II open-label randomized study of the triple combination of indinavir, zidovudine (ZDV) and didanosine (DDI) versus indinavir alone and zidovudine/didanosine in antiretroviral naïve patients. *3rd Conference on retroviruses and opportunistic infections* **1996**, 200.
91. Patick, A. K.; Boritzki, T. J.; Bloom, L. A. Activities of the human immunodeficiency virus type 1 (HIV-1) protease inhibitor nelfinavir mesylate in combination with reverse transcriptase and protease inhibitors against acute HIV-1 infection *in vitro. Antimicrob. Agents Chemother.* **1997**, *41*, 2159-2164.
92. Landers, M. B.; Fraser, V. J. Antiviral chemoprophylaxis after occupational exposure to human immunodeficiency virus: Why, when, where, and what. *Am. J. Ophthal.* **1997**, *124*, 234-239.
93. Notermans, D. W.; Jurriaans, S.; De Wolf, F.; Foudraine, N. A.; De Jong, J. J.; Cavert, W.; Schuwirth, C. M.; Kauffmann, R. H.; Meenhorst, P. L.; McDade, H.; Goodwin, C.; Leonard, J. M.; Goudsmit, J.; Danner, S. A. Decrease of HIV-1 RNA levels in lymphoid tissue and peripheral blood during treatment with ritonavir, lamivudine and zidovudine. *AIDS* **1998**, *12*, 167-173.
94. Wong, J. K.; Gunthard, H. F.; Havlir, D. V.; Zhang, Z. Q.; Haase, A. T.; Ignacio, C. C.; Kwok, S.; Emini, E.; Richman, D. D. Reduction of HIV-1 in blood and lymph nodes following potent antiretroviral therapy and the virologic correlates of treatment failure. *Proc. Natl. Acad. Sci. USA.* **1997**, *94*, 12574-12579.
95. Smith, R. A.; Remington, K. M.; Preston, B. D.; Schinazi, R. F.; North, T. W. A. A novel point mutation at position 156 of reverse transcriptase from feline immunodeficiency virus confers resistance to the combination of (-)-β-2',3'-dideoxy-3'-thiacytidine and 3'-azido-3'-deoxythymidine. *J. Virol.* **1998**, *72*, 2335-2340.
96. Kemp, S. D.; Shi, C.; Bloor, S.; Harrigan, P. R.; Mellors, J. W.; Larder, B. A. A novel polymorphism at codon 333 of human immunodeficiency virus type 1 reverse transcriptase can facilitate dual resistance to

zidovudine and L-2',3'-dideoxy-3'-thiacytidine. *J. Virol.* **1998**, *72*, 5093-5098.

97. Schinazi, R. F.; McMillan, A.; Cannon, D.; Mathis, R.; Lloyd, R. M.; Peck, A.; Sommadossi, J. -P.; St Clair, M.; Wilson, J.; Furman, P. A.; Painter, G.; Choi, W. B.; Liotta, D. C. Selective inhibition of human immunodeficiency viruses by racemates and enantiomers of *cis*-5-fluoro-1-[2-(hydroxymethyl)-1,3-oxathiolan-5-yl]cytosine. *Antimicrob. Agents Chemother.* **1992**, *36*, 2423-2431.
98. Rousseau, F.; Wang, L. H.; Delehanty, J.; Hulett, L.; McCreedy, B.; Sommadossi, J. -P.; Barry, D. W. Intracellular FTC-triphosphate levels correlate with the clinical antiviral activity of FTC. *AIDS* **1998**, 12, Suppl. 4, OP75.
99. Wilson, J. E.; Martin, J. L.; Borroto-Esoda, K.; Hopkins, S.; Painter, G.; Liotta, D. C.; Furman, P. A. The 5'-triphosphates of the (-) and (+) enantiomers of cis-5-fluoro-1-[2-(hydroxymethyl)-1,3-oxathiolane-5-yl]cytosine equally inhibit human immunodeficiency virus type 1 reverse transcriptase. *Antimicrob. Agents Chemother.* **1993**, *37*, 1720-1722.
100. Shewach, D. S.; Liotta, D. C.; Schinazi, R. F. Affinity of the antiviral enantiomers of oxathiolane cytosine nucleosides for human 2'-deoxycytidine kinase. *Biochem. Pharmacol.* **1993**, *45*, 1540-1543.
101. Cui, L. X.; Schinazi, R. F.; Gosselin, G.; Imbach, J. -L.; Chu, C. K.; Rando, R. F.; Revankar, G. R.; Sommadossi, J. -P. Effect of beta-enantiomeric and racemic nucleoside analogs on mitochondrial functions in HepG2 cells - Implications for predicting drug hepatotoxicity. *Biochem. Pharmacol.* **1996**, *52*, 1577-1584.
102. Frick, L. W.; Lambe, C. U.; St John, L.; Taylor, L. C.; Nelson, D. J. Pharmacokinetics, oral bioavailability, and metabolism in mice and cynomolgus monkeys of (2'*R*, 5'*S*-)-*cis*-5-fluoro-1-[2-(hydroxymethyl)-1,3-oxathiolan-5-yl]cytosine, an agent active against human immunodeficiency virus and human hepatitis B virus. *Antimicrob, Agents Chemother.* **1994**, *38*, 2722-2729.
103. Schinazi, R. F.; Boudinot, F. D.; Ibrahim, S. S.; Manning, C.; McClure, H. M.; Liotta, D. C. Pharmacokinetics and metabolism of racemic 2',3'-dideoxy-5-fluoro-3'-thiacytidine in rhesus monkeys. *Antimicrob. Agents Chemother.* **1992**, *36*, 2432-2438.
104. Furman, P. A.; Davis, M.; Liotta, D. C.; Paff, M.; Frick, L. W.; Nelson, D. J.; Dornsife, R. E.; Wyrster, J. A.; Wilson, L. J.; Fyfe, J. A.; Tuttle, J. V.; Miller, W. H.; Condreay, L.; Averett, D. R.; Schinazi, R. F.; Painter, G. R. The anti-hepatitis B virus activities, cytotoxicities, and anabolic profiles of the (-) and (+) enantiomers of *cis*-5-fluoro-1-[2-(hydroxymethyl)-1,3-oxathiolan-5-yl]cytosine. *Antimicrob. Agents Chemother.* **1992**, *36*, 2686-2692.
105. Condreay, L. D.; Jansen, R. W.; Powdrill, T. F.; Johnson, L. C.; Selleseth, D. W.; Paff, M. T.; Daluge, S. M.; Painter, G. R.; Furman, P. A.; Ellis, M. N.; Averett, D. R. Evaluation of the potent anti-hepatitis B virus agent (-)-*cis*-5-fluoro-1-[2-(hydroxymethyl)-1,3-oxathiolan-5-yl]cytosine in a novel *in vivo* model. *Antimicrob. Agents Chemother.* **1994**, *38*, 616-619.
106. Fourel, I.; Cullen, J. M.; Saputelli, J.; Aldrich, C. E.; Schaffer, P.; Averett, D. R.; Pugh, J.; Mason, W. S. Evidence that hepatocyte turnover is required for rapid clearance of duck hepatitis B virus during antiviral therapy of chronically infected ducks. *J. Virol.* **1994**, *68*, 8321-8330.
107. Cullen, J. M.; Smith, S. L.; Davis, M. G.; Dunn, S. E.; Botteron, C.; Cecchi, A.; Linsey, D.; Linzey, D.; Frick, L.; Paff, M. T.; Goulding, A.; Biron, K. *In vivo* antiviral activity and pharmacokinetics of (-)-cis-5-fluoro-1-[2-(hydroxymethyl)-1,3-oxathiolan-5-yl]cytosine in woodchuck hepatitis virus-infected woodchucks. *Antimicrob. Agents Chemother,* **1997**, *41*, 2076-2082.
108. Gish, R. G.; Leung, N. W. Y.; Wright, T. L.; Trinh, H.; Lang, W.; Kessler, H. A.; Fang, L.; Wang, L. H.; Delehanty, J.; Rigney, A.; Mondou, E.; Snow, A.; Rousseau, F. Dose range study of pharmacokinetics, safety, and preliminary antiviral activity of emtricitabine in adults with hepatitis B virus infection. *Antimicrob. Agents Chemother.* **2002**, *46*, 1734-1740.

109. Smith, R. A.; Remington, K. M.; Lloyd Jr, R. M.; Schinazi, R. F.; North, T. W. A novel Met-to-Thr mutation in the YMDD motif of reverse transcriptase from feline immunodeficiency virus confers resistance to oxathiolane nucleosides. *J. Virol.* **1997**, *71*, 2357-2362.
110. Bednarski, K.; Dixit, D. M.; Wang, W.; Evans, C. A.; Jin, H.; Yuen, L.; Mansour, T. S. Inhibitory activities of herpes simplex viruses type 1 and 2 and human cytomegalovirus by stereoisomers of 2'-deoxy-3'-oxa-5-(*E*)-(2-bromovinyl)uridines and their 4'-thioanalogs. *Bioorg. Med. Chem. Lett.* **1994**, *4*, 2667-2672.
111. De Muys, J.-M.; Gourdeau, H.; Nguyen-Ba, N.; Taylor, D. L.; Ahmed, P. S.; Mansour, T.; Locas, C.; Richard, N.; Wainberg, M. A.; Rando, R. F. Anti-human immunodeficiency virus type 1 activity, intracellular metabolism, and pharmacokinetic evaluation of 2'-deoxy-3'-oxa-4'-thiocytidine. *Antimicrob. Agents Chemother.* **1999**, *43*, 1835-1844.
112. Norbeck, D.W., Spanton, S., Broder, S., Mitsuya, H. (±)-Dioxolane-T ((±)-1-[(2β,4β)-2-(hydroxymethyl-4-dioxolanyl]thymine). A new 2',3'-dideoxy nucleoside prototype with in vitro activity against HIV. *Tetrahedron Lett.* **1989**, *30*, 6263-6266.
113. Wilson, L. J.; Choi, W. B.; Spurling, T., Liotta, D. C.; Schinazi, R. F.; Cannon, D.; Painter, G. R.; St Clair, M.; Furman, P. A. The synthesis and anti-HIV activity of pyrimidine dioxolanyl nucleosides. *Bioorg. Med. Chem. Lett.* **1993**, *3*, 169-174.
114. Jin, H. L.; Tse, H. L. A.; Evans, C. A.; Mansour, T. S.; Beels, C. M.; Ravenscroft, P.; Humber, D. C.; Jones, M. F.; Payne, J. J.; Ramsay, M. V. J. Unexpected effects of Lewis acids in the synthesis of optically pure 2'-deoxy-3'-oxacytidine nucleoside analogs. *Tetrahedron: Asymmetry* **1993**, *4*, 211-214.
115. Kim, H. O.; Ahn, S. K.; Alves, A. J.; Beach, J. W.; Jeong, L. S.; Choi, B. G.; Van Roey, P.; Schinazi, R. F.; Chu, C. K. Asymmetric synthesis of 1,3-dioxolane pyrimidine nucleosides and their anti-HIV activity. *J. Med. Chem.* **1992**, *35*, 1987-1995.
116. Kim, H. O.; Schinazi, R. F.; Nampalli, S.; Shanmuganathan, K.; Cannon, D. L.; Alves, A. J.; Jeong, L. S.; Beach, J. W.; Chu, C. K. 1,3-Dioxolanylpurine nucleosides (2R,4R) and (2R,4S) with selective anti-HIV-1 activity in human lymphocytes. *J. Med. Chem.* **1993**, *36*, 30-37.
117. Kim, H. O.; Shanmuganathan, K. Alves, A. J.; Shanmuganathan, K.; Cannon, D. L.; Alves, A. J.; Jeong, L. S.; Beach, J. W.; Chu, C. K. Potent anti-HIV and anti-HBV activities of (-)-L-β-dioxolane-C and (+)-L-β-dioxolane-T and their asymmetric synthesis. *Tetrahedron Lett.* **1992**, *33*, 6899-6902.
118. Kim, H. O.; Schinazi, R. F.; Shanmuganathan, K.; Jeong, L. S.; Beach, J. W.; Nampalli, S.; Cannon, D. L.; Chu, C. K. L-β-(2S,4S)- and L-α-(2S,4R)-Dioxolanyl nucleosides as potential anti-HIV agents: asymmetric synthesis and structure-activity relationships. *J. Med. Chem.* **1993**, *36*, 519-528.
119. Abushanab, E.; Bessodes, M.; Antonakis, K. Practical enantiospecific syntheses of (+)-*erythro*-9-(2*S*-hydroxy-3*R*-nonyl)adenine. *Tetrahedron Lett.* **1984**, *25*, 3841-3844.
120. Abushanab, E.; Vemishetti, P.; Leiby, R. W.; Singh, H. K.; Mikkilineni, A. B.; Wu, D. C. J.; Saibaba, R.; Panzica, R. P. The chemistry of L-ascorbic and D-isoascorbic acids. 1. The preparation of chiral butanetriols and butanetetrols. *J. Org. Chem.* **1988**, *53*, 2598-2602.
121. Belleau, B.; Evans, C. A.; Tse, H. L. A.; Jin, H. L.; Dixit, D. M.; Mansour, T. S. Oxidative degradation of L-ascorbic acid acetals to 2',3'-dideoxy-3'-oxaribofuranosides – Synthesis of enantiomerically pure 2',3'-dideoxy-3'-oxacytidine stereoisomers as potential antiviral agents. *Tetrahedron Lett.* **1992**, *33*, 6949-6952.
122. Belleau, B., Dixit, D.M., Ba, N.N., Brown, W.L., Gulini, U., Lafleur, D., Cameron, J.M. Synthesis and anti-HIV activity of (±)-1,3-dioxolane nucleoside analogs and their enzymatic resolution. *Abstr. Papers Am. Chem. Soc.* **1992**, *204*, MEDI 138.
123. Evans, C. A.; Dixit, D. M.; Siddiqui, M. A.; Jin, H.; Tse, H. L. A.; Cimpoia, A.; Bednarski, K.; Breining,

T.; Mansour, T. S. Divergent asymmetric syntheses of dioxolane nucleoside analogs. *Tetrahedron: Asymmetry* **1993**, *4*, 2319-2322.

124. Siddiqui, M. A.; Brown, W. L.; Nguyen-Ba, N.; Dixit, D. M.; Mansour, T. S.; Hooker, E.; Viner, K. C.; Cameron, J. M. Antiviral optically pure dioxolane purine nucleosides analogs. *Bioorg. Med. Chem. Lett.* **1993**, *3*, 1543-1546.
125. Efimtseva, E. V.; Mikhailov, S. N.; Meshkov, S. V.; Lönnberg, H. Synthesis and physicochemical properties of dioxolane nucleoside analogs. *Acta. Chem. Scand.* **1992**, *46*, 1122-1126.
126. Mikhailov, S. N.; Efimtseva, E. V.; Meshkov, S. V.; Kern, E. R. Nucleoside analogs on the basis of 4(*R*), 5(*R*)-dihydroxymethyl-2-methyl-1,3-dioxolane. *Nucleosides Nucleotides* **1994**, *13*, 615-623.
127. Efimtseva, E. V.; Mikhailov, S. N.; Victorova, L. S.; Rosovskaya, T. A.; Beabealashvilli, R. S. Synthesis of dioxolane analogs of dideoxynucleotides and their substrate properties in DNA synthesis reactions. *Nucleosides Nucleotides* **1995**, *14*, 727-729.
128. Du, J.; Qu, F.; Lee, D.; Newton, M. G.; Chu, C.K. 1,3-Dioxolane *C*-nucleosides – Asymmetric synthesis of 4 stereoisomers of 2-[2-(hydroxymethyl)-1,3-dioxolan-5-yl)]-1,3-thiazole-4-carboxamide. *Tetrahedron Lett.* **1995**, *36*, 8167-8170.
129. Qu, F.; Hong, J. H.; Du, J.; Newton, M. G.; Chu, C. K. Asymmetric synthesis of (2'*R*,4 '*R*) and (2'*S*,4'*S*)-1,3-dioxolanyl triazole *C*-nucleosides. *Tetrahedron* **1999**, *55*, 9073-9088.
130. Grove, K. L.; Guo, X.; Liu, S. H.; Kukhanova, M.; Chu, C. K.; Cheng, C. -Y. β-L-(-)-Dioxolane cytidine (β-L-(-)-OddC) as a potent compound for the treatment of cancer. *Nucleosides Nucleotides* **1997**, *16*, 1229-1233.
131. Grove, K. L.; Guo, X.; Liu, S. H.; Gao, Z. L.; Chu, C. K.; Cheng, Y. -C. Anticancer activity of β-L-dioxolane cytidine, a novel nucleoside analog with the unnatural L-configuration. *Cancer Res.* **1995**, *55*, 3008-3011.
132. Grove, K. L.; Cheng, Y. -C. Uptake and metabolism of the new anticancer compound β-L-dioxolane cytidine in human prostate carcinoma DU-145 cells. *Cancer Res.* **1996**, *56*, 4187-4191.
133. Moore, M.; Belanger, K.; Jolivet, J.; Baker, S.; Wainman, N. NCIC CTG IND 103: A phase I and pharmacokinetic (PK) study of the novel L-nucleoside analog troxacitabine (BCH-4556) given every 21 days. *Eur. J. Cancer* **1999**, *35*, S284-S284.
134. Schwartz, P. M.; Haggerty, J. G.; Cheng, Y. -C. β-L-1,3-Dioxolane-cytidine: A novel nucleoside that inhibits proliferation and induces differentiation of keratinocytes *in vitro*. *Skin Pharmacol. Applied. Skin Physiol.* **1998**, *11*, 207-213.
135. Mar, E. C.; Chu, C. K.; Lin, J. C. Some nucleoside analogs with anti-human immunodeficiency virus activity inhibit replication of Epstein-Barr virus. *Antiviral Res.* **1995**, *28*, 1-11.
136. Schinazi, R. F.; McClure, H. M.; Boudinot, F. D.; Xiang, Y. J.; Chu, C. K. Development of (-)-β-D-2,6-diaminopurine dioxolane as a potential antiviral agent. *Antiviral Res.* **1994**, *23*, suppl. 81.
137. Rajagopalan, P.; Boudinot, F. D.; Chu, C. K.; Tennant, B. C.; Baldwin, B. H.; Schinazi, R. F. Pharmacokinetics of (-)-β-D-2,6-diaminopurine dioxolane and its metabolite, dioxolane guanosine, in woodchucks (*Marmota monax*). *Antiviral Chem. Chemother.* **1996**, *7*, 65-70.
138. Chen, H.; Boudinot, F. D.; Chu, C. K.; McClure, H. M.; Schinazi, R. F. Pharmacokinetics of (-)-β-D-2-aminopurine dioxolane and (-)-β-D-2-amino-6-chloropurine dioxolane and their antiviral metabolite (-)-β-D-dioxolane guanine in rhesus monkeys. *Antimicrob. Agents Chemother.* **1996**, *40*, 2332-2336.
139. Furman, P. A.; Jeffrey, J.; Kiefer, L. L.; Feng, J. Y.; Anderson, K. S.; Borroto-Esoda, K.; Hill, E.; Copeland, W. C.; Chu, C. K.; Sommadossi, J. -P.; Liberman, I.; Schinazi, R. F.; Painter, G. R. Mechanism of action of 1-β-D-2,6-diaminopurine dioxolane, a prodrug of the human immunodeficiency virus type 1 inhibitor 1-β-D-dioxolane guanosine. *Antimicrob. Agents Chemother.* **2001**, *45*, 158-165.

140. Gu, Z.; Wainberg, M. A.; Nguyen-Ba, P.; L'Heureux, L.; de Muys, J. -M.; Rando, R. F. Anti-HIV activities of 1,3-dioxolane guanine and 2,6-diaminopurine dioxolane. *Nucleosides Nucleotides* **1999**, *18*, 891-892.

141. Richman, D. D.; Kessler, H.; Eron, J.; Thompson, M.; Raffi, F.; Jacobson, J.; Harris, J.; McCreedy, B.; Bigley, J.; Rousseau, F. Anti-HIV activity and tolerability of DAPD, a novel dioxolane guanosine RT inhibitor: Initial results of a phase I/II 14-day monotherapy clinical trial. *7th Conference on Retroviruses and Opportunistic Infections*. **2000**, Abstr. 668.

142. Mewshaw, J. P.; Myrick, F. T.; Wakefield, D. A. C. S.; Hooper, B. J.; Harris, J. L.; McCreedy, B.; Borroto-Esoda, K. Dioxolane guanosine, the active form of the prodrugs diaminopurine dioxolane, is a potent inhibitor of drug-resistant HIV-1 isolates from patients for whom standard nucleoside therapy fails. *J. Acquir. Immune Defic. Syndr.* **2002**, *29*, 11-20.

143. Ono, S. K.; Kato, N.; Shiratori, Y.; Kato, J.; Goto, T.; Schinazi, R. F.; Carrilho, F. J.; Omata, M. The polymerase L528M mutation cooperates with nucleotide binding-site mutations, increasing hepatitis B virus replication and drug resistance. *J. Clin. Inv.* **2001**, *107*, 449-455.

144. Chin, R.; Shaw, T.; Torresi, J.; Sozzi, V.; Trautwein, C.; Bock, T.; Manns, M.; Isom, H.; Furman, P. A.; Locarnini, S. *In vitro* susceptibilities of wild-type or drug-resistant hepatitis B virus to (-)-β-D-2,6-diaminopurine dioxolane and 2'-fluoro-5-methyl-β-L-arabinofuranosyluracil. *Antimicrob. Agents Chemother.* **2001**, *45*, 2495-2501.

145. Gu, Z. X.; Wainberg, M. A.; Nguyen-Ba, N.; L'Heureux, L.; de Muys, J. M.; Bowlin, T. L.; Rando, R. F. Mechanism of action and *in vitro* activity of 1 ',3 '-dioxolanylpurine nucleoside analogs against sensitive and drug-resistant human immunodeficiency virus type 1 variants. *Antimicrob. Agents Chemother.* **1999**, *43*, 2376-2382.

146. Bazmi, H. Z.; Hammond, J. L.; Cavalcanti, S. C. H.; Chu, C. K.; Schinazi, R. F.; Mellors, J. W. *In vitro* selection of mutations in the human immunodeficiency virus type 1 reverse transcriptase that decrease susceptibility to (-)-β-D-dioxolane-guanosine and suppress resistance to 3 '-azido-3 '-deoxythymidine. *Antimicr. Agents Chemother.* **2000**, *44*, 1783-1788.

147. Lin, J. -S.; Kira, T.; Gullen, E.; Choi, Y.; Qu, F.; Chu, C. K.; Cheng, Y. -C. Structure-activity relationships of L-dioxolane uracil nucleosides as anti-Epstein Barr virus agents. *J. Med. Chem.* **1999**, *42*, 2212-2217.

148. Kira, T.; Grill, S. P.; Dutschman, G. E.; Lin J. -S.; Qu, F.; Choi, Y.; Chu, C. K.; Cheng, Y. -C. Anti-Epstein-Barr virus (EBV) activity of -L-5-iododioxolane uracil is dependent on EBV thymidine kinase. *Antimicrob. Agents Chemother.* **2000**, *44*, 3278-3284.

149. Choi, Y.; Li, L.; Grill, S.; Gullen, E.; Lee, C. S.; Gumina, G.; Cheng, Y.-C.; Chu, C. K. Structure-activity relationships of (*E*)-5-(2-bromovinyl)uracil nucleosides and related analogs as anti-herpesvirus agents. *J. Med. Chem.* **2000**, *43*, 2538-2546.

150. (a) Li, L.; Dutschman, G. E.; Gullen, E. A.; Grill, S. P.; Choi, Y.; Chu, C. K.; Cheng, Y.-C. Inhibition of the growth of varicella-zoster virus by L-β-5-bromovinyl-(2-hydroxymehtyl)-1,3-dioxolanyl)uracil. *Abstracts of Papers*, 91th Annual Meeting of the American Assciation for Cancer Research, San Francisco, CA; American Association for Cancer Research: **2000**; A22. (b) Li, L.; Dutschman, G. E.; Gullen, E. A.; Tsujii, E.; Grill, S. P.; Choi, Y.; Chu, C. K.; Cheng, Y. -C. Metabolism and mode of inhibition of varicella-zoster virus by L-5-bromovinyl-(2-hydroxymethyl)-(1,3-dioxolanyl)uracil is dependent on viral thymidine kinase. *Mol. Pharmacol.* **2000**, *58*, 1109-1114.